Of the Human Heart

Benjamin Peirce, ca. 1859. Harvard University Archives, call # HUP [4b].

Of the Human Heart
A Biography of Benjamin Peirce

Edward R. Hogan

Lehigh
University
Press

Bethlehem: Lehigh University Press

Associated University Presses
2010 Eastpark Boulevard
Cranbury, NJ 08512

The paper used in this publication meets the requirements of the American National Standard for Permanence of Paper for Printed Library Materials Z39.48-1984.

Library of Congress Cataloging-in-Publication Data

Hogan, Edward R., 1942–
    Of the human heart : a biography of Benjamin Peirce / Edward R. Hogan.
        p.   cm.
    Includes bibliographical references and index.
    ISBN-13: 978-0-934223-93-5 (alk. paper)
    1. Peirce, Benjamin, 1809–1880.   2. Mathematicians—United States—Biography.   I. Title.

QA29.P66.H56   2008
510.92—dc22
[B]

2007015983

PRINTED IN THE UNITED STATES OF AMERICA

To my wife with my whole heart

Contents

Preface   9

Acknowledgments   11

1. Hurrah for Young America   15
2. The Father of American Geometry   28
3. The Finest Lady in Northampton   49
4. The Voice of God   63
5. The Feeling of Mutual Goodwill   95
6. A Prince of the Humbugs   122
7. As the Tree Grows Towards Heaven   142
8. No Idea of an Observatory   165
9. A Stranger in My Own Land   189
10. This Holy Cause   203
11. Every Drop of Blood   217
12. We Shall Always Be Boys Together   240
13. A Public Functionary   256
14. My Most Precious Pearls   281
15. My Best Love   306

Appendices   341

Notes   346

Bibliography   402

Index   419

Preface

Benjamin Peirce (pronounced "purse") was one of the principal contributors to nineteenth-century American science. His *Linear Associative Algebra* was the first important mathematical research done by an American; he was a key figure in the professionalization of American science; and, as superintendent of the United States Coast Survey, he was an effective scientific administrator. Peirce also played an important part in the education of many American scientists, including Peirce's son, Charles Saunders Peirce, and Simon Newcomb, the most widely honored and recognized American scientist of the generation after Peirce.

Peirce belonged to an impressive American family. The intellectual tradition in this family is apparent with Peirce's feminist mother. And his scholarly father wrote a history of Harvard College. The tradition finds its climax in Peirce's son Charles who Max Fisch described as "the most original and the most versatile intellect that the Americas have so far produced."[1]

As a man, Peirce felt things strongly and passionately. He lamented to a friend that despite the power of mathematics to describe the universe, mathematics could not describe a "single emotion of the human heart. . . . Poor mathematician! Is he not to be pitied?"[2] Peirce's extraordinarily strong feelings governed his life much more than dispassionate science.

In the spring of 1864, Benjamin Peirce feared that he was terminally ill. He wrote to his friend, Alexander Dallas Bache, "if I should be taken, dearest Chief, exert all your influence to save me from eulogistic biographers."[3] Until now Peirce has fared fairly well. For 125 years he has escaped a full length biography.

Partly due to Peirce's request and partly because I have been captivated by his correspondence, I have quoted extensively from his letters as well as from the letters of his correspondents and others, especially his sister, Charlotte Elizabeth Peirce. I feel it best to let Peirce tell his own story as much as possible rather than to attempt to tell it for him. Thus this work might be described as a documentary biography.

Unfortunately, I have not always found Peirce's hand legible. Peirce, himself, recognized that his writing was not always easy to read. He wrote to his wife, "I hope, my dear Sarah, that you do not find it such hard work to read my writing as I do myself. If you do, you will never undertake to do more than finish the first sentence, and therefore it will be quite unnecessary for me to write any more."[4] I'm sure that I have not always succeeded in deciphering Peirce's handwriting successfully, but hopefully I have conveyed his intended meaning.

Acknowledgments

M<small>Y WIFE, CONNIE, READ PARTS OF THE MANUSCRIPT MANY TIMES AND</small> read the entire manuscript at least once. My daughter, Sarah Flandro, helped me with the research, especially in chapter 2. My son, Daniel, gave me invaluable technical aid. LynnKay Brown, Russell Osmond, Deborah Prince, and Lawrence Squeri read the manuscript and all of them made suggestions for its improvement, as did the reviewer, David Zitarelli.

The entire East Stroudsburg University Mathematics Department helped me to decipher Peirce's letters. Mary Lou Smith, our departmental secretary, helped me well beyond the call of duty. Ralph Vitello translated the Latin and French phrases for me. Both Allison Walsh and Amy Ackerberg-Hastings shared pertinent sections of their PhD dissertations with me.

I am indebted to scores of librarians from many libraries for aiding me in finding and accessing material about Peirce. The staffs of the Houghton Library and the Harvard University Archives helped me many times over a period of many years, as did my colleagues at the Kemp Library at East Stroudsburg University. Many other libraries helped me find information about Peirce and provided me with copies of Peirce's correspondence.

Of the Human Heart

1

Hurrah for Young America

IN EARLY AUGUST 1847, A MAN IN HIS LATE THIRTIES BOARDED AN OMNIbus traveling from Cambridge to Boston. He was a good-looking man of just over average height, broad-chested, with a slender figure and an erect bearing. His hair reached to his shoulders and he wore a beard, but no mustache. He was dressed rather peculiarly, wearing striped sack trousers, but his appearance was not that of a haberdasher who typically wore such clothing, but as "a lion out of Judah." He had remarkably clear and piercing eyes; his countenance commanded attention.[1] As he stepped onto the omnibus, he was clutching a letter in his hand. When he recognized his friend and colleague on the Harvard faculty, the classicist Cornelius C. Felton, he called out, "Gauss says I am right."[2]

The man was Benjamin Peirce. On March 16 of the same year, Peirce had announced that the mathematics of the French astronomer, Leverrier, that led to the discovery of the planet Neptune was faulty. Peirce had made this startling announcement before the American Academy of Arts and Sciences, of which he was a member.[3] Most of those who had heard it found it to be reckless, even irresponsible. To them, it had been presumptuous for an American to criticize the work of a leading European scientist. Few of Peirce's friends and colleagues had supported him. Now he had received confirmation from none other than Karl Frederick Gauss, unquestionably the best mathematician of the nineteenth century, perhaps the best mathematician of all time.

$$\Phi^n$$

The discovery of Neptune had been unlike any previous astronomical discovery. Neptune had been discovered by mathematics. On the evening of September 23, 1846, the astronomer J. G. Galle of the Berlin Observatory had merely pointed his telescope to the spot where mathematical calculations had indicated the new planet would be.[4] The computations had been made by Urbain Jean Joseph Leverrier.[5] Over a year before, John Couch Adams had completed similar calculations. Furthermore, using Adams's calculations, James Challis, in Cambridge, England, had actually seen the new planet well before Galle did, but unfortunately had mistaken it for a star.[6]

Both Leverrier and Adams had computed the orbit of Neptune by considering the mathematical irregularities of the orbit of Uranus. If the solar system consisted only of the sun and a single planet, the planet would circle the sun in a perfectly elliptical or oval orbit. In reality, however, nearby planets exert a significant gravitational force on a given planet in addition to that of the sun. These gravitational forces pull the planet out of its perfectly elliptical orbit and create irregularities in the planet's path called perturbations. Leverrier's work was a seminal event in the history of astronomy because it was the first time a mathematical astronomer had successfully calculated the trajectory of an unknown planet by considering the perturbations of another planet. After his success, solving the "inverse perturbation problem" became an important astronomical tool.

News of Leverrier's discovery came to America on the steamship *Caledonia* on October 20, 1846. The next evening William Cranch Bond and his son George, astronomers at the Harvard Observatory, received the news and immediately began to search for the new planet.[7] After searching for two nights, they were confident that they had seen it. Other American observatories also made observations of the new planet. One of them was the Cincinnati Observatory, directed by Peirce's friend, Ormsby MacKnight Mitchel. Upon first seeing the planet, Mitchel's assistant cried out, "There it is! there's the planet! with a disc round, clear, and beautiful as that of Jupiter."[8]

The new planet sparked public as well as scientific interest, not only because of the novelty of a new planet, but because of the verification of mathematical theory by the planet's discovery. Peirce himself forwarded an article written by the Berlin astronomer, Johann Encke, to the *Boston Courier,* praising "the grand discovery of Leverrier."[9] The Washington *Union* extolled Leverrier's achievement as "the unequalled triumph of mathematical science."[10] Other papers were equally enthusiastic about the discovery.[11] In addition, the fierce controversy between Adams and Leverrier over the priority of their discoveries was also avidly followed in the United States.[12]

Once the planet was discovered, astronomers were quick to make calculations of its orbit. The first American to make an important mathematical study of the planet was Sears Cook Walker, who had been given the assignment by Matthew Maury, superintendent of the

Naval Observatory.[13] Walker was perhaps the best astronomer in the country at the time. A Harvard graduate and friend of Peirce, he had made himself well acquainted with German astronomical techniques while working as an actuary for a Philadelphia insurance company. He then became the director of the Philadelphia High School Observatory upon its opening in 1839. Despite the observatory's being connected with a high school, it was the best observatory in the country at the time and the only observatory in the Untied States during the early 1840s whose prime mission was research.[14] Walker's study formed the basis upon which Peirce did his work and upon which he based his later claims.

Only two sightings of a new planet are necessary to compute its orbit, but they must be significantly far apart. Since Neptune has an orbit of 164.79 earth years, astronomers would have had to wait a long time to observe the planet at two places that were far enough apart. To get around this, astronomers searched observatory records to find a previous sighting of Neptune that had not been identified as such. One promising observation was made by Louis François Wartmann of Geneva fifteen years before Neptune was discovered. Wartmann had observed what he thought to be a trans-Uranian planet roughly matching the orbit predicted by Leverrier and Adams. The planet had been 38.8 AU[15] from the sun, the distance predicted by Bode's Law, a sequence of numbers that gave the known planets' distances from the sun. It was derived empirically in the eighteenth century when there were only six known planets and had no theoretical underpinning. Although Bode's Law held considerable weight with astronomers until the discovery of Neptune, it was discredited when it failed to give the correct distances from the sun for both Neptune and Pluto.[16]

Although Edward C. Herrick was the librarian of Yale University, he was an excellent astronomical observer.[17] He suggested to both Peirce and Walker that Wartmann had seen Neptune. But both men were able to show that no orbit could include the positions recorded by Wartmann and the recent positions of the new planet.[18] Peirce wrote to Herrick, "As soon as I received your notice of Wartman's [sic] star, I investigated the observations sufficiently to be able to report to the American Academy . . . that this star could not be the same with Leverrier, alias Neptunus, alias Oceanus; and moreover that it could not be a proper planet moving about the sun in a nearly circular orbit."[19] Peirce concluded facetiously, "do you not think that it proves this star to be one of a new class of celestial bodies, which may perhaps have the same relation to the cluster of stars to which we belong, which the comets have to our solar system? or does it rather show the deficiency of Wartman's (sic) observations."[20] A few weeks later, Peirce wrote Herrick that his friend Agassiz, the Swiss naturalist, had told him that Wartmann had a poor reputation as an observer, "and we must not, therefore, indulge in over sanguine hopes of success in finding the star from his observations."[21]

When Wartmann's star failed to be Neptune, Walker examined old star catalogues and found a likely candidate in Lalande's well-known *Histoire céleste français.* Verification that the sighted star was no longer in the recorded position indicated that the object recorded as a star was probably a planet.[22] Walker wrote to Peirce, referring to the new planet as "Leverrier," in honor of its discoverer, "That the star is missing was a prediction readily fulfilled, that the star was Leverrier is offered as a hypothesis to be sustained or set aside by future discoveries."[23] Now using the recorded location, Walker made new computations which described a planet with eccentricity and mean distance from the sun that were significantly different than those for the planet predicted by Leverrier and Adams.[24]

Walker's work was not appreciated by Maury at the Naval Observatory, however. Maury was unhappy with Walker as an astronomical observer and fired him.[25] Maury, who did his most notable work in oceanography, not astronomy, was considered by Peirce and his scientific friends as a tragic choice to head the Naval Observatory. Peirce lamented to his friend Herrick: "I believe that Walker has just reason to complain of Maury for the miserable garbled manner in which he first presented Walker's labours to the public. It is a sad thing for the national observatory that its head has not training enough in it to appreciate the capacity of Walker."[26]

$$\Phi^n$$

Initially, Peirce had not doubted the validity of Leverrier's computations.[27] But inconsistencies between the description of the actual planet and the proposed planet of Leverrier and Adams began to appear. It was these inconsistencies that led Peirce to his controversial views. For one thing, Leverrier and Adams had both predicted a planet with a highly eccentric or elliptical orbit, but Walker's computations showed that the orbit of the new planet was almost circular.[28]

At a meeting of the American Academy of Arts and Sciences in Boston, Peirce reported on Walker's research on Neptune up to and including the discovery of Lalande's missing star. Although Walker's calculations were based on the fact that Lalande's star and Neptune were one and the same, by no means a certainty, Peirce publicly announced that the orbits calculated by Walker differed so significantly from the predictions of Leverrier and Adams that he had recomputed the orbit himself. He concurred with Walker that the orbit had a mean distance of 30 AU. George Bond made independent calculations, which agreed with those of Walker and Peirce. Peirce concluded that: "From these data, without any hypothesis in regard to the character of the orbit, . . . THE PLANET NEPTUNE IS NOT THE PLANET TO WHICH GEOMETRICAL ANALYSIS HAD DIRECTED THE TELESCOPE; that its orbit is not contained within the limits of space which have been explored by geometers searching for the source of the disturbances of Uranus; and that its discovery by Galle must be regarded as a happy accident."[29]

Peirce criticized two things in Leverrier's solution. First that Neptune's mean distance from the sun must be between 35 and 37.9 AU, and second that the planet's mean longitude on January 1, 1800, must have been between 243° and 252°.[30] Peirce stressed that although neither of these statements taken alone was inconsistent with the observations of Neptune since its discovery, the two combined were totally inconsistent with observation. Peirce did not rule out the possibility that the newly discovered planet may, in fact, account for the perturbations of Uranus, however.[31]

Pointing out that Walker's computations did not match the description of the predicted one was obvious. In reference to the shape of the orbit, rather than its radius, Leverrier, himself, remarked about two weeks later that: "We confine ourselves for the present to the remark that this smallness of the eccentricity, which would result from the calculations of M. Walker, would be incompatible with the nature of the perturbations of the planet of Herschel [Uranus]."[32]

Peirce's claim that the discovery of Neptune was an accident was rash, however, especially since Lalande's star had not been definitely identified as Neptune. Certainly, the members of the American Academy thought so. Edward Everett, president of Harvard, was present at the meeting where Peirce made his announcement. Everett addressed the academy pleading that the members would not allow such an improbable declaration to be issued to the world with the academy's sanction.

"It may be utterly improbable," Peirce retorted, "but one thing is more improbable still, that the law of gravitation and the truth of mathematical formulas should fail."[33] Despite his strong reservations, Everett did forward Peirce's paper to Europe's leading astronomical journal, the *Astronomische Nachrichten*.[34]

Peirce was gratified when the *Astronomische Nachrichten* reprinted his report to the American Academy, and later cited the article to bolster his position in exchanges with Leverrier. Leverrier may not have paid much notice to Peirce's criticisms if they had not been published in this prestigious journal.

Peirce drew criticism from scientists as well as laymen, such as Everett. Walker, himself, wrote, "You must excuse me if I cannot go to the full length of agreeing with you that the discovery of Neptune by Galle was accidental. It seems to me that Leverrier and Adams, though they failed in the limits of the elements, still were successful in the main point of pointing out in advance the present position . . . of the principal disturber of Uranus."[35] And the eminent American physicist, Joseph Henry, confided to Bache, "Were you not surprised by the announcement of the results obtained by Peirce; I fear he has been too hasty."[36]

$$\Phi^n$$

Another of Peirce's critics was Jared Sparks, the Harvard historian (and soon to be Harvard president). He suggested that Peirce "should present to the public a brief statement of the positions you have assumed in regard to the planet Neptune, before the next steamer sails for Europe. The hints that you have thrown out may mislead, & it is extremely important that the first impressions in Europe should be accurate, & founded on such explanations as this subject will admit."[37]

In response to Sparks's advice, Peirce released an abstract of his opinions to the press in order to clarify his position. In this statement he stressed that "I would not have you suppose that I am disposed to contest in the least the greatness of Leverrier's genius. I have studied his writings with infinite delight, and am ready to unite with the whole world in doing homage to him as the first geometer of the age, and as founder of a wholly new department of *Invisible Astronomy*."[38]

Peirce's praising Leverrier confused many of his associates. Mitchel, having just read an account of Peirce's announcement at the American Academy, was "quite astonished." Although Peirce rejected Leverrier's computations, Mitchel complained that "you join the world in his praise and liken him the first geometer of the age. This may all be light but I do not exactly see through it."[39]

Peirce even praised Leverrier at the meeting of the American Academy where he proclaimed that Leverrier's discovery had been a happy accident.[40] Peirce's doing this may simply have reflected his reluctance to criticize such an eminent scientist as Leverrier. But he expressed this dual opinion so pervasively, in private as well as public, that it is hard to attribute his praising Leverrier to diffidence. For example, he wrote to Professor Nichol, "I cannot believe that you so far underrate this masterly portion of Leverrier's investigation, a portion which is the highest glory of his mathematical reputation, as to regard it as detracting from his merit. Leverrier saw very clearly that he should not expect observatories to devote their time to the discovery of a planet unless he could give comparatively narrow limits to the necessary field of research."[41]

More than being merely equivocal, Peirce's paradoxical statements reflect his attitude towards science and mathematics. Mathematics was the work of God, and hence infallible. If Leverrier and Adams had done their mathematics correctly, the observed planet would have the same properties as the predicted one. One could admire a mathematician, such as Leverrier, and praise him for the fine work he had done on the inverse perturbation problem, even if it was only a partial success. But it was Peirce's responsibility as a scientist to point out the truth as he saw it. According to Benjamin Peirce's son Charles, his father and his father's scientific associates did not regard science as " 'systemized knowledge' . . . nor anything set down in a book; but, on the contrary, a mode of life; not knowledge, but the devoted, well considered life pursuit of knowledge; devotion to truth—not as one sees it, for that is no devotion to truth at all, but to party—no, far from that, devotion to the truth that man is not yet able to see but is striving to obtain."[42]

Peirce explained this attitude in a letter to the *National Intelligencer:* "It [Neptune being Leverrier's planet] cannot be included within this theory without so extravagant an extension of his limits, as to destroy all confidence in the elegant analysis, with which he so skillfully contracted them, and which has justly been the wonder and admiration of geometers and the chief glory of his memoir. . . . I am prepared to defend this most ingenious and profound theory, even against its illustrious author provided that forgetful, for a moment, of his true glory, he is ready to abandon the substance for the shadow, and seek popular applause at the expense of sound reputation."[43]

Peirce's former student, the astronomer Benjamin Apthorp Gould, stated the situation more succinctly, "The arguments, which tend to prove that Neptune is the planet of [Leverrier's and Adam's] theory, can only be based upon the supposition of error in that theory, a supposition I am unwilling to admit. Investigations conducted with the care and precision which characterized these must not be so lightly dealt with."[44]

Seven weeks later, on May 4, 1847, Peirce presented a second paper on Neptune to the American Academy in Boston. He now claimed, "The problem of the perturbations of Uranus admits of three solutions, which are decidedly different from each other, and from those of Leverrier and Adams, and equally complete with theirs."[45] Furthermore, in Peirce's mind, Walker's identification of Lalande's star with Neptune was "indisputably confirmed by an examination which Mr. Mauvais of the Paris Observatory had made into the original manuscripts of Lalande, at the request of Leverrier."[46] Mauvais had found that Lalande had observed the planet on May 8 and 10. Since it had changed position during the two days, Lalande had marked it as a doubtful observation. Both Peirce and Mauvais independently made calculations that confirmed that Neptune would have changed position exactly that amount in the two days.[47]

At first Peirce doubted that the irregularities of Uranus's path could all be attributed to Neptune, and that there must be another planet yet to be discovered. This question could not be resolved without knowing Neptune's mass, which could not be determined until a satellite of Neptune was discovered. Although William Lassell, the British amateur astronomer, strongly suspected that he had seen a satellite just a few days after the planet's discovery, he did not verify his suspicion until July 1847. Peirce used Lassell's initial inaccurate data to compute the mass of Neptune. This data gave an erroneous value for the mass for Neptune, which Leverrier severely criticized.[48] Peirce, however, immediately revised his estimate with updated data.[49] Since the computed mass of Neptune was much less than that predicted by Leverrier and Adams, Peirce continued to doubt that the planet was sufficient to account for the perturbations of Uranus. Then after making extensive calculations, he was convinced that Neptune did, in fact, fully account, for Uranus's perturbations.[50]

Peirce had originally based his "happy accident" thesis on two principal things: that the orbit of the planet was much smaller than the one predicted by Leverrier and Adams and that the new planet was unable to account for the perturbations of Uranus. Although one of his suppositions failed, he did not recant his views. As Peirce said: "The argument may now be concentrated into a single sentence. Neptune is five times the earth's [orbital] radius nearer the sun than the geometrical planet was to have been. Neptune's mean distance from the sun is five hundred millions of miles less than it was predicted by geometry that it possibly could be in order to produce the effect which it does produce. This inconsistency between the observed and theoretical mean distances is equal to one sixth part of Neptune's distance from the sun, and to one half the distance between the orbits of Neptune and Uranus."[51]

Φ^n

Although Peirce was a little-known American astronomer and mathematician, Leverrier did not take kindly to Peirce's comments. Ormsby Mitchel had published several articles giving Peirce's point of view in the *Sidereal Messenger,* a journal of popular astronomy, which Mitchel edited. When Leverrier was shown two of these articles, he was understandably annoyed and wrote to Matthew Maury at the Naval Observatory. Maury had Leverrier's letter published in the *National Intelligencer;* Leverrier stated: "I was resolved to be silent on the subject of the strange assertions in America with regard to the theory of Neptune. I cannot, however, permit the two numbers of the *Sidereal Messenger,* with which I was this morning furnished . . . to pass without a few words in reply."[52]

In response to Leverrier's letter, Peirce wrote a final reaffirmation of his position for the *National Intelligencer* and the *Sidereal Messenger.* He insisted that "*M. Le Verrier*" had "not touched the main point at issue." Neptune could not be Leverrier's planet without destroying "the theory to which it was attached." Peirce pointed out that whatever Leverrier might think of his views they "commanded sufficient respect in Europe to be republished at full length in the most eminent astronomical journal in the world, . . . *Schumacher's Astronomische Nachrichten.*" In no way backing down, Peirce declared, "I confidently reiterate my former assertion that '*the planet Neptune is not the planet to which geometrical analysis had directed the telescope.*'" After Pierce defended his computations based on early data regarding Neptune's satellite, he insisted that he was not in collusion with Mitchel. He was not the author of Mitchel's articles and that he had not "even read them, nor any parts of them, except the extracts in M. Leverrier's letter." Peirce reminded Leverrier that "although any foreigner may easily forget the vast extent of the American States and might pardonably overlook the fact that Cincinnati is as far from Cambridge as Paris from Rome."[53]

Leverrier's annoyance with Peirce and Mitchel did not mellow with time. George Bond, Peirce's colleague at the Harvard Observatory, visited Leverrier many years later and found him still outraged by Peirce's criticisms. Bond recorded that Leverrier "spoke of *Neptune* and Professor Peirce. That the remarks of the latter have irritated him to the last degree is plainly evident, and much to be regretted.

His impressions of Professor Peirce's position are, I infer, taken from the *Sidereal Messenger,* as he continually confounds what Professor Peirce has written with what Professor Mitchel has written, always mentioning the two names together."[54]

Φn

Some American scientists continued to be embarrassed by Peirce's challenging Leverrier. For example the Yale geologist and mineralogist, James Dwight Dana, wrote as late as the spring of 1848: "It is to be regarded, as a national calamity, we might almost say, that Prof. Peirce should have been so hasty in his conclusions, or rather in his publication of them. In his prominent position, and in a subject of so much general as well as scientific interest—the planet Neptune, it is bad to have to renounce such grand discoveries as error, especially as he had made himself a critic upon European astronomy. He is undoubtedly a man of great ability, but is too 'quick upon the trigger.' "[55]

Maury sided with Leverrier.[56] But since he was not an astronomer, Maury's actions only reinforced the opinion of Peirce and his scientific friends that Maury was unfit to hold his post as superintendent of the Naval Observatory.[57]

Some American scientists remained mystified, not knowing whom to believe. For instance, W. C. Bond recorded in his diary: "Leverrier holds to his claim as the discoverer, Mr Peirce denys [*sic*] it without either making out a clear case. This could not be if Leverrier and Peirce were both masters of the subject and gave their views and results clearly."[58]

Consequently, many American scientists sought safety in some sort of middle ground. They were pleased to laud Peirce, but reluctant to join him in challenging such a respected scientist as Leverrier. For example, Mitchel supported Peirce in the *Sidereal Messenger,* writing, "Prof Peirce's announcement that the planet Neptune was not, and is not, the planet of Le Verrier's analysis will in all probability be now confirmed."[59] Mitchel privately wrote to Peirce, "Permit me to offer you my most hearty congratulations upon your brilliant discoveries with reference to Neptune. Without flattery I regard your announcement as the boldest that has been made. Even more daring than those of Le Verrier or Adams and from all I can find out more truthful."[60] But in his *The Orbs of Heaven,* Mitchel heaped praise upon Leverrier, observing that, "To many minds, the resolution of such a problem may appear beyond the powers of human genius."[61] He did note Peirce's objections, but evaded taking sides, saying he was confident that "Time and observation will settle the differences of these distinguished geometers; and truth being the grand object of all research, its discovery will be hailed with equal enthusiasm by both of the disputants."[62]

But most American scientists were pleased with Peirce. Peirce's colleague at Harvard, the botanist Asa Gray, felt pride in American science when he wrote to Peirce: "When I read in the Daily, your letter, I was at the point of sitting down to write you a line and tell you that I think it a *perfect gem.* And the most beautiful contrast to the *Johnny Crapeau* vociferations of Le Verrier. I am perfectly charmed with its spirit, and all that I have heard speak of it have taken the same view. I am not alone, therefore, in the opinion that it does you the highest credit and it is just the style of reply calculated to place you at greatest advantage. As one zealous for the highest interests and character of American Science and American Savants, I thank you most sincerely."[63]

Sears Cook Walker had a quick change of heart and sided with Peirce as soon as Walker's computations based on the presumed sighting of Neptune appeared to be highly likely. He wrote to Peirce in May 1847, "I am pleased with your last article on Neptune in the *Boston Courier.*"[64]

Bache, the Superintendent of the Coast Survey, wrote: "I have just learned from S. Walker the very strong confirmation which your views of the "Neptunian theory" have rec'd from the discovery of the [twenty] days satellite, and must write a line of congratulations thereon. Hurrah for young America."[65]

$$\Phi^n$$

Since colonial times, American scientists had bemoaned their inferiority. After the Revolution, scientific and cultural independence became a matter of national pride. Peirce himself expressed this national pride when he wrote to his mentor, Bowditch: "Standing in the land of freedom, you have bounded your views by no narrow limits, but have shown that, though the genius of analysis must open the road to science, this road will remain untraveled if it be not leveled by the finished and masterly labours of the scientific patriot."[66]

Peirce was a scientific patriot. He saw the glory of America not in terms of Manifest Destiny or military might, but in terms of the nation's becoming a world leader in science and education. By the 1840s, many Americans, especially American scientists, were determined to make American science respected. No one worked harder to raise the level of American science during this period than Benjamin Peirce.

Alexander Bache and Joseph Henry were no less scientific patriots than Peirce, and they were particularly anxious that Peirce's statements about Neptune be seen in the best light. The astronomer and meteorologist, Elias Loomis, had consistently supported Peirce's criticisms of Leverrier and eventually wrote an article about the discovery of Neptune that appeared in the *American Journal of Science.*[67] Peirce wrote Loomis telling him that he had read his article "with great delight."[68] But when Loomis wrote a book[69] two years later, Bache and Henry felt that he wasn't sufficiently clear as to who had been right in the dispute between Peirce and Leverrier. They wanted to leave no question as to who was correct. Henry wrote to Loomis: "Your memoir on Astronomy was read sometime in April last by Profr Bache and myself, and . . . on the whole we did not fully accord with you as to what the article should be, and particularly the part relative to the Planet Neptune we thought would require recasting. On account of the delicacy of this part of the subject, and in order to get the opinions of others with reference to it, I requested Dr. B. A. Gould to furnish me with a sketch which would meet the approval of Peirce and the astronomers at Cambridge. This I intended to submit to you . . . to be incorporated in your report but the article was so full that I afterwards concluded to accept it as a separate report."[70]

In the history of Neptune[71] that Henry had Gould write, he, like Peirce, lavished praise on Leverrier. But he left no doubt that Peirce's pronouncement that the discovery of Neptune was a "happy accident" was correct. Gould's supporting Peirce is particularly significant. Gould spent three years in Europe, two of them in Germany, studying astronomy under leading European astronomers, including Friedrich Wilhelm Argelander, Johann F. Encke, Karl Friedrich Gauss, and Christian Schumacher. Gould's knowledge of astronomy and mathematics enabled him to fully understand the dispute over Neptune. Gould saw no reason to be embarrassed by his old teacher, who had first interested him in mathematics and astronomy. Gould praised Peirce's "clear and convincing" arguments, and stated that Peirce's conclusion, "paradoxical as it might at first have appeared to many, was announced with a candor and moral courage only equaled by that of Le Verrier in his original prediction of the planet's place. The reasoning by which Peirce defended his position deserves . . . the most careful consideration."[72]

Despite the controversy involved in the dispute, Peirce had taken on the giants of science and come out intact. To the surprise of the world, including his own scientific countrymen, he had shown that

American scientists and American science could no longer be ignored. Far from becoming a national embarrassment, as Dana had feared, Peirce was now widely believed to be correct in his controversy with Leverrier and had become a national hero.[73] Whether or not Peirce had been right, Americans generally believed he was right.[74] A decade after the incident one newspaper writer praised him: "This [Prof. Peirce] is the great American Geometer . . . who, while England and France disputed for the honor of the discovery of the new planet Neptune, stepped in and bore away the crown from both nations; the man who while poets, statesmen, and kings were striving to out do each other in honoring Le Verrrier had the courage to declare and the ability to demonstrate that the planet actually discovered was not the one computed either by Le Verrier or by Adams. As a mathematician Prof. Peirce has no living superiors and his authority has been repeatedly sought as final on the highest questions in that department."[75]

But Peirce had managed to do more than impress partisan American scientists, although as we have seen even this was difficult. In addition to Gauss, at least two other European scientists took a stance similar to Peirce's. The French physicist, Jacques Babinet, proposed searching for another planet before the Académie des Sciences in Paris, which sparked a heated response from Leverrier.[76] And the British astronomer, George Biddell Airy, remarked that " 'much was owing to chance' in the discovery of Neptune."[77]

More importantly, beyond the question of whether or not the discovery of Neptune was a happy accident, Peirce's remarkably accurate work on the perturbations of Neptune and Uranus won him international respect.[78] In 1850, four years after Leverrier's announcement of the discovery of Neptune, Peirce was elected an associate member of the Royal Astronomical Society of London. And in 1852 he was elected to the Royal Society of London.[79] Peirce was the first American to be elected since the election of his mentor, Nathaniel Bowditch in 1818, the two of them being the only members elected since colonial times. This was an honor that eluded such eminent American scientists as Alexander Dallas Bache, the superintendent of the Coast Survey, and Joseph Henry, the first secretary of the Smithsonian. Despite the international recognition that Henry had gained for his work in electricity and magnetism, the Royal Society turned him down on each of the two occasions that it considered him for membership.[80]

2

The Father of American Geometry

Benjamin Peirce was born into a wealthy Salem family on April 4, 1809. At the time of Peirce's birth, Salem was not a sleepy suburb of Boston, but a bustling seaport engaged in worldwide commerce. Boston itself was only a small city of sixty thousand. Salem's inhabitants were wealthy and well educated. Its Philosophical Library Company boasted one of the best scientific libraries in the country.[1] It was also home to one of the best mathematicians in the country, Nathaniel Bowditch, who was to have a profound influence on young Benjamin.

Jerathmiel Peirce, Benjamin's paternal grandfather, began his career as a leather merchant, but became involved in the East India trade, in which he made a substantial fortune. Benjamin's other grandfather, Ichabod Nichols, was the captain of several successful voyages to the East Indies, although in his later years he abandoned the sea to become a farmer. The two grandfathers became brothers-in-law; Jerathmiel marrying Sarah Ropes, and Ichabod marrying her sister Lydia.[2]

Benjamin's parents, Benjamin Peirce Senior,[3] and Lydia Ropes Nichols, were cousins. They began courting sometime before Benjamin Senior left to go to Harvard. He must have cut quite a figure. One of Lydia's friends described him: "I have <u>at length seen</u> & <u>only seen votre bon ami,</u> he is certainly a very handsome man as to form, his face I could not <u>discover,</u> though <u>that,</u> is of no consequence, <u>grace, symmetry,</u> & <u>height</u> are certainly <u>enough,</u> & if this <u>francois</u> has as captivating a <u>mind</u> as <u>person</u>—good heavens preserve me!"[4]

Upon arriving in Cambridge Benjamin Senior wrote Lydia: "In beginning a letter to you I presume, I need not make any ceremonies. Our near intimacy does or ought to forbid all stiff & formal introductions. It is true, this is the first of my ever writing to you; but it is not the first of our *conversations*—and what are letters among friends but conversations? Friend, I say & you'll indulge me in that expression for I assure you I feel a *secret* pleasure in the name. I am not so wedded to Cambridge as not to cast my thoughts home sometimes;

and the pleasure I take in *your* company hold not the least part of them."[5]

Lydia responded most congenially: "You must know I am an enemy to what the world calls fine letters, so you have not anything to expect from me but honest plain 'How do ye's. . . . Your epithet of 'friend' does so perfectly accord with my own feelings that it could not fail but of being agreeable to me."[6]

Their letters reflected this same mutual warmth and affection for some time, although Benjamin Senior soon recognized Lydia's propensity to take offense where none was intended. He confided to his sister, Sarah: "I wrote a letter to Lydia sometime ago, but something or other has prevented an answer. Her creative imagination, I suppose, has found something in the letter offensive. . . . I wish it had been in my power to have kept up a correspondence with Lydia. Her qualifications are such that I flattered myself with the hopes of deriving much pleasure & information from her."[7]

Benjamin Senior and Lydia did continue their correspondence, but Lydia's fiery nature, which matched her red hair, soon surfaced, and was perhaps a source of her son's quick temper and excessive sensibility. Lydia became vexed with Benjamin Senior for paying too much attention to his studies and too little attention to her. From that point on, their courtship was a rocky one. She wrote to Benjamin Senior:

> I am not going to be accountable for either the grammar or goodness of this letter. My head is indisposed, & I have no ideas, yet I must and will thank you for your letter, it being what I did not expect; that you would leave your *philosophy*, rather *exert* it, to have written again this term; Yet you say you like to write me, and I believe it, for it gives me pleasure to think so, and as the sources of happiness are so few in this World, should we not make the most of them?—Why must you Benjamin stay at Cambridge the next vacation, you will not be any wiser for it—I don't think—I would not tarry if I were you, you will enjoy yourself almost as well at Salem as there. We will exert ourselves to make you pass your time agreeably, and I am afraid you will lose your health by attending so closely to your studies, but just as you please Mr. Philosopher. I must bear the disappointment, and learn to be a Stoic *likewise*. I'll not have any friends, for they have caused me more pain than pleasure. Adieu. I feel very sick, and cannot tell you anything about my Latin, only that it is tedious stuff!!
> "I thank you Student for
> the little you have given
> me—Farewell my Cousin—Lydia N.[8]

At his father's insistence, Benjamin Senior did go home that vacation, but his romance with Lydia cooled off.[9] In part because, at this

time, Lydia acquired another suitor, Joseph Story. Story had graduated from Harvard in 1798, had read the law in his native town of
Marblehead, and had then come to Salem to practice law. Lydia's
family did not approve of him; he was a Republican (later called
Democrat) and a Unitarian. Although Lydia was attracted to Story,
she discouraged marriage:

> Tremblingly alive to the enchanting sound of friendship, I have never
> sighed for the more refined enjoyments of love. But why do I hesitate to
> declare I am acquainted with love. The importance of matrimony may
> prevent my wishing or consenting, to be engaged to you by any other ties
> than those of amity—But so necessary is your society to my felicity, I am
> convinced my heart is more than simply interested—Can we never love
> Henry [Joseph Story] unless we agree to be united in the bands of Hymen?
> A thousand circumstances may conspire to prevent such a union. . . . Mar-
> riage it is said is the grave of love. . . .
>
> You hinted in a former letter that interest could have some influence
> in my decisions—know my friend I possess a soul that is above pecuniary
> motives—were you master of millions you would not be more estimable
> to my view than at present—so infinitely does your merit eclipse property,
> it is impossible to consider that one when reflecting on the other—you
> have talents Henry for acquiring a living & that is all I seek for in the man
> of my choice—Injure me not again by so degrading an observation—it
> was almost sufficient to have destroyed my growing affection.[10]

Lydia added that she could get lost in Story's conversation and forget others, "But this is not love my friend for when in company with
Miss Sawyer I have frequently experienced the same feelings."[11]

Despite Lydia's reluctance to become engaged, her conservative
father feared that his daughter might marry Story. In his sight, Story
was an infidel and a radical, certainly not a suitable mate for his
daughter:[12]

> With respect to the difficulties between you and cousin Benjamin, your
> mama informs me that there was such, but if I rightly understood her, I
> considered them not of a weighty kind. If I am not mistaken in your
> cousin, he is a man of virtue and honor of strong & good natural capac-
> ity, and with those qualities united with a good education I can not doubt
> but that he would make any woman happy who may be a partner of his
> choice, with whom he may be united. With respect to Mr. Story I must con-
> fess that I have nothing of him, but this I will observe, that I believe it is
> very rare that the great are known to talk as much as he does, and who
> show themselves to be so much delighted with their own conversation. . . .
> as a kind friend and Parent I must my dear request you not to hastily com-

mit yourself to him, or any one else, but to use your Dear Brothers (sic) words 'to stand a while at ease and support your own disarmament. In a concern of so much magnitude as that of a Partner for Life, for one false step might make you unhappy to the end of time, and perhaps to eternity. May you not on further consideration palliate, in some measure, Cousin Benjamin's conduct when you consider he is a student in College. No kind of property at his command, at present, and if I am not much mistaken he is naturally of a diffident make. When I consider these things I can in some measure excuse him, for not naming his intentions to me, as I have not the least doubt but that they are Honorable—let me advise that you may not come to a fixed conclusion until you see Cousin Benjamin once more, for I am fully persuaded that he has an affection for you, and will be disposed if he has the opportunity to make you happy.[13]

To her father's relief, Lydia still had no intention of rushing into a marriage with Story. She responded: "As to Cousin Benjamin I will say with you, he is a man of Virtue and Honour. But he discovered a degree of indifference toward me in not being willing to make the small sacrifice of his feelings I requested him to that I think rendered separation necessary. . . . Mama misunderstood in supposing I wished to give Story a decisive answer at present, I am not sufficiently acquainted with his character myself."[14]

When Story asked Ichabod for his daughter's hand, he did not refuse him, but did not give his consent either.[15] Ichabod then wrote to Benjamin inquiring if he and Lydia were "under the least engagement to each other."[16] Benjamin replied immediately, showing that he did, indeed, have feelings for Lydia, despite her protestations of his indifference:

I do not, sir, consider Lydia under any engagements to me; but I consider myself engaged to her. That is, Lydia is at liberty to do with herself just as she pleases; and if at any future time she should say that she is perfectly satisfied with me, I shall be at her disposal. Lydia, sir, will, as it respects me offend against no principle of honor, or virtue, by marrying any person whatever; but a positive promise deprives me of that liberty. This may appear to you singular, Sir, but I have searched my own mind, I have found it to be a true statement.

Considering our mutual fondness (for of this I am certain) I have but little doubt but that at some future period we should, if agreeable to you & aunt, be engaged to each other.[17]

Shortly after this, Benjamin Senior was chosen to be the valedictorian of his class at Harvard. In this he bettered Story, his rival in love, who finished second in his class.[18] Lydia, who had sworn off writing

Benjamin, relented for this occasion: "Though I did say I should not write to you this term I cannot help breaking my resolution to congratulate you on having the highest part at Commencement. . . . I am far from being happy. Few things can interest me and still fewer interest me agreeably. I endeavor to be happy because . . . I am not formed to bear the mortifications of this life with indifference. Do come home as soon as you can. I long to see you."[19]

Despite the mutual affection expressed in the above letters, Lydia and Benjamin broke off their courtship. Benjamin, now fearing that Lydia's affections were with Story, and that she consented to marry him only to please her parents, refused to enter into a marriage or even an engagement. He wrote:

> I must then relinquish the object of my fondest attachment! But necessity demands it; and whatever pains it may cost me, I will not with my eyes open rush into a situation, which, I foresee, would be an unhappy one both to myself, and her whom I love more tenderly than any other human being.—Yes, Lydia, we will always love one another; we will always be friends; we will see one another, are you afraid of such intercourse?
>
> Oh Lydia I shall always love you too well! Let us be friends in the future, with out being lovers—I shall visit your house as usual and expect you will meet me with the smile of familiarity. . . .
>
> But though, Lydia, I have bid adieu to Cheerfulness; I am determined that my spirits shall not be breaking, that my energies shall not be repressed.[20]

An irate Lydia responded:

> Perhaps you can account for your actions on very <u>prudent</u> reasons. I must confess however let them be ever so praise worthy[,] the indifference they discover is by no [means] consistent with my wishes. I write now to prevent you from taking the trouble of visiting me to morrow evening as I am engaged. I shall also be engaged every evening this long time so it is quite unnecessary to attempt seeing me.
>
> You are ever making me miserable. You have been the greatest cause of my unhappiness—and my mind is no longer equal to support your caprice. I write with warmth. I feel so—and have reason to. If I ever marry it shall be from ambition—and then want of it in my friends. I know my character entitles me to a return of affection and I will not so far descend from the dignity of my situation, as to love a man—who feels not, what love is.[21]

To which Benjamin protested:

You observe the other day, Lydia, that we had not sufficient confidence in each other—How is it possible that I should have confidence in you? I am convinced that I am not the person whom you love. Nor is it in your power to love me. Another object has possession of your heart. I will not be exploited with any longer. My situation is not only painful but mortifying. I am therefore determined to adhere to the agreement we made the other morning. I will not give you an opportunity of being connected with me from motives of convenience or pity.

I therefore declare to you that it is my sincere wish that we treat one another in future with no more attention than to avoid notice both from friends and acquaintances—if this does not succeed I am determined I will not visit at your house—If this does not succeed, I am determined I will cross the ocean, and continue to India and Europe, till I have recovered complete possession of my heart.[22]

Benjamin's questioning the sincerity of Lydia's affection for him incensed her further; she responded:

Though I was prepared to receive an unkind letter—yours still surprised me—I will be silent on the subject of my interest in Mr. Story. You know what I have formerly said and the strongest proof I ever gave you of confidence was in telling you my real situation. I have perceived too late there are no friendships in this world so perfect as will in every instance admit of undisguised frankness. Hypocrisy must be practiced or social feelings must be given up—As you wish me to assist you in keeping to your agreement I cannot consistently with the present difficulties. I will only make a few observations on your letter and then be agreed to act any part you say.
   If I love Mr. Story—and should make myself yours in compliance with the requests of my parents—is it not going too far in virtue to sacrifice a life of happiness to a sense of filial obedience? But if I even erred on this hand, is it a weakness that would not rob me of the esteem of any just person—and had my heart ought to be called deceitful for doing what will make no one miserable but myself? . . .

I feel degraded you who have been so long an intimate friend, should have formed so contemptible an opinion of my heart. "Love cannot exist without esteem" has ever been my maxim, so you will not now be under the necessity according to this to cross the water to gain your complete possessions of your heart. . . .

If called upon by my father I request of you the liberty to inform him it is your proposal the connection should be forever dissolved. We cannot be friends—we cannot converse—a passing word shall and must be our only intercourse—if more than this I cannot be accountable for discovering my regard for you in assisting you to keep to your resolution.[23]

It took some time for love to run its true course. Lydia did not break things off with Story for another year.[24] Lydia and Benjamin married a year later, in 1803.[25] Benjamin went into his father's thriving shipping business, and he and Lydia shared a large house with another pair of cousins who had married, Lydia's brother George and Benjamin's sister, Sarah.[26]

The spurned Joseph Story had a distinguished legal career. After serving in the Massachusetts state legislature, he was made a justice of the United States Supreme Court at the age of thirty-four, the youngest person who had been appointed to the court at that time. The energetic and ambitious Story was not content to merely sit on the nation's highest court, but wrote extensively, and in 1829 joined the faculty of the Harvard Law School, an institution upon which he had a profound effect.[27] Benjamin and Lydia had moved to Cambridge by this time, where the three of them renewed their friendship.[28]

Lydia's first choice was not lacking in political activity, however. For one thing he could give a stirring political speech, as he did one Fourth of July: "An elegant and patriotic oration was . . . pronounced by the Hon. Benjamin Peirce; in which he took a brief view of our public affairs under the former and present administrations. The mind of the hearer was, by a clear and forcible display of the leading measures of the several administrations, irrefutably led to the conclusion, which it was apparent the Orator had formed in his own mind, that the unexampled prosperity which our country experienced under Washington was owing to an inviolable adherence to that policy which we are now proud to call the Washington Policy; and that our present degraded and miserable condition has been the natural consequences of a willful departure from Washington's principles, and unaccountable attachment to a policy of foreign growth."[29] More significantly, Benjamin Senior served in the Massachusetts Senate at an unusually young age and later served in the Massachusetts House of Representatives.[30]

Φ^n

The Peirces' impressive intellectual tradition was certainly alive with the generation of Benjamin's parents. In addition to finishing first in his class at Harvard, Benjamin Senior expressed his love for knowledge in a letter to his sister, Sarah: "The field of science is of immense extent—the most we can gain is but a little corner in it. Yet one who has once attain'd to that little corner, has the whole open (as it were) before him, & enjoys a free air, unconfined within the narrow limits, which ignorance (with its usual attendants), vanity, self-conceit & a thousand foolish impertinences would necessarily prescribe."[31]

Scholarly excellence wasn't limited to Benjamin Senior alone. His sister Elizabeth wrote her brother that she was almost always the highest in school.[32] At ten she wrote a charming letter, which reflected her intellectual curiosity: "The silkworm is a curious animal. The thread that one of them spins is said to be 300 yards long. I used to think people killed the silkworms and took the skins off them to make silk of. Sally has been reading about the Zoophites. She says that people doubt whether they are animals or vegetables. They may be divided in two or three parts, some cut in a thousand parts, turned inside out & yet in time they will become perfect animals."[33]

$$\Phi^n$$

In addition to the Peirces' high regard for learning and scholarship, a family tradition of feminism goes back at least as far as Benjamin Senior and Lydia. They exchanged feminist views early in their correspondence.[34] Lydia wrote to Benjamin: "I regret that custom does not authorize an education equally liberal to both sexes & were I to allow myself to reflect one moment on that truly unhappy circumstance, it would forever prevent my putting pen to paper & as you will inevitably make a comparison between your own & my compositions I hope you will make those deductions which I have reason to expect from your candour."[35]

Benjamin was sympathetic to Lydia's feminist viewpoint; he responded: "I presume you did not intend to class yourself with the generality of your sex, and I know if you were 'to allow yourself to reflect one moment' you would discover yourself too well to think so—Therefore I cannot account for your not 'putting pen to paper' but from your compassion to others; tho' I think this ought to be rather a stimulus for you to retrieve the honor of your sex—I am sensible of the neglect in their education in general, and heartily join with you in regretting it."[36]

Lydia's feminist views were much broader than equality of education. She expressed them in another letter to Benjamin Senior: "The world has established rules of right & wrong[.] [W]e cannot deviate from them without incurring its censure & ridicule; but must I because you are of different sex deny myself the gratification of saying I esteem your virtues? Must I practice that cautious & affected reserve to gain their respect? In short my good friend must I play the coquette to continually remind you of your superiority?—No, my soul despises this servility & want of independence, though a delicacy inherent in the mind renders it unwilling to disregard the established opinion & customs of Mankind: yet I think my deviation is allowable in this instance."[37]

During this same time, Benjamin wrote from college, encouraging his sister's intellectual growth: "I hope your pleasures are equal to your <u>industry,</u> and that a little more of this is given to books, & less to drudgery. At least don't keep Betsy so light at it. It is impossible that study & reading can be estimated too highly, if one wishes to make any figure in society, or, what is more, to enjoy pleasures <u>substantial</u> & <u>refined.</u>"[38] Benjamin then noted that he was sending his sister five shirts, four collars, two pocket-handkerchiefs, two napkins, and three pairs of stockings, which Sarah was to launder, and "which you'll please to send next Saturday—by all means! Send also that other pair of stockings."[39]

When Sarah found her brother's advice, not his laundry, offensive, he wrote again:

> I am sorry that any thing should escape my pen, which might need an apology. The word, drudgery, I am sensible, sounds too harsh, and was really misapplied. But you ought to consider the state of my mind at that instant, that it was warmed with an <u>honest indignation</u> (if I may be allowed the expression) at what ever should oppose itself to the improvement of the most <u>valuable</u> part of the most <u>amiable</u> of sisters. And you know, Sally, that needle work & every part of housewifery contribute nothing to the cultivation of the mind. These are certainly laudable & should be attended to by every Miss whatever. Yes "not always in the parlour, but sometimes in the kitchen"—is one of the maxims in common life. One who has to "earn her bread by the sweat of her brow," is <u>obliged</u> to give herself up wholly to those manual employments; but one, whose situation in life is easy, should give at least a part of her attention to the refinement of study. What can be a more delightful object than a lady, with an improved taste, good understanding and amiable disposition, filling her particular sphere of life with that <u>dignity</u> which proceeds from a mind stocked with noble & exalted sentiments? While I beg your pardon for any offence I may have given you, I must repeat my request that you would devote considerable of your time every day, to reading. Words would fail me to express to you the desire I have to see you enriched with learning, and become the honour of your sex. Nature has formed you, my amiable sister, with every quality which you need to beautify your person—you only want "The skillful cultivator's fostering hand."[40]

Lydia's feminist viewpoint was influenced by her friend Mrs. A. D. Rogers, and a group of young women that met together for intellectual stimulation. Sometimes they discussed the eighteenth-century

English feminist, Mary Wollstonecraft. Wollstonecraft is best known for her book, *A Vindication of the Rights of Women,* which first appeared in 1792.[41] Her work was original, provocative, and popular. After becoming pregnant with his child, she married William Goodwin, the leading British radical of his time. She died as a result of childbirth in 1797, at thirty-six. Her daughter,[42] Mary Goodwin, survived, however, to marry the poet and radical, Percy Shelley, and to write the novel *Frankenstein.*

Soon after Wollstonecraft's death, her husband wrote a memoir of her, which included such details of her personal life as her having lovers and her twice attempting suicide. Although several of her books were popular, *A Vindication of the Rights of Women* was especially so, immediately going into a second edition in England; American, Paris, and Dublin editions soon followed. Initially her publications were fairly well received, since people had been clamoring for women's education for at least a century. But they were soon vehemently attacked in England, France, and America. Those who opposed her ideas often pointed to what they considered a dissolute life as evidence of the folly of her opinions.

Mary Wollstonecraft's ideas were also attacked in a Boston newspaper, the *Columbia Centinel.* This assault began with a column ironically named Latitudinarian, and was followed by a series of letters and additional articles that blossomed into a full-fledged newspaper war.[43] Although the articles began with a general discussion of the lot of women in the United States and the pernicious ideas of Wollstonecraft and her husband, Goodwin, a letter soon appeared targeting the wayward young ladies in Salem who met to discuss Wollstonecraft's ideas:

> Trusting to your candor and impartiality, *a friend to her sex* begs leave to offer to your attention a subject for your next number. There is a society of Young Ladies residing not more than *twenty miles* from *Boston,* who throwing aside all *that delicacy* which is *so natural and essential* to the female character, have presumed to come forth in defiance of order and authority and boldly assume what *they stile*[44] the *"rights of women."* Led on by those pernicious principles they dress out vice in the most alluring colours; and would fein obliterate forever the fair form of virtue. Not content with the corruption of their own minds, they spare no enticement—they regard no pains which can possibly tend to contaminate those who are so unfortunate as to fall in their way. And with their *satanic arts* they but too often succeed. They think that every girl who is not furnished with the ideas from *Mary Wollstonecraft* is wanting in understanding—and is destitute of every requisite that should entitle them to an admission into their *society;* which they denominate the *"Social Group"*—the *"Moschetto Fleet,"* and sev-

eral other imaginations equally ridiculous and absurd. Point out to them how destructive the sentiments which they possess are to the well being of society. Ask them if after reading the life of *Miss Wollstonecraft* they did not blush to own she was a woman.[45]

After reading this article, Benjamin Senior wrote Lydia, assuring her that "a friend to her sex" had overstated her case, and done little damage to the young women of Salem: "Malice never inhabited a soul more worthy of her residence. But such an outrageous piece must, I conceive, prove beneficial to the objects of it. A moderate one might have gained some degree of credit; but no one will be stupid enough to believe this. What resources then have your enemies? It is too late to go back; and surely they won't have the effrontery to repeat such a violent attack. What an instance of the blindness of zeal! . . . Hathorn, I am told, wrote 'One of Many.' I am glad of it, for he certainly can't be esteemed in Salem now less, than he deserves."[46]

A few days later a letter signed "One of Many" defended the young women: "For the information of those, who are unacquainted with the characters, and families of these ladies, I declare that the charges are *false, malicious,* and *without foundation.* There is not a town in the Union, where religion and morality are more respected by the old, and enforced on the young, than in *Salem;* where these precepts are more regarded; nor where the female sex are more improved, pure and unblemished."[47]

Several people suspected "One of Many" to be Joseph Story, and the author of the Latitudinarian referred to the author as a "*pretended lover*" of one of the young ladies. Story wrote to the *Centinel* denying authorship, although agreeing with the author's sentiments.[48]

The response of "One of Many" did not impress the author of the Latitudinarian. He wrote another scathing article, pointing out that friends of the worst of people will praise their associates' virtue. He included another anti-Wollstonecraft letter.[49] This column was followed by a similar one a few days later.[50]

Lydia was devastated by the attack; she wrote to Benjamin:

> What am I to do in this situation! Must I sink under these calumnies or endeavor to support them with all the philosophy of a Wollstonecraftian? The latter pride enables me to wear the appearance of but the suffering, I endure in secret, plainly showing my enemies have gained their point, in giving a mortal blow to my happiness—Oh cousin this is indeed more than I am able to endure with fortitude—to be in a moment thus unexpectedly robbed of a reputation—which was far dearer than life, friends, or fortune—to gain which I would have sacrificed everything—but principle and integrity.

I am now truly miserable. . . . Yet I have one comfort. That religion they accuse me of being destitute of, consoles me with an assurance this is only a state of probation. . . .

And do you think Peirce—my feelings would permit me to consider myself the first object of your heart now? NO—Know I should despise my conduct, could I consent to become the partner of a man, if the world thought me his inferior. It is a tax I can never lay on my friends, to ask them for their society in misery—or disgrace. So until my reputation is again established, on the slender basis of public opinion you must consider me only as a common acquaintance.

I am extremely sorry any notice was taken by the pen of the friend of her sex—the author of the Latitudinarian—is an obstinate unprincipled fellow and even was he convinced of his error he would sooner sacrifice the Universe than confess it. And to have my character or rather the circle of ladies with whom I associate made the subject of a Newspaper war is unfortunate were we all as free from error as purity's self. The reputation of a woman is too sacred and too delicate to be offered to the eyes of a curious prejudiced world. We all in common with the rest of mankind have our imprudences—what the uncandid and envious can misrepresent as the greatest crimes. You know me as I really am. If it is irreligion and Wolstonecraftian to cultivate our minds and innocently enjoy the blessings of Providence—I am one—But as to wives ruling their husband with a rod of iron. I have no idea of it.[51]

In another letter, Lydia complained to Benjamin: "You think I view these attacks in too serious a light. Indeed I do not—I rather think from some reports on the situation, I quite err on the other hand. From their first appearance they have deeply affected my sensibility. I did not suppose they would be generally credited—but the accounts which we have from Boston—inform us—we are not only as bad as there represented— but we are considered as characters so *vile*, we can not sink lower—and this to girls whose ambition has taught them to aspire to immortality in fame—is a stroke sufficient to destroy all happiness."[52]

Despite the pain of being the object of an attack in a newspaper war, Lydia enjoyed her association with Mrs. Rogers and her husband. She wrote, "frequently when conversing with her [Mrs. Rogers] & Mr. Rogers they discover so much wisdom that I imagine myself in company with all the great men & women of the age."[53]

$$\Phi^n$$

Thus Lydia was an important contributor to the distinguished intellectual tradition of the Peirce family, although its most evident beginning may be Benjamin Peirce Senior's graduating from Harvard in 1801 as valedictorian. He wrote his senior thesis in mathematics, as his son would twenty-eight years later.[54] Benjamin Peirce Junior also had several uncles who distinguished themselves as scholars, most notably Ichabod Nichols, his mother's brother. Ichabod, who was in the same class at Harvard as Benjamin Senior, was a tutor in mathematics at Harvard before becoming a minister. He had the reputation of being an eminent mathematician.[55]

$$\Phi^n$$

Benjamin Senior shared a large house with Lydia's brother George, and his wife Sarah, who was Benjamin Senior's sister. Benjamin and Lydia had four children, of whom Benjamin was the third.[56] Young Benjamin was sent to the Salem Grammar School, where he was under the tutelage of John Walsh, an accomplished scholar, whose father had written a popular arithmetic text.[57] One of Benjamin's classmates, Ingersoll Bowditch, showed Benjamin a solution to a problem the boys had been working on, which Ingersoll's father, Nathaniel, had prepared. When the elder Bowditch, then one of the two best mathematicians in the country,[58] was told that Benjamin had found a mistake in his solution, he said that he would like to meet the young man who corrected his mathematics.[59] Bowditch became Peirce's mentor, a debt that Peirce repeatedly acknowledged. For example, he dedicated his book on mechanics: "To the cherished and revered memory of my master in science, Nathaniel Bowditch, the father of American geometry."[60]

This same title was given to Peirce, himself. Sir William Thomson (Lord Kelvin) referred to Benjamin Peirce as "the Founder of Higher Mathematics in America," and Arthur Cayley called Peirce "the Father of American Mathematics."[61]

As the time neared to send Benjamin to college, his mother fretted about sending him away to boarding school; the newly founded Round Hill sounded attractive, but it would only take boys under twelve and Benjamin was now fourteen. Benjamin's father favored placing him with "Mr. Putnam," in North Andover, Massachusetts.[62] So in the fall of 1824, Benjamin's father took him there to attend the Franklin Academy. Despite Mr. Putnam's reputation for being a tyrant, young Benjamin quickly reported that he was content with the school.[63]

Benjamin's father wrote to him at North Andover, already recognizing that sometimes his son's feelings were too strong and his temper too hot: "I cannot help writing a few lines to repeat to you my most earnest request that you will take good care of your <u>health</u> &

your <u>habits;</u>—your health & habits, not only of body, but of mind and heart. Govern your thoughts, your feelings, your temper, your passions. Aim perpetually at advancement in that acquisition, which includes all valuable acquisitions, 'mens sasa in corpes sano.' Verbem sat."[64]

Peirce entered Harvard in 1825 at sixteen years of age. Writing to his wife, Benjamin's father discussed young Benjamin's prospects at college: "If Ben will keep himself cool & not be impatient nor too ambitious, I have no doubt he will get along very well. . . . I can't think he will find it very difficult. Nobody will expect so much from him now, as they will 10 years hence."[65]

$$\Phi^n$$

Benjamin's mathematical attainments were already impressive by the time he entered Harvard. Andrew Peabody, a senior when Peirce became a freshman, and who shared, with Peirce, the duties of teaching mathematics at Harvard in 1832–33, mentioned that "even in our senior year we listened, not without wonder, to the reports that came up to our elevated platform of this wonderful freshman, who was going to carry off the highest mathematical honors of the University."[66]

Once at Harvard, Peirce did not lose his reputation for mathematical excellence. One of his classmates, James Freeman Clarke, recalled that: "Each class had one day a week in which to take books from the college Library; and I recollect that Peirce, instead of selecting novels, poetry, history, biography, or travels, as most of us did, brought back under his arm large quarto volumes of pure mathematics. When we came to recite in the Calculus or Conic Sections, it was observed that the tutor never put any question to Peirce, but having set him going, let him talk as long as he chose without interruption. It was shrewdly suspected that this was done from fear lest the respective *rôles* should be reversed, and the examiner might become the examinee."[67]

The mathematics professor at Harvard at this time, John Farrar, was one of the best teachers of his time.[68] George Emerson, who graduated from Harvard in 1817, and was subsequently a mathematics tutor there, praised Farrar, "whose lectures on natural philosophy and astronomy I have never known surpassed or equaled. I have seen day after day, a whole class so charmed by one of his lectures as to forget the approach of Commons hour, and to leave with reluctance, to go to dinner, though the lecture had gone on more than half an hour beyond the time allotted to it."[69]

In addition to being a popular instructor, Farrar was the first American to translate French mathematical texts for the use of American students.[70] The first of Farrar's books, *An Introduction to the Elements of Algebra,* was published in 1818. Farrar translated seven other mathematical textbooks, and additional textbooks in physics and astronomy.[71] This was an important contribution to mathematical education in the United States, for these books were far superior to the English texts they replaced. Yet despite the relatively favorable position of Harvard in teaching mathematics, Bowditch's guidance, not Farrar's instruction, was crucial to Peirce's mathematical education. Farrar probably knew little mathematics beyond the scope of his texts.[72] Furthermore, the largely prescribed curriculum at Harvard offered little opportunity for specialization in mathematics.

Nathaniel Bowditch[73] grew up in a poor Salem family, and, as a boy, was apprenticed as a clerk to a ship-chandler. As a young man he went to sea, soon mastering the elements of seamanship and the principles of navigation. Bowditch had a much broader interest in mathematics than its applications to practical navigation. He taught himself Latin in order to read Newton's *Principia;* he also learned French and German in order to read mathematical works in those languages. Bowditch completed five voyages, one to Europe and four to the East Indies. On his last voyage he was master and part owner. After a successful career as a sailor, he became an insurance company executive and then an actuary. In 1806 he turned down the professorship in mathematics and natural philosophy at Harvard. He also declined academic positions at West Point and the University of Virginia.

Bowditch did two things that were truly remarkable mathematically. One was his writing of *The American Practical Navigator.* The other was his translating Laplace's *Traite de mécanique céleste* into English. When Bowditch first went to sea, navigation tables were notoriously inaccurate; these inaccuracies had caused shipwrecks. Bowditch began to correct the tables in John Hamilton Moore's *The Practical Navigator.* After issuing several editions under Moore's name, he felt he had changed the work sufficiently to warrant issuing it under his own name and under a new title, *The American Practical Navigator.* Bowditch's *Navigator,* published in 1802, explained the fundamentals of navigation in a way that was highly accessible to the typical seaman. For example, assuming a knowledge of only addition and subtraction, Bowditch gave instructions on how to use log tables for multiplication and division, thus making mathematical computation accessible to seamen who had not learned their times and division tables.

Computing accurate tables for navigation is not a thrilling mathematical accomplishment; indeed, it is a task of tedious drudgery. It was an extremely important thing to do, and no one else had done it. Bowditch made scientific navigation a reality. Sailors all over the world used a "Bowditch." Revisions of *The American Practical Navigator* are still published under Bowditch's name.

Bowditch's translating the *Mécanique céleste* into English, his other significant mathematical accomplishment, required a far greater mathematical knowledge and ability. Bowditch provided much more than a translation. He added copious footnotes, which included demonstrations that Laplace skimmed over. There were few people alive in Bowditch's day who had the mathematical expertise to understand Laplace's work. While still an undergraduate, Peirce read the proof sheets of Bowditch's translation of the *Mécanique céleste,* which would require considerably more mathematical knowledge than he could have gleaned from his Harvard classes.[74] Indeed, Peirce may have known more mathematics as an undergraduate than Farrar himself.[75]

Peirce remains enigmatically quiet about Farrar, never mentioning that he benefited by his instruction.[76] He never explicitly criticized him either, although he did give a scathing recounting of his undergraduate experience studying mathematics at Harvard:

> When I was in college and in one of the lower classes, I was asked to furnish a mathematical question to a college periodical. I gave one, simple enough to excite the mirth of any one who has recently graduated with mathematical honors, so simple indeed that I was prevented by shame from answering it myself. One of the higher classes . . . complained to me of its extraordinary difficulty, and it remains an unanswered and an unanswerable indication of the science of the college. About the same period, I was called upon by the instructor to explain the spinning of a top. I replied that I did not understand the explanation given in the mechanics which we then studied, & was therefore most unmercifully screwed by the teacher & laughed at by the class. I now know of myself what I was told the evening of that unhappy day by Dr. Bowditch that the pretended explanation was utterly false & the most unintelligible nonsense ever written by one who had crossed over the Pons Asinorum. These bricks of old edifice are, in my opinion, fair specimens of the unsound instruction. When I entered upon the office of instructor it was with the solemn resolution that I would never hesitate to confess my ignorance to my pupils, & that above all I would always present to them the real difficulties of the science without attempting to disguise them under forms of expression which are only calculated to deceive, & not in the least to simplify.[77]

Peirce's everyday routine at Harvard was busy, but generally reward-ing; he described his college routine, when writing home:

Dear Mother,

I have very little time now to write to you, as I have to study a great deal. In the morning I first go to prayers, then either to recitation in Greek or to my room and study; at common bell I go to breakfast. At study bell half an hour after commons bell at eight I recite or study till 10 or 11 (my hour for mathematical recitation) after which I study till 1 o'clock when I go to dinner; after dinner I kick foot-ball till study bell at 2. From that time till prayers I study or am at recitations; after prayers come commons. Then I exercise till eight, the evening study bell. My time is occupied in this way every day except Sunday & Saturday (when I exercise) from 10 1/2 o'clock, the rest of the day. On Sunday Grotius[78] is studied.

I wish you boil me a good lot of chestnuts and send them up.[79]

Benjamin was by no means exaggerating the rigors of a Harvard education.[80] The hours were long and the living conditions spartan. Morning prayers were at six in the summer and a half hour before dawn in the winter in an unheated chapel. There were two periods of recitations between chapel and breakfast. Breakfast consisted of only coffee, hot rolls, and butter, unless a student had succeeded in pinning meat from the previous day's dinner to the underside of a table with a two-pronged fork. Dinner, which was at half past twelve, was ample in quantity, but deficient in quality. After evening prayers at about six, came supper, which consisted of coarse cold bread, "the consistency of wool," and butter, this time served with tea.

Rooms were plainly furnished, with a pine bedstead, washstand, table, desk, and assortment of cheap chairs. Only the more affluent Southern seniors had carpets—and threadbare ones at that. The rooms were heated with open fireplaces. Matches had not yet been invented, so coals were carefully buried in ashes overnight to rekindle the fire the next morning. When weather was warm enough to do without a fire, the lamps were lighted "by the awk-ward, and often baffling, process of 'striking fire' with flint, steel, and tinderbox."

Understandably, many students sought room and board off campus, where conditions were more comfortable and food tastier, if slightly more expensive. Benjamin was one of them. When he returned to school during his sophomore year, he found a friend had taken lodg-ings off campus and wished him to join him. He wrote home very apprehensively: "I pray you not to be angry but wait till you have read to the end of my letter before you form an opinion. I have changed my lodgings." After explaining the rather complicated arrangements made so he could make the move, he then described the room, which cost only five cents more a year than the dormitory: "Here we have a room where we live to which is attached a closet, large enough to hold a bed. This is on the lower floor a fine square room. Upstairs is a bed-chamber with a closet in which are drawers. We have meat twice or three times a day by ourselves, a boy to make a fire & brush our boots & also to go any errands we have in the village. The man of the house is very much praised for his good feelings by every one that knows him & has just married a wife, who used to live as a servant at doctor Holmes's[,] whose son told me [that] they thought the finest servant they ever had."[81]

In addition to problems with lodging and boarding, relations between students and faculty were far from congenial. "The students considered the faculty as their natural enemies. There existed between the two parties very little of kindly intercourse, and that little generally secret. If a student went unsummoned to a teacher's room, it was almost always by night."[82]

Just two years before Peirce went to college, one of the worst student riots in Harvard's history occurred. Forty three of the seventy seniors in the class of 1823 were expelled, in what is called the great rebellion of 1823.[83] Students remained rowdy when Peirce was a student. Recalling his college days later in life he wrote: "It is many years since Harvard has been visited with a student—tornado. But when I was myself an undergraduate, rebellion was in habit of raging every three or four years—and my own class especially notorious for its contumacy."[84]

$$\Phi^n$$

Bowditch had further impact upon Peirce's Harvard education, as well as on that of his classmates. From Kirkland's inauguration in 1810, until 1826, the corporation gave the immensely popular President Kirkland a free hand in running the university.[85] During the 1825–26 academic year four of the five fellows of the corporation were replaced. One of the new members was Nathaniel Bowditch, although he had already been connected with the university for some time as a member of the board of overseers.[86]

By the spring of 1827, Harvard College had fallen into serious financial problems; the college was spending more than it was taking in. Indicative of Kirkland's casual fiscal policies, Kirkland had never received his regular salary. He had just drawn money from the steward whenever he needed funds. By 1827, he had overdrawn $1700. The college's treasurer had been so sloppy with the college's accounts that it later took an auditor six months to straighten them out. Kirkland had magnanimously waived tuition and fees for needy students, without regard to fiscal consequences. Money owed the college had not been collected.

Typically, Kirkland congenially accepted suggestions and criticism when offered, but if he did not like them, he did nothing about them. He was not successful in handling Bowditch in this manner, however; once Bowditch had uncovered the school's fiscal problems, he was neither timid nor tactful in demanding reforms. He attacked Kirkland on many fronts. The salaries of the president and of the faculty were reduced. Some of the faculty were required to do additional teaching, so that tutors could be eliminated. In particular, Bowditch insisted that Farrar take on additional teaching duties and that the mathematics tutor, James Hayward, be eliminated. When both Kirkland and Farrar objected, Bowditch pointed out that Farrar had been making two thousand dollars a year teaching one hour a day; if he was not willing to do his share, he could resign.[87] Kirkland then pleaded to retain Hayward, pointing out that he "had by choice prepared himself for this occupation." Bowditch replied that Hayward had sought to be a minister but had taken this position because he had failed to get a call. Furthermore he had little taste for the sciences. Kirkland countered that Hayward was "however a very good mathematician" to which Bowditch answered, "There is nothing very great about him, for at this moment [Benjamin] Peirce of the Sophomore Class knows more of pure mathematics than he does."[88]

By 1828, Bowditch became convinced that the university would never be properly administered with Kirkland at the helm. After a final attack from Bowditch at a meeting of the corporation, Kirkland resigned.[89]

$$\Phi^n$$

In addition to Bowditch, the *Mathematical Diary*[90] had an important influence on Peirce's early mathematical development. The *Diary* was founded and first edited by Robert Adrain, who, with Bowditch, was one of the two leading mathematicians in the country. Adrain was an Irish immigrant, and, like Bowditch, essentially self-taught. He contributed copiously to the *Mathematical Correspondent,* the country's first mathematics journal, and when the journal ceased publication, attempted to replace it with the *Analyst,* which he edited. Although the *Analyst* was short-lived, Adrain himself published two proofs of the experimental law of error in the journal, as well as other impressive results.[91]

Then in 1825, Adrain began publishing the *Mathematical Diary*. In the first issue of the *Diary*, Adrain proclaimed that its principal object was "to excite the genius and industry of those who have a taste for mathematical studies, by affording them an opportunity of laying their speculations before the public."[92] However well it may have done with the general population, the *Diary* definitely succeeded in exciting Peirce's genius and industry.

The *Diary*, like other early American mathematical journals, published problems and, in a subsequent issue, their solutions, both of which were submitted by the journal's readers. The *Diary* occasionally published articles as well. Peirce began contributing to the *Diary* when he was in college. In the eighth number of volume I, the *Diary* published three of Peirce's solutions, his first published work. One of them was the solution to the question: "If through a given cone, the diameter of the base being 5 feet and perpendicular altitude 15, an equilateral triangle, whose side is 3 feet, move, having its surface parallel with the base of the cone, and its centre of gravity coinciding with the axis: What will be the solid content of the remaining part?"[93]

The *Diary* became the focus of mathematical activity for the young Peirce. His senior thesis, which he dedicated to Bowditch, consisted of the solutions to all of the questions in the November 1828 issue of the *Diary*.[94] These twenty-four questions are an eclectic collection of problems on algebra, number theory, geometry, applied calculus, and mechanics. They reflect the interests and expertise of American mathematicians of this time. The *Diary* published six of Peirce's solutions and cited a seventh as being similar to another one that was published.[95] Since the *Diary* usually published multiple solutions to a given problem, Peirce's work did not dominate the issue to the extent it might appear. It was, however, an impressive contribution by a senior at Harvard in 1829.

Two of the problems are especially worthy of comment. The twenty-second problem is: "A straight line and an ellipsoid of revolution being given in any manner in space, to draw a plane through the line cutting the surface so that the area of the plane surface common to both may be given." In his solution to this problem, Peirce demonstrated his familiarity with Gaspard Monge's *Application de l'analyse à la géométrie*. The last problem in the set is: "Determine the motion of a ball rolling down the surface of a hemisphere, the base of which slides by virtue of the force of the body on a perfectly smooth horizontal plane." The solution to this problem shows Peirce was acquainted with Joseph-Louis Lagrange's *Mécanique analytique.* In citing these works, Peirce exhibited a mastery of mathematics that was well beyond the Harvard curriculum for 1829.

After Peirce graduated from Harvard, he continued to contribute to the *Diary* while he taught at the Round Hill School.[96]

3

The Finest Lady in Northampton

IN 1826, WHILE PEIRCE WAS STILL A STUDENT AT HARVARD, THE FAMILY business met with sudden failure.[1] The Peirces met their loss of fortune with remarkably good spirits. In three generations of Peirces, it may well have been their finest hour. Lydia and her daughter Charlotte Elizabeth, or Lizzie as she was called,[2] moved to Cambridge to open a boardinghouse. Although the Peirces needed the income the boardinghouse would generate, many Cambridge families took in Harvard students as boarders.[3] Benjamin Sr. stayed on in Salem to finish up his business affairs. He hoped to become Harvard's new librarian, a position that had recently become vacant.[4]

At this time the town of Cambridge consisted of three separate villages. The village that included Harvard Square was known as Old Cambridge, which was largely isolated by surrounding salt marshes. Only a few roads and canals connected Old Cambridge with other communities.[5] Lydia and Elizabeth took possession of the Gannet House there, and began the job of making the structure suitable for boarders. Elizabeth described their work:

> Aunt Saunders & Mother & the girls & I are cleaning etc. The other day Aunt & Ma & Uncle Harry cleaned out the stable & wood house & this was the dirtiest job that has been done for the dirt flew & the cobwebs came down & when they had finished they had nothing but hard pump water to wash in. This by the way—being obliged to use pump water is the greatest inconvenience here, so you may conjecture that we have not many troubles. The failing business is no trouble to me & why should it be? For I should distain to be as foolish as to depend at all for happiness upon riches. . . .

> Can you believe that it is quite a common thing for people to take 20 [boarders] & some have been known to take 30 & so keep two tables. Don't you wish you were going to be as busy as perhaps we shall be—for happiness after all dear aunt depends upon our being well and constantly occupied. The chief thing that I regret in leaving Salem is leaving my friends, you & Grandpa and Grandma & to tell the truth this is about the

49

only thing, but you must come to Cambridge & I will go to Salem to make visits. . . .

Have you heard what sort of house we live in? It is a very good one & the worst thing about it is a stable which is apt to be offensive on warm damp days. It consists of an excellent cellar, three good parlours, a kitchen & pantry on the first floor & two good chambers in the garret. . . .

Tell Grandpa that if Pa is Librarian, he will have a nice time if he comes down here for he will have the best books in the world to read—& that Ben and I will read to him if he wants us to. The library here is said I believe to be the best in the United States. . . .

I am going to keep the accounts.[6]

Lydia, too, kept an optimistic attitude; she wrote to her husband: "She [Mrs. Lee] said when she heard of our trouble she told Mr. Schuyler—she knew me too well to believe I should droop—that loss of property would never make a spirit like mine, whine & hang its head. I think I shall be more pleasantly situated than I ever was & associate with the first society in the country—She says your great fault has been that you were too much of a book worm to lounge in Salem offices—but now you will be in your element . . . and may devour all the books in the Library if you wish it. . . . We both congratulated ourselves that things had turned out as they had done—That instead of our burying ourselves in Salem we should now be brought forward to show the world what wonders we were. She told me something about what Franklin said on riches—do get the essay & bring it to me."[7]

The Peirces were involved with manufacturing as well as the sea trade.[8] Benjamin wrote to Lydia, describing their failing business, and the provision made for their son, Benjamin:

When I got to Salem . . . I found there had been more failures. . . ; There will be more. . . .

Ben has gone to Andover with Mr. Pickering. . . . Mr. Pickering speaks highly of Ben. He says he is a fine boy, & that it is a pleasure to teach such a one. I told Mr. P that I felt under great obligation to him for his attention to Ben.[9]

Lydia wrote to her husband extolling the joys of running a boardinghouse: "We like our boarders exceedingly. If we find them all so & these continue as they are now, I shall have a very pleasant time here—they are all attention to Elizabeth and myself & we pass the time we are together, which is several hours everyday in very interesting pleasant conversation—they are young men who have seen a good deal of the world & perfectly polite & amiable in their manners. Our conversation, as you may imagine, is quite literary & it really makes the Salem society appear flat."[10]

One of the Peirce's boarders was Timothy Walker, who became a distinguished lawyer and legal educator. He graduated from Harvard in 1826, and then taught mathematics for a short time before returning to study law at Harvard under Lydia's former suitor, Joseph Story. He boarded with the Peirces while in law school. Soon after leaving, he wrote to Benjamin Senior: "I have not heard who fills my place, but I guess it is an amiable interloper, who will more than make it good. I hope Mrs. Peirce kept a record of our puns and jokes and aphorisms. If she did, and will send it on to me, I will have them all over again at the table to which I now belong. . . . I have not a dignified professor opposite, but I have a talkative lady from Mississippi, who has seen more of the world than I ever expect to. Above all, I have not a kind, intellectual, orthodox Mrs. Peirce at the head, but I have a Methodist old maid, who has read many dry books."[11]

Needless to say, not everything about taking in boarders was to the Peirces' liking. Years later Elizabeth wrote, "She [a Mrs. Johnson] has let all her rooms but one—she told me that she hated to keep boarders—& I said that even the Angel Gabriel would not be pleasant as a boarder."[12]

Benjamin Senior, who was still in Salem waiting to hear if he had got the position of librarian, wrote:

> I should prefer not to go to Cambridge till the Corporation have had their meeting, for personal reasons. In the first place I have a good deal to do here which it is very important that I do. . . . & then I will have to speak to some persons about giving my office in the Mill Dam—to Harry, if I resign. So that you must be patient and get along as well as you can without me. . . . Do be cheerful—You & Elizabeth must not work too hard. If we do our part, we may safely trust to Providence for the rest. . . .
>
> Ben is reading Mrs. Edgewerth's novels, which he says are extremely interesting—he thinks them superior to Scott's, leaving out the historical parts in the latter. He wants to stay in Salem as long as he can.[13]

Benjamin Senior did become Harvard's librarian, and being librarian suited his taste much better than being in business.[14] But it paid only the salary of a tutor at Harvard, six hundred dollars, a small salary even in 1826. Before Peirce assumed the duties of librarian, it was considered as a part-time job, which justified the low compensation. Peirce Senior pointed out to Nathaniel Bowditch, now looking after Harvard's financial affairs, that the custom of having a part-time librarian had been detrimental to the college in the past and that he was now spending all his time working there, even during the vacations. He complained repeatedly that he deserved and needed a larger salary.[15]

Φ^n

As graduation approached, young Peirce set about looking for a teaching job.[16] The change in the family's circumstances may explain why Peirce did not go to Europe for further study.[17] He certainly knew of men who had done this; many of the younger Harvard faculty members, such as Cogswell, Bancroft, and Ticknor, had already studied in Europe by the time Peirce was an undergraduate. Given Peirce's scientific ambitions, it seems likely that he would have gone to Europe to study if his family's means had made it possible. He did greatly value Bowditch's guidance, and may have felt his education under Bowditch was sufficient. He wrote his mentor shortly after graduation: "You have, for a long time, been to me a father. You have placed me in possession of advantages, such as no young person in this country ever could have enjoyed; perhaps such as no one ever yet enjoyed. A kind Providence has bestowed upon me an excellent constitution. I have never been visited by the stroke of adversity. How shamefully negligent must I then be or how extremely dull, if I do not acquire more than common acquaintance with the science of mathematics!"[18]

Peirce graduated from Harvard in 1829, having been elected to Phi Beta Kappa.[19] He found a position at the Round Hill School in Northhampton, Massachusetts, an exceptional prep school.[20] The school's faculty was as outstanding as that of any college of the day. The school was founded by George Bancroft, who later became a distinguished man of letters and a diplomat, and Joseph Cogswell, who became the first librarian of the Astor Library. Both men were teaching at Harvard when they decided to found the school. Their experience there had convinced them that reforms in American education must begin below the college level. After graduating from Harvard, Bancroft received a PhD from Göttingen in 1820. Cogswell had traveled widely in Europe, studying educational institutions there, and was especially impressed with the school of Fellenberg, at Hofwyl, near Berne. Consequently, Cogswell and Bancroft founded a school that was patterned closely after the German Gymnasium and the French Collége.

Modern languages were not only emphasized, a radical innovation at the time, but French was taught by a graduate of the University of Paris, whom George Ticknor[21] called the best tutor in French that Harvard College had ever employed. A native German scholar was also employed to teach that language. In addition to teaching modern languages, physical education, an equally radical addition to the curriculum, was stressed. Instructors in gymnastics and dance were employed by the school, and horses were kept for the school's equestrian program. In addition, the school was beautifully situated overlooking the Connecticut River, and one Round Hill student recalled late in his life that the school provided an excellent table.[22]

Benjamin enjoyed the beauty of his surroundings: "I wish that you could be here now and see how very beautiful N[orthampton] looks. The blossoms are fallen from the trees and everything looks green and flourishing. I, yesterday, took a walk into the woods and N[orthampton] seems to exceed in the beauty and profusion of its flowers all other places that I ever visited. From the tops of all our hills you enjoy a distant and most grand prospect, at their feet are valleys romantic enough to deserve the pen of a Waverly."[23]

Despite the beauty of Northampton, Peirce found the Round Hill School to be less than a utopia; he wrote home about the plight of one of his colleagues: "Poor Hillard is very sick. Yesterday he began to starve himself well, and went without dinner. This morning he took scarcely any breakfast. He is afraid that he shall be compelled to leave Northampton. He suffers a great deal from want of attention. Mr. Bancroft and Mr. Cogswell have not the least feeling for a sick person. They seem to think that if a man is so ill, as to be unable to perform his duties, that he is only endeavoring to get rid of them. It is a heartless place on this hill, I can tell you."[24]

The duties of the faculty, including those of Cogswell and Bancroft, were arduous. Peirce complained to Bowditch: "You was (sic) so kind, the last time I saw you, as to desire me to write and inform you respecting my situation, and my hopes of success in the business of Instructor. Had I had any time to myself, a month would not have passed by without my taking advantage of your request. But except Saturday and Sunday, I am employed on an average of ten hours a day, sometimes $10\frac{1}{2}$, never less than $9\frac{1}{2}$. I exercise an hour and a half, and my meals occupy about three quarters of an hour; so that from seven in the morning till nine at night when I go to bed, completely exhausted, I have but about two hours of leisure time, i. e. of time not employed in some regular manner, when I can read and write as I please. Of course, I am obliged to neglect my own studies."[25]

Peirce then goes on to describe his classes, most of which he was pleased with; one exception was:

A very dull class, and the course, that Mr. Cogswell has put them on, would for the brightest boys be almost too difficult. He wishes them to become acquainted only with the theory of arithmetic. He wishes them to become acquainted with the abstract principles of quantity, and they are too young to understand them. Indeed, they have already been through the arithmetic with himself once or twice, and I should not think, from their appearance at recitation, that they had ever seen it before. Mr. Cogswell believes that they do understand it, but that they are lazy, and they are punished for not getting what it is beyond their strength to learn. Yet I am determined to layout my whole powers to carry them on. I will set them short lessons, and what they cannot arrive at by analysis, I will endeavor to lead them to by induction. I will repeat a lesson again and again, till I am confident that they understand it. But, perhaps, then at the end of the term Mr. Cogswell will complain that they have not advanced more rapidly; for I cannot hope to persuade him that it was necessary to go slow because of the difficulty of the subject. What, then, ought I to do in this case?[26]

Peirce wrote to his father of the same class: "I feel very much interested in my duties. Many of the boys are quite bright, and there are but few that I despair of. And these few I believe I could teach something to, were it not that Mr. Cogswell has put them on a course of study far above their comprehension. But as there is but one class, of which this can be said, I am determined to lay out my whole strength in the endeavor to make a set of block-heads understand the fundamental abstract principles of arithmetic; why we carry one for every ten addition; what, in subtraction, is the nature and effect of borrowing and the like. These are subjects which are not understood by many grown persons; even by those that have received a liberal education."[27]

Peirce did like being a teacher. He wrote his father, "How glad I am that I am not to be engaged in trade!"[28] He also took his duties at the school seriously. He explained to his father that his responsibilities left him with neither time to prepare solutions for the current issue of the *Diary*, nor time to send letters to friends in Cambridge: "Some of the Round Hill boys are going to College in Spring and . . . I must give them as much extra time as possible."[29]

Despite his heavy duties at the Round Hill School, Peirce found time to teach some of the young women in Northampton. "I went down to Mrs. Judge Lyman's to instruct her daughter, Annie Jean, and Miss Wilson a young Lady on a visit to her, in mathematics. They are both possessed of talent and I should much rather teach a school of girls than one of boys."[30]

There were other problems at the Round Hill School besides the heavy workload. Cogswell had hoped that the school might run smoothly without the severe discipline used in many schools. One of the things that pleased Cogswell most while traveling in Europe and visiting schools was finding one where there was no harsh discipline. He was as impressed with Fellenberg's school near Berne in this regard as in other aspects of the school; he said: "More heartfelt joy I never witnessed in my life, not, as it seemed to me, because they were about to relax from their labor, but because they had the happiness to be placed for their education in a school, the head of which was rather a father than a master to them. I saw a thousand proofs of the sentiments they entertain toward each other, and nothing could resemble more a tender and solicitous parent, surrounded by a family of obedient and affectionate children. There was the greatest equality and at the same time the greatest respect, a respect of the heart I mean, not of fear; instructors and pupils walked arm in arm together, played together, ate at the same table, and all without any danger to their reciprocal rights; how delightful it must be to govern, where love is the principle of obedience!"[31]

This became Cogswell's aim for administering the school. To some extent he succeeded. He did take the boys on jolly excursions, and students and masters took their meals together, apparently in harmony, but there were problems with discipline. Samuel Ward, a student there shortly before Peirce arrived and later a friend of Peirce, wrote home to his father about problems with discipline.[32] And young Peirce also experienced serious discipline problems. He wrote to his father: "Our school has, since Mr. Cogswell left for Washington, fallen into a terrible state. The boys have been nearly in open rebellion, and our persons have even been in danger. There are in the school several very bad boys, and we shall not be able to get along till they leave. Everyone is satisfied of this from their present behavior. They play cards on the Sunday, drink much, and own daggers. Now, when Mr. C[ogswell] returns he will probably only make a few speeches to the boys on their bad behavior and everything will go on as before. If this is the case the instructors will be, continually, liable to insults. What ought we to do in this situation? I should like to have you consult with Dr. B[owditch]. I feel confident that the boys have great respect for me. I have seen this plainly. . . . Yet with the present discipline of the school, this can be of no avail to me."[33]

Peirce did seem to get the hang of handling a class, however, and developed a presence in the classroom. He wrote his mother: "You do not know how grand I feel when I get out with my cane. Tell George H. that if he will become a school-master, he can wear his sword-cane, for the continual danger that he is in of being stabbed or bruised will be a sufficient excuse. Besides it gives a certain commanding air to the figure which makes the boys very submissive."[34]

Peirce took this presence with him to Harvard. He was at once a solid disciplinarian and a popular member of the Harvard faculty, if not a popular teacher. One of his students, Colonel Henry Lee, recalled: "Why we should have given him the diminutive name of 'Benny' I cannot say, unless as a mark of endearment because he could fling the iron bar upon the Delta[35] farther than any undergraduate; or perhaps because he always thought the bonfire or disturbance outside the college grounds, and not inside, and conducted himself accordingly. His softly lisped 'Sufficient' brought the blunderer down from the blackboard with a consciousness of failure as overwhelming as the severest reprimand. There was a delightful abstraction about this absorbed mathematician which endeared him to the students, who hate and torment a tutor always on the watch for offenses, and which confirmed the belief in his peculiar genius."[36]

Peirce also demanded an aura of authority outside of the classroom:

> Jenny Lind's last concert of the original series, given under the auspices of Phineas T. Barnum, was given at the hall over the Fitchburg Railroad Station. Tickets were sold without limit,—many more than the hall could hold,—and there was every prospect of a riot. Barnum had taken the precaution to leave for New York. I got about one third up the main aisle, but could get no further. Just ahead of me was Professor Peirce. The alarm was increasing. The floor seemed to have no support underneath, but to hang over the railroad track by steel braces from the rafters above. Would it hold? The air was stifling and windows were broken, with much noisy crashing of glass, in order to get breath. Women were getting uneasy. And there was no possibility of escape from a mass of human beings so packed together. We knew, from the conductor's baton that the orchestra was playing, but no musical sound reached us. Professor Peirce mounted a chair. Perfect silence ensued as soon as he made himself seen. He stated, very calmly, certain views at which he had arrived after a careful study of the situation. The trouble was at once allayed. Jenny Lind recovered her voice and the concert went on to its conclusion.[37]

$$\Phi^n$$

Despite Peirce's complaints about Cogswell, Peirce eventually developed a deep respect for him as a disciplinarian and a schoolmaster. He confided to Bowditch:

Mr. Cogswell has just issued an "Outline of the System of Education at the Round-Hill School." One of his objects, I suppose, is to let the world know that Mr. Bancroft has no longer a connection with the school; and the friends of the school must rejoice at this, for Mr. B, though undoubtedly a man of superior talent and acquirement, is by no means fitted for an instructor, so little that, with all his brilliancy, he had been of far greater injury than benefit to the establishment. Mr. Cogswell is, I venture to say, unequalled in the power of conciliating the warmest affections of his pupils, and at the same time of preserving his dignity and a rigid discipline throughout the establishment. All respect yet love him. His characteristics are simplicity, independence, and energy. The calmness of demeanor that never deserts him and I have seen him tried. The mild and authoritative tone which the most outrageous scoundrel and most impudent blockhead never have had the hardihood to withstand show him to be above the common level.[38]

$$\Phi^n$$

Peirce did not spend all of his Northampton years struggling to teach the boys of rich men. His letters during his two years at the Round Hill School show his wide range of interests. For one thing, he found sufficient time to write to his brother, Henry, giving him advice and three large pages of algebra problems. He admonished him: "The last sums are so very hard my brother that I do not expect you to be able to solve many of them in a day, particularly if you do not neglect your other duties. Night before last I received your letter or poem, or doggerel rhyme. Not very good rhyme in my opinion. . . . Glad that you are done with Herodotus, Thucidides and Xenathon, the three best historians of old Greece! I am amazed. You've learnt two problems in Geometry! I am amazed. I am amazed. I can hardly go on from my astonishment. Two whole problems! So little taste for the classics! I did not expect this abounding intelligence. . . . Study hard my lazy brother. . . . you ought to tell Pa and Ma that they must write me oftener."[39]

Peirce also wrote of studying history, of religion, and of growing dahlias. Neither was Peirce without social interests; at the beginning of his second year he explained to his father that he had little time to work on the problems in the *Diary*, not only because of his work at school, but because he was spending much more time visiting than he used to.[40] Indeed, Peirce's social life took a turn for the better. As an aid to his social exploits, the school's dancing master gave him free lessons in the waltz. One of the young ladies Peirce visited was Sarah Mills. He wrote: "I made a call this eve at Mrs. Mills's. S[arah] Mills does not please me so much as I expected, and though some call her

Sarah Mills Peirce. By permission of the Houghton Library, Harvard University, MS CSP 1643.

the finest lady in N[orthampton], I do not think her. But perhaps I judge too hastily and in a week I may be writing a letter filled with the most extravagant praises of her. Fickle mortals we are dear Mother!"[41]

Benjamin was right. He soon became smitten with Sarah Mills and began courting her. Soon he was happy to report to his father that she had promised to make him a waistcoat.[42]

Sarah belonged to a distinguished family. Her father, Elijah Hunt Mills, was a graduate of Williams College and an eminent lawyer in Northampton. He served in the United States House of Representatives from 1815–19 and in the United States Senate from 1820–27. Upon his retirement, his seat was taken by Daniel Webster, a family friend. In 1829, after retiring from public life, Elijah Hunt and two other men opened a law school in Northampton. Sarah's paternal grandfather, a Yale graduate, was a member of the Provincial Congress and the Massachusetts General Court.[43]

Despite the beauty of Northampton and being in love with a woman there, in the spring of 1831, Peirce accepted a position as a mathematics tutor at Harvard for the following fall. He was pleased about the competence of his replacement. He wrote Bowditch: "Mr. [French] will be my successor, and, his favorite occupation being to instruct, is in some degree a proof that he succeeds in it. He is uncommonly well read in elementary works on mathematics, and is far from the usual cast of the would be mathematicians of this country. He seems attached to no particular system and has made some progress in the higher branches."[44]

Although Peirce was happy to be going to Harvard, he was having such a grand time in Northampton that he was reluctant to leave. He asked his brother, Charles, to contact President Quincy to learn when he would have to assume his responsibilities at Harvard, as he wanted to stay in Northampton "till the very last hour possible."[45] He also wished to share these delights with his brother. Peirce urged his brother to visit him in Northampton: "I will see that your visit shall be an agreeable one—indeed when I think of the beauty and fascination of some of the Northampton young ladies I think that their charms joined to the easy style of manners universal here will render it a delightful one."[46]

Benjamin proposed marriage to Miss Mills, apparently before leaving Northampton to return to Cambridge. But Sarah Mills did not accept him. Benjamin wrote to her soon after returning to Cambridge, saying that his love for her, even though not reciprocated, had fundamentally changed him:

> What was it before I felt true love that inspired me to such laborious efforts? I tremble to answer. Ambition. I was resolved that I would make

myself known throughout the world, and that I would bow to no living man as my superior in science. And what was my chance of success? Altered as I am, I may speak of what I was as of a different being. I do it without boasting, except that I have roused now from the dream; for I now know the inferiority of intellectual to moral greatness. My hopes were, then, founded on my deeds. Again and again I had surpassed the highest anticipation of the most powerful mathematician of the present day.[47] The work of what he thought months, I had finished in a week. I had gone through calculations in my head, for which he thought it necessary to resort to his pencil; and on one occasion when he he (sic) had mentioned to me an elegant demonstration that had just made its appearance, I immediately proposed another that he judged in several respects to be the better one. But I will leave this disagreeable subject and glance at my present state. My passions are no longer chained in the chilling dungeons of philosophy. The poet's fire can now warm me into something like enthusiasm. The voice of nature is not now that of mere reason. Since I have come to love you, I feel that I have risen in the scale of being, my soul is elevated in its moral reflections; it is more than gratitude that now inspires my devotions to the god of love; to the Creator of so beautiful, so magnetic, so lovely a being. God grant that I may be animated by this most ennobling principle. Neglect shall not chill me. I am too proud for that and I feel that what would render me most worthy of your disdain is ceasing to love you.

The vanities of this life never appeared so clear to me, though they never troubled me less. I am soon going into the very market place of worldliness, but I do not fear contamination. Alone in the world, my wants will be small and easily gratified. Had I been successful in my suit, how different it would have been? With such an associate, I should not have been less safe from the frivolities of ambition. Still I could never have rested till I had placed you in the only situation worthy of your dignity and character. I could not have endured that any other woman should take the station which you are so admirably fitted to adorn; and which would prove to be the most splendid and useful that a lady could fill.

But is it so certain that I must forever be single?

"Amore duce, nil desperadum"[48]

Sarah had a fairly quick change of heart and accepted Benjamin, although there were some rough moments during their engagement. A mutual friend, Charles Phelps Huntington, described the following incident when Peirce was visiting the Mills' home:

Ben, Sally, and Elizabeth Cabot came a week last night. Ben and Mrs. Mills had a family scene Sunday p. m. Mrs. Mills reproached Peirce with not going to her church, with offering an indignity to Sally thereby, telling him it was the worst thing she had ever known him do, and that she should mark it down against him. Ben said it must be so marked then. Mrs. Mills withdrew. Ben walked the room once or twice and then rushed out of the house in a grand passion. Sally thereupon blew up at her mother, saying she had made her engage herself to him, and that now she should treat him civilly and not insult him while a guest in her own house. Thereupon we walked off to a third service at the Episcopal, Sally talking rather loud on the way and to the hearing of divers good citizens of this commonwealth. Ben returned soon after to the front Parlor. Mrs. Mills made a sort of apology and he cried and so the matter ended without a violation of betrothment.[49]

There were much happier moments during their engagement. Benjamin describes his fiancée's entrance to a ball: "Sally was dressed a la dignité con simplicita vigorosa. Her gown white lace worked over white satin black shoes, a head of false hair, a black necklace[,] flowers in her hair[,] & smiles on her lips. When she first entered the room she cast her eyes round & in a moment all was silence. She moved forward one step & before her lay a heap of vanquished lovers. She waved her hand & they darted to her feet. She smiled & all again was gaiety."[50]

Sarah and Benjamin Peirce. By permission of the Houghton Library, Harvard University, MS CSP 1643.

Shortly before Benjamin and Sarah married, his mother, now widowed and living in Rhinebeck, New York, wrote, advising him about buying a house in Cambridge. She gave other counsel and offered assistance, adding that she hoped to move to Cambridge herself:

> Unless you keep several servants—which by the way are the greatest plagues in housekeeping—If I was only at liberty, I should go on to Cambridge and take you and Sarah to board—then take students enough to enable me to get along . . . it would require considerable of exertion but then my mind would be much easier to be near you & the other children. If you have a family you will want <u>old Mrs. Peirce</u> more than you think for—Still I don't think seriously of leaving my present situation—though I should be willing to give up many of my present comforts to be near my children. . . .

> I have many things in the housekeeping line in Salem—& some things also that Sarah may like in her preparations for an expected event—It would be well for you to go to Salem and see about them.[51]

In May 1833 Peirce was appointed to a new professorship of mathematics & natural philosophy;[52] the following July he and Sarah Hunt Mills were married.[53]

4

The Voice of God

BENJAMIN'S FATHER DID NOT LIVE TO SEE HIS SON MARRY. HIS DAYS AS Harvard's librarian were productive, but short. He compiled the first complete catalogue of the library ever printed.[1] He also wrote a history of Harvard up to the time of the Revolution. It was still in manuscript form when he died in the summer of 1831, shortly before Benjamin moved back to Cambridge to become a tutor at Harvard.

Although Peirce's father was only fifty-three when he died, Peirce took his father's death stoically.[2] He wrote to his brother Charles Henry that he was delighted, "to see . . . that you all view our loss in so happy a light, and we have every reason to be grateful for an affliction that has given us such a realizing sense of the existence of a father in heaven, and let us so improve this chastening of God's love, that we may ever look back to it with joy, unmingled with regret except it be for the loss of so agreeable society as that of our dear parent."[3]

Peirce also wrote a gracious letter to his father's physician, George C. Shattuck, thanking him for caring for his father during his illness. He concluded, "Had he lived? Would he have been happier? Thank God, no. He is in heaven, and I will not regret that no human power could ward off that fatal blow."[4]

Peirce took his responsibilities as the oldest child in the family seriously, and wrote to his uncle asking him for his support and advice.[5] But Peirce's modest salary of six hundred dollars was hardly sufficient to support his mother and siblings. Peirce's sister, Elizabeth, had been working as a governess in New Bedford before her father's death, and she continued to do so.[6] She liked teaching the children, who learned quickly; she also shared some of her brother's aptitude for mathematics: "I like Smith's arithmetic. With their former instructress more attention was paid to the outside of things than I think good—She seems to have been satisfied with the children's appearing to understand a thing—which is not a good plan as if they do not really understand what they study, they never can make any great pro-

63

ficiency—they can go to a certain extent & then they stop—I do not
believe she knew much herself—She had never studied geometry &
yet she attempted to teach it to the children—the consequence was,
that they are quite satisfied with knowing that things are so & so—but
they are unwilling to learn how to prove that they must be—Tell Ben
I do not like Colburn—but I cannot give my reasons exactly—I wish
he would tell me <u>why</u> Smith's is the best & what are the objections to
Colburn."[7]

Peirce's mother, Lydia, was anxious to have her husband's manuscript history of Harvard published. In fact, nine months after her
husband's death, she was still more concerned about his book than
her own situation. She wrote to Benjamin: "As to my plans, I have
none, but I think on the whole [as] it will require so little effort to
maintain myself I shall lead a pretty easy life for the present. Elizabeth & myself can do this in many different ways—I want to know how
you get on with the History of the College—and what has been done
about it since you wrote me?"[8] A family friend, John Pickering, was
editing the manuscript and overseeing its publication.[9] It was published the next year in 1833.[10]

Lydia soon found a position as a governess and housekeeper for
Judge Samuel Jones in New York City. She was happy with the position:

> I am truly most delightfully situated and have every thing but my dear chil-
> dren and friends to make me happy—I find myself as completely the head
> of our establishment as though it were my own—the house and furniture
> are about as handsome as Mrs. Leis—We have excellent domestics—a
> cook—chamber maid—nurse, and man servant—I feel quite at home.
> The children seem as fond of me as though they were my own. . . .
> My time is almost entirely devoted to the children.[11]

Lydia was still pleased with her position a year later, when she wrote
that "I am very pleasantly & elegantly situated in Judge Jones Family—
. . . though there is great care and responsibility in taking care of
so many little children."[12] Nevertheless, she was not sure she would
remain with Judge Jones. That depended on what would be best for
her youngest child, Charles. He was studying medicine and still needed
her assistance, so seeing him well established was her top priority. She
wrote, "There is no sacrifice I am not willing to make for my children—
and until Charles is fixed in business, I intend to do all in my power to
make his situation in life an eligible one."[13] Although she considered
"taking a school in Boston with Elizabeth," she never acted on the idea.

Despite being well situated, Lydia was homesick; she wrote to her
parents:

But N[ew] England people are not made to live in N[ew] York—everything here is for show—everyone is selfish & regards no ones feelings but their own—it is a principle with them to feel so—& not to do it they consider a proof of weakness & folly—I do not mean I have experienced any unkindness—far from—Had you been on the spot yourself you could not have wished me to have been treated more fairly—But if it seems to be the will of Providence for me to remain where I am a much longer time—(as desirous as I am of living in the midst of my friends) I shall feel it proper to practice the most useful lesson in life to be contented with my situation—This is a lesson so difficult to learn that necessity alone can teach it to us—It seems strange it should be so when we can remain for so short a time in this life—you have no reasonable grounds my dear parents to have any anxiety about me & my children—Ben is well established in Life & ready & [willing] to do for me & his brother & sister all we can wish him to—but he is young & has to work more I fear than is good for his health—so I am determined to make every exertion that he may be at no more expense for Charles than is absolutely necessary—Charles in two or three years at farthest will be so situated as to get a respectable living.[14]

Lydia did not remain in New York City long, but she did stay in the employ of Judge Jones, moving to Rhinebeck, New York, where she had the responsibility for looking after his farm as well as his children. Elizabeth joined her mother in New York, just before she moved to Rhinebeck.[15] Elizabeth described her life in Rhinebeck: "It seems to me that I cannot write anything about Rhinebeck which will interest you much. One day passes almost just like the other—I go on just as I used to in Cambridge, washing cups, ironing, sewing & besides I keep school for the children—I am always busy with something. We have had sleighing for a fortnight & I have enjoyed it exceedingly."[16]

Judge Jones may have had more of an interest in Lydia than that of an employer. He wrote: "It is so long since I have written to any of you at Rhinebeck that I almost fear that you may begin to think that I have forgotten you. But I do assure you . . . I think of you often, very often, so often indeed that I may truly say you are never out of my mind."[17] If Jones had a romantic interest in his employee, nothing came of it. Lydia remained a widow.

Lydia's mother died in 1837. Since Peirce's brother, Henry, had begun his medical practice, Lydia and Elizabeth were free to move to Salem to keep house for Lydia's father, Ichabod Nichols, who died two years later.[18] After Ichabod's death, Lydia and Elizabeth moved to Cambridge to live with Lydia's sister, Charlotte, and her husband, Charles Saunders.[19]

Φⁿ

During this time, Peirce's career had progressed smoothly at Harvard. His initial position as tutor in mathematics was created because of the poor state of Professor Farrar's health.[20] Peirce, with the aid of a proctor, took over Farrar's recitations and the responsibilities for teaching mathematics.[21] Hoping to regain his health, Farrar took an eight-month leave of absence in 1831 to travel to England with his wife. When he returned, he delivered all of his lectures for that academic year in its four remaining months. Although Farrar still continued to lecture, he relinquished all of his duties in the recitation room.

Farrar's health continued to decline and, after Peirce had been a tutor at Harvard for one year, Farrar secured a leave of absence for the next year, leaving Peirce in charge of the department.[22] Since Farrar continued to be unable to perform most of his duties, the corporation created the University Professor of Mathematics & Natural Philosophy, and appointed Peirce to the position in 1833. As holder of the professorship, he was responsible for overseeing all mathematical instruction at the university.[23] In 1842 the Perkins Professorship of Astronomy and Mathematics was established, and Peirce was elected to the post, at a salary of two thousand dollars per year.[24]

Peirce also received the Master of Arts from Harvard in 1833. There were no requirements for this degree other than allowing three years to pass after the earning of a bachelor's degree and paying a fee of five dollars.[25] Later, when Harvard was struggling to introduce real graduate work, Peirce lamented: "The proposed extension of the collegiate course for a fifth year and the examination for the second degree appear to me to be most desirable improvements. Judging however, from a conversation which I had a few years ago with one of our own graduates, I fear that they will meet with the opposition of an uncompromising conservatism, which is disposed to claim for every blockhead, who is a graduate of three years standing, a vested right to the title of master of arts."[26]

Φⁿ

When Peirce was an undergraduate and tutor at Harvard, colleges differed significantly from modern ones. The students were somewhat younger; the typical age for entering freshmen was fifteen.[27] Moreover, the curriculum and mode of instruction were markedly different.

College education in America during the first half of the nineteenth century consisted largely of a prescribed curriculum consisting primarily of Latin, Greek, and mathematics. The primary means of instruction was the classroom recitation; students were assigned a lesson to study and each student was examined orally in class to determine if he had mastered it. Explanations and lectures were seldom given. The tutors, who often presided over the recitations, were young men like Peirce who had recently graduated from college. George Ticknor, a professor at Harvard while Peirce was an undergraduate, described the procedure: "The most that an instructor now undertakes in our colleges is to ascertain, from day to day, whether the young men who are assembled in his presence have probably studied the lesson prescribed to them. There his duty stops. If the lesson have been learnt, it is well; if it have not, nothing remains but punishment, after a sufficient number of such offenses shall have been accumulated to demand it; and then it comes halting after the delinquent, he hardly knows why. The idea of a thorough commentary on the lesson; the idea of making the explanations and illustrations of the teacher of as much consequence as the recitation of the book, or even more, is substantially unknown in this country."[28]

Andrew Peabody, who graduated from Harvard three years ahead of Peirce and shared the tutoring duties in mathematics with him in the academic year 1832–33, described the classes at Harvard: "The recitations were mere hearings of lessons, without comment or collateral instruction. They were generally heard in quarter-sections of a class, the entire class containing from fifty to sixty members. The custom was to call on every student in the section at every recitation."[29]

There were also lectures at Harvard when Peabody was an undergraduate, which he found to be excellent.[30] These lectures were not required, however. A particular professor could insist that his students attend a series of lectures, but he was prohibited from examining them on their content.[31]

Not only was most of the instruction by means of recitation, but mathematical instruction at this time often stressed rote memorization, with little or no attempt at understanding the underlying principles. A student might be called upon to recant the steps in a demonstration, but as an exercise in memorization, with little attention to the students actually following the argument. One commentator on mathematical education observed: "After much drilling he [the student] is able to make a kind of mechanical application of the rule to several problems which the sagacity of the author, not his own, has placed in its train. It is quite a mistake, however, to suppose that he has made any considerable acquisition in the science of numbers. Of these relations he knows little more than at first. His adoption of the rule is purely an exercise in faith, with which his understanding has no concern, and he learns to confide in the authority of the book and the instructor, without expecting to see for himself, the reason of the process which he performs, and consequently without making any considerable effort to understand the relations upon which the rule is founded."[32]

Shortly before Peirce attended Harvard, a small cadre of Americans —George Bancroft, Joseph Cogswell, Edward Everett, and George Ticknor—had gone to Europe for advanced education and returned to teach at Harvard, full of enthusiasm for reform. Their efforts for change met with firm resistance. Two of them, Cogswell and Bancroft, quickly left Harvard to found the Round Hill School in Northampton, Massachusetts. Everett remained five years before he left to pursue a successful political career. He returned as president, but discovered that he had no taste for the job and left again after only three years. Of the four, only Ticknor stayed an appreciable length of time, himself leaving after sixteen years, in 1835.[33] But Ticknor probably stayed as long as he did only because he was exempt from many of the burdensome responsibilities of a Harvard professor in the 1820s. His Smith professorship exempted him from Harvard's residency requirement and its concomitant disciplining of students. He also was free of hearing daily recitations. Ticknor lived in Boston and arrived on campus only to deliver a formal lecture or supervise the study of modern languages.[34]

$$\Phi^n$$

This was the academic environment that Peirce entered as a student and, a few years later, reentered as a professor. Unhappy with this situation, Peirce was deeply committed to transforming American higher education in general and Harvard in particular. Wishing to advance scientific research and education in the United States by establishing a university patterned after those in Germany, Peirce continually proposed and supported innovations that he felt would help transform Harvard into such an institution. Beyond Harvard, Peirce was an important player in reforming American higher education.

Unlike Peirce, Harvard's President, Josiah Quincy, was not progressive, but he was concerned about the quality of mathematical instruction at Harvard. In the spring of Peirce's first year as a tutor, Peirce received a letter from Quincy expressing concerns about the mathematical textbooks then being used at Harvard. Quincy gave specific criticisms of these textbooks and suggested other textbooks as substitutes. He stressed that he was not asking merely out of curiosity, but felt he needed this information to perform his duties as president competently. He asked the same questions of Farrar and the proctor, Mr. Giles.[35]

Quincy apparently requested Peirce to write mathematical textbooks for Harvard. The next fall Peirce wrote President Quincy, saying that he was willing, if unenthusiastic about the project. Peirce, unlike most other contemporary American mathematics professors, had greater scientific ambitions than writing textbooks:

In about a week I shall probably be at leisure to begin the preparation of some mathematical works for instruction in the College, provided the Corporation desire it. But I cannot undertake a task that must engross so much time and is so elementary in its nature and so unworthy of one that aspires to anything higher in science, without the certainty that no circumstance will retard the publication and use of the works when prepared but their defects. Let them be submitted to the most rigid examination. If they do not satisfy it, I shall destroy them without regret, or any feeling of disappointment. I dare not promise myself success where men of genius have failed and no one has yet succeeded. I would not even make an attempt, did I not think that the difficulty and the humble nature of the undertaking have hitherto deterred those that were not more desirous of money than fame. I must therefore beg you to lay this subject before the Corporation at their next meeting.[36]

The corporation responded positively to Peirce's letter, and he commenced writing a textbook on trigonometry.[37] Textbook production was a principal activity of American mathematicians of this period. As in the case of Farrar, these textbooks were often not original, but translations of French textbooks or merely American editions of English textbooks. Beyond the obvious desire to generate extra personal income, producing textbooks was a far more accessible activity than doing research for most American mathematicians of this period, most of whom lacked both the training and the time to attempt mathematical research. Furthermore, the English textbooks used in this country through the beginning of the nineteenth century were deplorable, generally giving rules with few explanations, so textbook reform was an important contribution to mathematical education as well.[38]

Despite his reluctance, Peirce wrote at least six different textbooks. His first, *Elementary Treatise on Plane Trigonometry*, was published in 1835. His last was the second volume of *An Elementary Treatise on Curves, Functions, and Forces*, which appeared in 1846. He also wrote a book on analytical mechanics, which appeared in 1855, but this was not so much a textbook as a mathematical treatise. Peirce's books were certainly not the most popular mathematical textbooks of their day, but they were adopted at other colleges and several went through a number of editions.[39]

Peirce was glib about some of his textbooks merely being the rehashing of other people's material, and hence, no more than the editing of a compilation of European works.[40] Yet many of Peirce's textbooks contained innovations, which, although sound, were no doubt one of the reasons for their lack of popularity. For example, Peirce followed Ferdinand Rudolph Hassler and Charles Bonnycastle[41] in abandoning the line system in trigonometry. Although this approach was eventually universally adopted and is the one currently used, it met with great resistance.[42] It was not widely adopted in the United States until the publication of William Chauvenet's work on trigonometry in 1850, fifteen years after Peirce published his book.[43] Peirce also used calculus techniques, called infinitesimals, in his geometry book, again an innovation.[44]

Many people complained that Peirce's textbooks were too condensed for the average student. Thomas Hill, who had been Peirce's student and was to be a president of Harvard, defended them, however:

> They were so full of novelties that they never became widely popular, except, perhaps, the trigonometry; but they have had a permanent influence upon mathematical teaching in this country; most of their novelties have now become common-places in all text-books. The introduction of infinitesimals or of limits into elementary books; the recognition of direction as a fundamental idea; the use of Hassler's definition of a sine as an arithmetical quotient free from entangling alliance with the size of the triangle; the similar deliverance of the expression of derivative functions and differential co-efficients from the superfluous introduction of infinitesimals; the fearless and avowed introduction of new axioms, when confinement to Euclid's made a demonstration long and tedious — in one or two of these points European writers moved simultaneously with Peirce, but in all he was an independent inventor, and nearly all are now generally adopted.[45]

Although Hill, as well as others,[46] liked Peirce's textbooks, they were not popular; they met with criticism even at Harvard, Peirce's own domain. Some of his students were known to write in their books, "He who steals my Peirce steals trash."[47] More serious objections were raised to Peirce's textbooks, however. Each year a committee supervised the examinations at the end of the year at Harvard. In 1848, the committee for examination in mathematics was concerned that so few people were taking mathematics there. The committee suggested that the reason for the general lack of popularity of mathematics was Peirce's textbooks. They were abstract and difficult, and few could understand them without extensive explanation. The committee suggested replacing them with textbooks by Bourdon, J. R. Young, and Hutton.[48]

Thomas Hill and J. Gill filed a minority report defending Peirce's textbooks: "Your minority of the committee believe that these textbooks, by their beauty and compactness of symbols, by their terseness and simplicity of style, by their vigor and originality of thought, and by their happy selection of lines of investigation, offer to the student a beautiful model of mathematical reasoning, and lead him by the most direct route to the higher regions of the calculus. For those students who intend to go farther than the every-day applications of trigonometry, this series of books is, in the minority, by far the best series now in use."[49]

Peirce also defended his textbooks. He wrote to President Everett, "It is admitted, I believe, that my whole course is universally well adapted to develop strong mathematical powers and . . . my Geometry and Trigonometry are easy enough for all college purposes. The complaints appear to be exclusively directed against portions of the algebra." Peirce then explained why those parts of the algebra were valuable and not excessively difficult for most students. He pointed out, "At the last examination of the Freshmen class, indeed, many of the students were rigidly questioned upon these very portions of the Algebra and exhibited the most thorough mastery of them, and the committee were, I believe, satisfied that they were not too difficult for a majority of the class, and that the complaints upon this point had very little foundation. I shall rest my defense here, for I cannot fear that under such circumstances, the government of the college will see sufficient reason for taking the extraordinary step of interfering between the head of a department [Peirce] and his textbook." He concluded by thanking Everett, "and be pleased, sir, to accept my grateful acknowledgements for your kindness and delicacy through out this unpleasant transaction."[50]

While Peirce's textbooks were more challenging than many of their contemporaries, the committee's recommendation for Hutton[51] as a replacement is telling. It was an old and obsolete textbook, belonging to the early class of British textbooks that were being replaced by much better books not only in the United States, but in England. That the committee would suggest such a book puts their general credibility into question. At any rate, Peirce and the minority won the day; his textbooks continued to be used at Harvard for many years.[52]

Peirce's students must have liked at least one thing about his textbooks. "When a new book came off the press, Peirce would distribute proof sheets among the students, and accept the discovery of a misprint in place of a recitation."[53]

Φⁿ

Before Josiah Quincy took over the helm at Harvard in 1829, he prepared himself by making a five-week tour of nearby colleges. He was particularly impressed with his visit to Yale and its president, Jeremiah Day. In response to criticisms by the Connecticut legislature, Day, with the aid of the Greek professor, James L. Kingsley, had just issued The Yale Report of 1828. It was an eloquent rationale for the status quo, pointing out the advantages of studying a largely fixed curriculum of Latin, Greek, and mathematics. It impressed many others besides Quincy and was extremely influential on the course of American higher education for the next several decades.[54]

The Report's principal argument was that the current college curriculum was ideally suited to achieve what were widely held to be the two principal goals of a college education. As Day put it: "The two great points to be gained in intellectual culture, are the *discipline* and *furniture* of the mind; expanding its powers, and storing it with knowledge." Of these two goals, Day allowed that the former was perhaps the more important.[55] Peirce recognized the value of mental discipline, but his main objective was to reform Harvard's mathematical curriculum in order to make advanced instruction available to capable pupils.[56] In Peirce's mind, a highly desirable consequence of this action would be to free him of the drudgery of teaching introductory courses.[57]

Day had convinced Quincy of the importance of studying the classical languages. Rather than replace some of the Latin and Greek courses with modern languages, as Ticknor had proposed, Quincy announced that students at Harvard would be spending more time, not less, on the classical languages. The entire freshman year would be devoted solely to Latin, Greek, and mathematics.[58]

Although Quincy wanted Harvard students to study mathematics, he was not averse to making changes, and continued to take an interest in the mathematical curriculum at Harvard and the students' problems with it. He asked Peirce if it might not be better to delay all mathematical instruction until the beginning of the sophomore year.[59] Peirce gave a tactful, although negative, reply to President Quincy's suggestion. Geometry was the first college course at that time, and Peirce pointed out that the novelty of mathematical proofs would not be any less so a year later. He feared the students would do even worse in their sophomore year, pointing out that the older a person is the more difficult it is to begin a new study. "The increase of age is, indeed in the present case only a year, but it is at a time when the habits are rapidly forming, and the mind is hardening into maturity; so that the oldest students in a class are rarely good mathematicians."[60]

Peirce pointed out that freshmen were more enthusiastic and industrious than sophomores. Furthermore, the mathematical knowledge that students brought with them to college would "lie idle and decay for a whole year; time enough for a boy to lose everything he had ever acquired."[61]

Not caring for Quincy's reforms, Peirce lost little time in suggesting changes of his own. Several months later he suggested teaching natural philosophy (physics) earlier to sophomores, using more apparatus in the recitation room, and delaying "the more difficult parts of Pure Mathematics into the Junior Year."[62]

The next fall, Peirce suggested the more radical change of making much of the mathematical curriculum optional. But first he felt compelled to defend the value of mathematics; he wrote to Quincy: "It is thought by some, who are better situated to judge impartially on this subject than I can be, that Mathematics occupies too large a space in the College course of instruction. I shall not undertake the task of showing the incorrectness of this opinion, although I cannot but think that it arises in some degree from a narrow view of the objects of education. It seems to be founded on the fact that Mathematics is of little use to a professional man, and it assumes that practical utility is the only criterion by which to judge of the value of a study."[63]

Peirce agreed that if judged by such a test, very little mathematics need be taught, as few people use it in everyday life. He defended the use of mathematics because of its connection to the physical sciences, and even more importantly for its spiritual value: "Were it only for its connection with the physical sciences, mathematics must hold a high place in every liberal course of education; but when we view it in the light in which it has been viewed by one of the most enlightened and most truly Christian Philosophers of the day, and reflect that its truths are so abstract, so unlimited by time and space that they cannot be lost to the soul even when the soul leaves the body, but that they will go with it into eternity; viewed in this light, does not Mathematics deserve its lofty reputation? And can any attention, which is paid to it, be too great? Or is the time wasted, which is devoted to it? But I will not trouble you farther with my ill-digested reflections."[64]

After defending the value of mathematics, Peirce admitted: "However exalted mathematical truth may be, I am aware that much time is wasted in College pretending to study it." Then Peirce made what was for the time the radical suggestion that only freshmen be required to take mathematics.[65]

A week later Peirce sent Quincy more details of his plan. His chief reason for proposing the plan was that a student could acquire in the freshman year a competent knowledge of "Geometry, Algebra, and Trigonometry," that would be sufficient for an understanding of physics as it was currently taught. Peirce stressed that he did not wish to discourage the study of mathematics among those who liked it: "But still I trust that every inducement will be held out to those who have any taste for Mathematics to continue to attend to it, and that they will not be permitted to desert it for an easier branch of study and yet maintain their rank as scholars; and may it not be forgotten that Mathematics is a far more difficult and therefore a less attractive department than almost any other."[66]

Peirce's proposal did not meet with success. The classicists at Harvard, particularly Charles Beck, the professor of Latin, feared that relaxing the requirements in mathematics would lead to taking the same action with the classics. This might lead to a mass exodus of students from Latin and Greek. Quincy was aware of the radical nature of Peirce's proposal, and realized that his aim was not to excuse indolent and incompetent students from mathematics, but to allow those with mathematical ability and interests to pursue them. Nevertheless, Quincy was not ready for such a sweeping change and he tabled it.[67]

Yet despite a conservative president, time was ripe for reform at Harvard. In the past, change had been opposed not so much by the administration, but by the conventional faculty. By 1835 many of the faculty were new, young, and progressive. Although Beck had opposed Peirce's plan, he was just as committed to Harvard's developing into a true university as Peirce was. He merely had a different agenda to achieve this goal. In fact, one of the reasons Beck had opposed Peirce's proposal was that he thought that if Harvard were to develop into a true university, requirements and standards needed to be raised, not lowered.[68]

Undaunted by his failure, Peirce continued to push for mathematical electives. In January 1838, he wrote President Quincy again.[69] This time Peirce added an ingenious twist to his proposal that would meet the needs of those not mathematically inclined as well as the mathematically gifted. He assured Quincy that by the end of his freshman year a student "will have acquired all the mathematics for which he will [probably] have any use during his life, and will have satisfied the general opinion as to the advantages of these studies in disciplining the mind." For the sophomores, Peirce proposed three options, should they wish to take additional mathematics. One class would be available for those who wished "to become better conversant with practical mathematics." There would be a second class for those who wished "to become qualified instructors," and a third class for "those who have the ability and inclination to become mathematicians, and to be qualified for mathematical professorships." The first two classes would be concluded at the end of the sophomore year, whereas the third would continue until graduation.

Peirce anticipated that the third class "will be a very small one and sometimes be entirely wanting, and yet if successfully instructed, will add more to the reputation of the College than either of the others."

After having been referred to a committee, Peirce's plan was adopted in May 1838.[70] The Harvard *Catalogue* for 1839 stated that after the successful completion of freshman mathematics, a student "may discontinue the study of Mathematics . . . at the written request of his parent or guardian."[71]

Peirce reported to President Quincy in August 1839 that surprisingly few sophomores opted to abandon mathematics at the end of their freshman year, and a remarkable number chose to pursue advanced mathematical study with Peirce. Of fifty-five sophomores, forty-seven took mathematics their sophomore year, "of whom 32 were in the first course, 3 in the second course, and 12 in the third or highest course; during the second term, out of 48, 42 took mathematics, of whom 27 were in the first course, 4 in the second course, and 11 in the third course."[72]

Peirce complained that the principal difficulty in teaching the first course was the want of a textbook for what was then an innovative course. Overall, he was pleased with the results: "With regard to the success of the experiment, I regard it as complete and proving most decidedly the superiority of the volunteer system, and the practicability of adapting different courses of instruction to different classes of students; and this superiority will, I think, become more obvious with each new trial and the introduction of improvements into the details of the instruction."[73]

Peirce added, "My only <u>selfish</u> reason for wishing to have it put into practice, is that my labours may not be of so little avail as they needs must be by the present course."[74]

When students complained that the mathematics course during their freshman year was too demanding, the committee on studies concluded that the time freshmen must spend on mathematics was detracting from their other studies and extended the required mathematical study through the middle of the sophomore year.[75]

In making mathematics an elective after the freshman year, and offering choices to upper classmen, Peirce was treading on relatively new ground. Not until the 1820s did colleges, including Harvard, begin to introduce a limited number of electives. In 1824, the year before Peirce's freshman year, juniors at Harvard could choose a substitute for Hebrew and seniors could choose between chemistry and calculus.[76]

Quincy, seeing the brilliance of Peirce's proposal, became a staunch advocate of electives. Beck, too, had a change of heart; although he considered Peirce's proposal a radical one, he wrote: "I am more and more inclined to think it will be a beneficial one. The college would still furnish, to say the least, the same means of improvement to the students; but it would free itself from the responsibility of obliging the students to avail themselves of them. The public would lose nothing, but the college would gain much."[77] Fully agreeing, Quincy now convinced the corporation to adopt Peirce's proposal to have electives in the mathematical curriculum on a two-year trial basis.[78]

Quincy soon won over most of the faculty to electives. Cornelius Felton, another classicist, happily stated in August 1839, "It begins to appear that the age of scholastic uniformity and conformity has gone."[79] Quincy recommended that the trial program in the Mathematics Department be made permanent, and that it be extended to the Latin and Greek Departments as well. In making these changes, Quincy was deliberately moving towards a university. As one fellow later said, "We regard this question as precisely the question whether Harvard College shall or shall not become a University. . . . In no institution intended to answer the purposes of a University, and to be called by that name is it attempted to carry all the scholars to the same degree of advancement in all departments of study. . . . We hope, therefore, that our college may become, in fact, a University."[80]

By 1843 essentially all courses beyond the freshman year at Harvard were electives. But Harvard had taken too radical a step for the innovations to become permanent. When Quincy was replaced by Edward Everett in 1846, electives were eliminated for sophomores. At this point, Peirce himself had become ambivalent about them: "The practical difficulties of the elective system are so great, and there is so little hope of materially diminishing them that I would gladly persuade myself of the inexpediency of its continuance, and am almost too ready to listen to anything which can be said against it. I think that I perceive it to be accompanied with serious evils, but its advantages are so strong and evident that I dare not yield to my inclinations, but find myself wavering in regard to the question of its abandonment."[81]

Peirce went on to discuss pros and cons of electives. He pointed out that young students, themselves, do not know what course of study is most appropriate for them and that a professor has the responsibility of selecting a course "which they should pursue for the proper discipline of their minds and thorough acquisition of knowledge." On the other hand, Peirce had no doubt that students were much more interested in courses they selected themselves. He also observed that electives allowed a student to obtain specialization in a field, and increased the "quantity and quality of high-scholarship"; he was "fearful, however, that this system tends to a distorted development of some of the mental powers at the expense of the rest." But perhaps the decisive argument for electives was that even "under a prescribed course, a large proportion of students will neglect those studies which they dislike; and thereby make a virtual selection of studies in the worst possible way."[82]

Despite their advantages, more elective courses were withdrawn. Under President Jared Sparks, options were eliminated for juniors. Not until 1883, fourteen years into Charles Eliot's administration, did the curriculum at Harvard become as flexible as it had been under Quincy.[83]

Φ[n]

When Peirce corresponded with President Quincy about splitting the sophomores into three sections, he included the general principles of engineering in the curriculum for the applied mathematics class. He pointed out that he would "be able to give but little instruction upon this subject, for the greater part of Engineering, such as the making of roads, canals, &c, the making of boundaries &c is a trade and not a science; it is not to be acquired at school or even in the office of an engineer, except in actual service."[84] Despite the negative comment, Peirce's including engineering in the applied course for sophomores marked not only the beginning of engineering at Harvard, but also the dawning of the Lawrence Scientific School.

After talking with an engineer, Peirce had a complete change of heart about collegiate engineering instruction. Recognizing that an increasing number of young men wished to become engineers, Peirce pointed out the benefits of collegiate education in engineering and proposed establishing an engineering school at Harvard on a plan similar to Harvard's "eminently successful" law school. While Peirce feared that many practical men would object to college training for engineers, he was "of the belief that it would meet with great favour from the public, and would be much frequented, and would do much good. Even the most practical admit that a thorough mathematical education gives a young engineer great command over the details of his profession."[85] The corporation referred Peirce's suggestion to teach civil engineering at Harvard to a committee consisting of the president, Judge Story, and Judge Shaw.[86]

Almost a year later, Peirce wrote President Quincy, saying that on further consideration his original plan for an engineering school might require too many of the university's resources, and proposed instead that students be allowed to enter Harvard who would not be required to complete the entire curriculum, but "who may join the college, shall have the liberty of studying any branch they may desire to study, and shall be entitled to certificates upon the same principles as the regular students of the College."[87] By doing this a student would be able to specialize in a particular field and receive far more advanced instruction in it than he would by following the usual degree curriculum. Professors would benefit from the plan by being able to work with more advanced students, although apparently without additional compensation or reduction of their existing workload.

Discussion of an engineering school continued, and in the spring of 1840, Peirce reported that the plans "for the formation of a School of Engineers" had been delayed by "want of information regarding the Polytechnic School of France." But the information was now available, and Peirce was now ready to present his ideas to the corporation.[88]

Peirce proposed a school to be patterned after the Royal Polytechnic School of France. As with the Law School, the complete course of the school would take two years. Students entering the program would either have completed the usual collegiate scientific education or be given the opportunity of devoting a year of study at Harvard before beginning the engineering curriculum. Instruction in writing and languages would be dispensed with to allow time to study civil engineering.[89]

The committee that reviewed Peirce's proposal felt that an engineering school should be established, "provided it can be done without the appointment of any other Professors or incurring any important additional expense to the college."[90] They agreed that the school should be patterned after the Law School and recommended that the matter should be referred to the scientific faculty of the college—

Peirce, Treadwell, Webster, and Lovering—since the success of the school would depend on their efforts.[91] Since additional endowment was clearly needed to support the school, Peirce's plans received the blessing of the corporation, but languished for lack of money.

Even without an engineering program, Harvard in 1845 could boast considerable scientific distinction, although it lagged behind some other American universities, most notably Yale.[92] Harvard now had a fine observatory ably manned by William Bond and his son, George. It also could boast of two professors who were the leading Americans in their fields: Benjamin Peirce in mathematics and Asa Gray in botany. Gray's appointment to Harvard in 1842 was a bellwether for science there. Gray was chosen over several men who had closer connections to Harvard, but lacked specialization and excellence in research. President Quincy had agreed that the newly founded professorship should emphasize research as much as teaching.[93]

Gray and Peirce had much in common; both were outstanding in their field and both were dedicated to the advancement of American science. But they had different temperaments and were unfortunately usually at odds. The two scientists worked together briefly on phyllotaxy, the study of the arrangement of leaves on a stem, in 1849, but the matter-of-fact Gray thought Peirce "ran off into the sky, dementedly."[94] In addition to having different scientific temperaments, Peirce resented Gray's leadership in the American Academy of Arts and Sciences and often told his friend, Alexander Dallas Bache, that Gray was showing the "cloven foot."[95]

Despite Harvard's scientific strengths, instruction at the college was still largely by recitation and financial support for science was intermittent. For example, Gray had insufficient funds for the botanical garden, and there was no endowment for a salary for the director of the observatory. But suddenly the future for science at Harvard looked much brighter. Three things happened in 1846 that boded well for science at Harvard: the Rumford Chair became vacant, Edward Everett became its new president, and the Swiss naturalist, Louis Agassiz, joined the faculty.

The Rumford Chair had been endowed by Benjamin Thompson (1753–1814), later Count Rumford, who came from a family of modest New England farmers. He married into wealth, however, and since he was a Tory, he left America for London during the American Revolution. Thompson did important scientific work on the nature of heat and was one of the first great scientists born in America, although he did all his work abroad. He was knighted by George III and given the title of Count of the Holy Roman Empire by the Elector of Bavaria.[96]

Although an expatriate, Thompson remembered his native land with several bequests. One of them was to Harvard University to endow a professorship "of the application of science to the art of living."[97] Although by 1846 the endowment for this chair provided only the modest salary of about seventeen hundred dollars, the position had only nominal teaching duties. Thus, it was one of the very few, probably the only, research professorships in the country.[98] "My own desire," Peirce wrote to Bache, "and one I am sure of being able to accomplish if the right man can be found, is to make this professorship be head and center of a new school to be attached to the University for instruction in practical science; and upon this school I hope to concentrate all the scientific ability attached to the University. You see enough to understand what I am, for a word is sufficient to the wise man. Oh that we had a Bache to undertake it!"[99]

The leading candidate for the position was the geologist, Henry Darwin Rogers. Rogers had strong support from the Boston area, including that of G. B. Emerson, president of the Boston Society of Natural History. However, two years previously Rogers had endorsed the doctrines espoused in the *Vestiges of Creation*.[100] This book was an anonymous forerunner of Darwin's theory of evolution, and Rogers's accepting it made him controversial in scientific circles. Despite the fact that he was the only American to have held a faculty position in a European university up to this time, Peirce thought his science was "deficient in accuracy" and opposed him.[101] Peirce's efforts to counter his appointment were sidetracked, because of his wife's illness. But, uncharacteristically, Asa Gray sided with Peirce in opposing Rogers, and they were successful in blocking his appointment.

Although Rogers's election was at least temporarily blocked, no obvious alternative appeared. Peirce favored the naturalist, Jeffries Wyman,[102] but there were other candidates. John Webster, Harvard's professor of chemistry and mineralogy, was lobbying for his former student, Eben Horsford, who was studying chemistry at Giessen under Justus Liebig, one of Europe's leading organic chemists. With Horsford on the faculty, Webster would no longer be required to teach chemistry, but could concentrate on geology and mineralogy, which he much preferred to teach. But Webster had even stronger reasons to oppose Rogers than did Peirce. Rogers and G. B. Emerson had attempted to have Webster dismissed on charges of incompetence and replace him with Rogers. Understandably Webster was now adamantly opposed to Rogers's election.[103] For a year things remained at an impasse.

Even though Peirce said he wished he could find a Bache to fill the position, he had no desire to take his friend away from his work on the Coast Survey. Luring Joseph Henry from Princeton to Cambridge was an entirely different matter. Henry was not only a friend of Peirce, but an old and good friend of Asa Gray. Gray wrote to Henry telling him that only the small salary had prevented his name from being brought before the corporation. "If you would come," he wrote, "we would all be delighted and you would be elected by acclamation."[104] Gray wrote to Henry again at the end of the year, urging him not to take the position of secretary of the Smithsonian unless everything there was to his liking. He now had reason to believe that the Rumford professorship would carry a better salary.[105] Henry declined the Rumford professorship and became secretary of the Smithsonian.

With the Rumford Chair still vacant, Harvard got a new president, Edward Everett. After Everett had graduated first in his class from Harvard in 1811, he became minister of the Brattle Street Church in Cambridge, one of the leading Unitarian pulpits in the country. He was soon asked to serve as Harvard's first Eliot Professor of Greek Literature. Everett accepted on the condition that he be allowed to go to Germany for advanced study. Armed with the first German PhD to be earned by an American, he assumed his short-lived duties as professor of Greek in 1819. At this time he also became editor of the *North American Review*.[106]

Everett excelled as an orator even more than as a scholar. Unhappy at Harvard, he was urged by his friends, including Daniel Webster, to pursue a political career.[107] Before returning to Harvard as its president, Everett had had a distinguished career in public service. He had served in the United States House of Representatives from 1825 to 1835 and as governor of Massachusetts from 1836 to 1839. Before becoming president of Harvard, Everett had been appointed by his friend, Daniel Webster, then secretary of state, to be ambassador to Great Britain. When Quincy resigned as president of Harvard, Everett was a popular choice for his replacement. Despite the general enthusiasm for his being president, he was reluctant to return to academia. The duties of a college president at that time included an enormous amount of minutia; he was also personally responsible for student discipline. When contemplating the job, he wrote, "No parent who has children of his own needs be told what it must be to be the head of a family of two hundred boys, all at the most troublesome and anxious age."[108] Everett reluctantly accepted the post, but he found many of the duties to be degrading. After repeated threats to resign, he finally did so three years later. He then did work for Daniel Webster, who was secretary of state, and upon Webster's death became secretary of state himself. In 1853 the Massachusetts legislature elected him to the United States Senate.[109]

Despite his short tenure and lack of enthusiasm, Everett made a significant contribution to Harvard, and Peirce found him to be a key ally in his quest for a scientific school there. On December 27, 1845, shortly before Everett took the helm, the corporation appointed a committee to study how science at Harvard could be improved. The subject was discussed at a meeting the following January,[110] but conditions were now favorable for change.[111]

Early in 1846, Peirce presented a new plan for a "School of Practical and Theoretical Science." The plan, much like Peirce's previous one for the engineering school, proposed a course of study that was to be completed in two years, although a student would be allowed to begin his course of study at any point and elect the course he wished to pursue. Engineering could be studied at this school, but so could "any branch of science." The school would be capable of turning out architects and astronomical observers as well as civil engineers.[112]

Peirce also saw the new school as a vehicle for reorganizing science and scientific instruction at Harvard: "The object of this school is to bring the University into more immediate and intimate connection with the community to which it belongs, by supplying the public demand for scientific education of various kinds. It is intended to subserve the purpose of organizing the different scientific professorships, and enabling the professors to cooperate in an efficient system of instruction, distinct from that given in the present academical course. It is not to be regarded as an institution for general liberal education, but as designed to furnish one of a particular and professional character."[113]

In his inaugural address in April 1846, Everett called not only for a scientific school at Harvard to furnish "a supply of skillful engineers," and the establishment of a "philosophical faculty, in which the various branches of science and literature should be cultivated, beyond the limits of an academical course."[114] Everett was aware that his plans called for much more than a mere reorganization at Harvard and that they would require new funds.

$$\Phi^n$$

As Peirce told Bache, finding the right person for the Rumford Chair was a key factor in establishing the scientific school. Horsford, Webster's candidate for the Rumford Chair, was still in Giessen and poorly informed about the status of the position. He arrived in New York in January 1847, only to receive a letter from Webster telling him that all was lost. Although no one had yet been selected for the post,

Webster was sure that Horsford was no longer a candidate. Horsford, nevertheless, went on to Cambridge, if only to see what he had missed. It was a wise decision; after spending only two days there talking with faculty and university officials, he convinced everyone that he was the man for the job. The corporation unanimously elected him at the end of the month.[115]

Later that year, Abbott Lawrence, a New England industrialist, answered Edward Everett's call for money to support academic reforms at Harvard. Lawrence was frustrated by the lack of trained engineers and scientists available to support New England industry and wished to found a school "for the teaching of the practical sciences." He was pleased when he "learned . . . that the government of Harvard University had determined to establish such a school in Cambridge and that a professor had been appointed who is eminent in the science of Chemistry and who is supported by the munificence of the late Count Rumford."[116]

Lawrence donated fifty thousand dollars to endow a scientific school.[117] He suggested that the school should have three permanent professors, professors of engineering and geology, as well as the already appointed professor of chemistry. Lawrence described what he envisioned for the new school: "I deem it of the highest importance; and in fact essential, that none but first rate men should occupy the Professors['] chairs in this School—Its success depends upon the character of the Instructors—They should be men of comprehensive views, and acknowledged talent, possessing industry and integrity with an enthusiastic devotion to the great interests of science—they should love their profession, and work in it day by day—such teachers will soon gather around them a large number of pupils."[118]

Unfortunately, Lawrence's donation of fifty thousand dollars was insufficient to finance all of these plans. Horsford had hoped to have that much endowment for his chemical laboratory alone, and the money had to be divided between the laboratory and the endowment for the other professorships. Nevertheless, using what money was available, Horsford managed to build the best chemical laboratory in the country. Yet he failed to become a leading research chemist, contributing little to the advancement of scientific knowledge.[119]

A third thing that proved beneficial for science at Harvard in 1846 was the coming of Louis Agassiz to the United States.[120] Agassiz's coming also decisively shaped the Lawrence School. Born in French Switzerland in 1807, Agassiz pursued a degree in medicine to please his family, but his passion was zoology. After completing both an MD and PhD at Munich, he began a successful career as a naturalist by completing an outstanding study of Brazilian fishes. Blessed with a charming personality, he soon won the patronage of Georges Cuvier, the French paleontologist, and Alexander von Humboldt, the Prussian naturalist. Humboldt obtained a position for Agassiz at a museum in Neuchâtel. There he established himself as a leading scientist by completing an outstanding five-volume work, *Researches on the Fossil Fishes*. Agassiz's best-known work, however, was on glacial theory; he was one of the formulators of the concept of the Ice Age.

Agassiz had an insatiable appetite for scientific work and for collecting zoological specimens. By 1846, he had overextended himself and had fallen deeply in debt from a scientific publishing business. His wife, frustrated with his publishing losses, troubles with his scientific assistants, and his lack of attention, took the children and moved back to live with her parents. Agassiz was looking for a fresh start. He turned to his friend Humboldt, who was able to secure for him a grant from the King of Prussia to study the natural history of North America. Seeking additional income, he arranged through another friend, the British geologist Charles Lyell, to give a series of lectures at the Lowell Institute in Boston. Although Agassiz announced that he would stay in America for only two years, he concluded his affairs in Europe with a thoroughness that implied an indefinite stay.

Agassiz charmed Boston; he charmed the American scientific community; he charmed American society; he charmed America. He chose to deliver a series of lectures on the "Plan of Creation in the Animal Kingdom," for the Lowell Institute. Agassiz had an uncanny gift for making science fascinating to the most disinterested layman. The response to his lectures was overwhelming; some five thousand people wanted to hear them and they had to be repeated twice to accommodate the demand.

Abbott Lawrence was as charmed by Agassiz as anyone. Although Lawrence had originally wanted to train engineers for his mills, with the availability of Agassiz, Lawrence's plans changed. He wanted Agassiz at his scientific school. Although Agassiz's accomplishments in geology were substantial, he was primarily a zoologist, and he was certainly not an applied scientist. It is unclear whether Lawrence made his bequest to Harvard before or after talking with Agassiz about taking the professorship in geology. Late in his life, however, Agassiz said that he had talked to Lawrence before the donation was made, and that one of the reasons he accepted the position was to ensure Lawrence's carrying out his plans.[121]

Soon after Agassiz came to America, his wife died of tuberculosis, and two years later Agassiz married Elizabeth Cabot Cary, the daughter of a wealthy, socially prominent Boston lawyer. Elizabeth was sympathetic to her husband's voracious zeal for collecting zoological

Louis Agassiz. By permission of the Harvard University Archives, cal # HUP [17]

specimens, and founded a school for girls in order to help finance his work. The school was the forerunner of Radcliff College, of which Elizabeth became the first president.

Peirce found in Agassiz a staunch ally in his quest to build American science, and the two of them became close friends. Agassiz and his wife bought a house near the Peirce's home, and Peirce's son Charles recalled, "Agassiz came in every day without ringing, and standing in the large hall, would call 'Ben!' "[122] Agassiz's personality, charm, social standing, and European reputation as a scientist added greatly to American science. In an age when the American public did not believe that science was worthy of support, Agassiz made it very much so.[123]

Φⁿ

Hiring Agassiz for the Lawrence School produced an institution that was significantly different than Lawrence's original concept. But engineering was still to be taught in the school, although finding a professor of civil engineering proved difficult. An army engineer, Captain Henry W. Halleck, was offered the position in September 1848, but he turned it down. The position wasn't filled until a year later by a man who had graduated from both West Point and Harvard, Henry Lawrence Eustis. Consequently, engineering instruction did not begin until three years after the school was launched.[124]

The founding of the Lawrence School and making mathematics an elective for upper classmen allowed Peirce to offer a course of mathematics that was far more advanced than any given in an American university until the founding of Johns Hopkins in 1876. By 1848 students could study the works of Cauchy, Monge, and Hamilton, as well as take advanced courses in analytic and celestial mechanics, and optics.[125]

Φⁿ

In 1848 there were apparently criticisms of Peirce's light workload and of his teaching; perhaps these accompanied the criticisms of his textbooks. Peirce responded to them by describing his duties and his mode of instruction:

> I have no fear that I shall be called upon to work harder than I now do, for my whole time is occupied, & I shall always give the preference to that work which is most likely to be useful and to promote the best interests of the institution whose servant I am in heart and soul. Others may boast of

their love of our alma mater as loudly as they please, I challenge any one of them to manifest his affection by more earnest & self-denying efforts to contribute to her permanent welfare and her true reputation. . . .

The whole of the mathematical instruction is performed by me without the aid of a tutor. I occupy as many hours in the recitation room as the others of my departments will grant me, and I do not think that I have any right to complain of their illiberality to me in this respect. I deliver about two hundred and fifty lectures in a year every one of which is extemporaneous, and some of them embrace discussions upon the most difficult points of mathematical science. Some of the students desire to become mathematicians, and I consider it my duty to do all in my power to aid them in this object, and to let them into the very work shop of the mathematician. I am well aware that some regard me as aiming too high, and overshooting the intellect of my pupils; as a reply to this charge, I appeal with unhesitating confidence, to the reports of the mathematical committees, and to the mathematical state of the course as it was before I had charge of this department.[126]

Peirce did care about his students as individuals. He stressed their importance before the American Association for the Advancement of Education: "I begin to think that even in our Common School system, excellent as it is, there is one great defect. As it is administered at present in my own State, Massachusetts, I am sure that there is. It partakes too much of the character of a sort of manufactory, in which masses of educated men are to be turned out as if they were screws or pins. This is no way to educate men. Men have individual characters. Their Deity has made them with specialty, and we can not unmake them."[127]

$$\Phi^n$$

As a teacher, Peirce received mixed reviews from his students, many of whom left descriptions of his teaching.[128] Perhaps the most charming one was written by a twelve-year-old girl, Lillie Greenough. Miss Greenough was a student in Mrs. Agassiz's school:

You say in your last letter, "Do tell me something about your school." If I only had the time, I could write you volumes about my school, and especially about my teachers.

To begin with, Professor Agassiz gives us lectures on zoology, geology, and all the other ologies, and draws pictures on the blackboard of trilobites and different fossils, which is very amusing. We call him "Father Nature," and we all adore him and try to imitate his funny Swiss accent.

Professor Pierce, who is, you know, the greatest mathematician in the world, teaches us mathematics and has an awful time of it; we must be very

An early portrait of Benjamin Peirce, ca. 1845. By permission of the Harvard University Archives, call # HUP [1a].

stupid, for the more he explains, the less we seem to understand, and when he gets on the rule of three we almost faint from dizziness. If he would only explain the rule of one! The Harvard students say that his book on mathematics is so intricate that not one of them can solve the problems. . . .

Cousin James Lowell replaces Mr. Longfellow the days he can't come. He reads selections of "literary treasures," as he calls them, and on which he discourses at length. He seems very dull and solemn when he is in school; not at all as he is at home. When he comes in of an afternoon and reads his poems to aunty and to an admiring circle of cousins and sisters-in-law, they all roar with laughter, particularly when he reads them with a Yankee accent. He has such a rippling little giggle while reading, that it is impossible not to laugh. . . .

There are about fourteen pupils now; we go every morning at nine o'clock. We climb up the three stories in the Agassiz house and wait for our teachers, who never are on time. Sometimes school does not begin for half an hour.

Mrs. Agassiz comes in, and we all get up to say good morning to her. As there is nothing else left for her to teach, she teaches us manners. She looks us over, and holds up a warning finger smilingly. She is so sweet and gentle.

I don't wonder that you think it extraordinary that all these fine teachers, who are the best in Harvard College, should teach us; but the reason is, that the Agassiz's (sic) have built a new house and find it difficult to pay for it, so their friends have promised to help them to start this school, and by lending their names they have put it on its legs, so to speak.[129]

Lillie was more than a girl who could write entertaining letters. Peirce's sister, Elizabeth described her as being "as beautiful as the dawn."[130] She also had a beautiful voice. Lillie wrote to her mother about the following incident:

The other day I was awfully mortified. Mr. Longfellow, who teaches us literature, explained all about rhythm, measures, and the feet used in poetry. The idea of poetry having feet seemed so ridiculous that I thought out a beautiful joke, which I expected would amuse the school immensely; so when he said to me in the lesson, "Miss Greenough, can you tell me what blank verse is?" I answered promptly and boldly, "Blank verse is like a blank-book; there is nothing in it, not even feet," and looked around for admiration, but only saw disapproval written everywhere, and Mr. Longfellow, looking grave, passed on to the next girl. I never felt so ashamed in my life.

Mr. Longfellow, on passing our house, told aunty that he was coming in the afternoon to speak to me; aunty was worried and so was I, but when he came I happened to be singing Schubert's "Dein ist mein Herz," one of aunty's songs, and he said, "Go on. Please don't stop." When I had finished he said:

"I came to scold you for flippancy this morning, but you have only to sing to take the words out of my mouth, and to be forgiven."

"And I hope you will forget," I said, penitently.

"I have already forgotten," he answered, affectionately. "How can one be angry with a dear little bird?" But don't try again to be so witty."

"Never again, I promise you."

"That's the dear girl you are, and 'Dein ist mein Herz'!" He swooped down and kissed me.

I burst into tears, and kissed his hand. This is to show you what a dear, kind man Mr. Longfellow is.[131]

Three years later, when Lillie was fifteen, she was the first serious love of Peirce's son Charles. But her mother whisked her off to Europe to avoid the match. The Peirces were close friends with the Greenoughs, but they didn't have enough money.[132] Lillie studied voice in London under Manuel Garcia, the most celebrated European voice teacher of the nineteenth century. Among his pupils were Jennie Lind, the Swedish Nightingale, and the operatic soprano, Mathilde Marchesi. At seventeen, Lillie married Charles Moulton, the son of an exceptionally wealthy American banker in Paris. She was a favorite at the Court of Napoleon III, and became friends with much of the nobility of Europe, as she did with the musicians Wagner, Liszt, Auber, Gounod, and Rossini. After her first husband's death, she married the Danish diplomat, Johan Henrik de Hegermann-Lindencorne. Peirce's son Charles subsequently married Lillie's cousin, Melusina Fay, or Zina. The marriage was not happy, but not because Charlie still cared for Lillie.[133]

Although Lillie found Peirce baffling, as a teacher, she had fond memories of him. Years later she wrote Zina, "And the dear, dear professor how I sh'd like to see him."[134] Most of Peirce's students had a remarkably similar experience with Peirce as a teacher as did Lillie. He was not only capable of confounding twelve-year-old girls; he was capable of confounding all of his students. Yet, like Lillie, most remembered him with affection.

Several writers have maintained that Peirce was not a good teacher for the average student. Perhaps so. A host of anecdotes have survived in which Peirce is described as highly eccentric and horribly incompetent in the classroom, but they all end with the caveat, "but don't misunderstand, Peirce was a great teacher."[135] One such description is by Charles W. Eliot, later president of Harvard:

His method was that of the lecture or monologue, his students never being invited to become active themselves in the lecture room. He would stand on a platform raised two steps above the floor of the room, and chalk in hand cover the slates which filled the whole side of the room with figures, as he slowly passed along the platform; but his scanty talk was hardly addressed to the students who sat below trying to take notes of what he said and wrote on the slates. . . .

When I entered College in 1849 Professor Peirce had ceased to have to do with the elementary courses in Mathematics. He taught only students who had been through the two years of prescribed Mathematics and had elected to attend his courses, which were given three times a week

throughout the junior and senior years. Two or three times in the course
of the hour, Professor Peirce would stop for a moment or two to give
opportunity for the members of the class to ask questions or seek explanations; and these opportunities were utilized by all the members who
really wanted to learn. If a question interested him, he would praise the
questioner, and answer it in a way, giving his own interpretation to the
question. If he didn't like the form of the student's question, or the manner in which it was asked, he would not answer it at all, but sometimes
would address an admonition to the student himself which went home. . . .

   In spite of the defects of his method of teaching Benjamin Peirce was
a very inspiring and stimulating teacher. He dealt with great subjects and
pursued abstract themes before his students in a way they could not grasp
or follow, but nevertheless filled them with admiration and reverence. His
example was much more than his word.[136]

Another President of Harvard, A. Lawrence Lowell, who was Peirce's
student, described him similarly: "Looking back over the space of fifty
years since I entered Harvard College, Benjamin Peirce still impresses
me as having the most massive intellect with which I have ever come
into close contact, and as being the most profoundly inspiring teacher
that I ever had. His personal appearance, his powerful frame, and his
majestic head seemed in harmony with his brain."[137]

As to the details of his teaching, Lowell noted:

   As soon as he had finished the problem or filled the blackboard he
   would rub everything out and begin again. He was impatient of detail, and
   sometimes the result would not come out right; but instead of going over
   his work to find the error, he would rub it out, saying that he had made a
   mistake in a sign somewhere, and that we should find it when we went
   over our notes.
      Described in this way it may seem strange that such a method of teach-
   ing should be inspiring; yet to us it was so in the highest degree. We were
   carried along by the rush of his thought, by the ease and grasp of his intellectual movement. The inspiration came, I think, partly from his treating
   us as highly competent pupils, capable of following his line of thought
   even through errors in transformations; partly from his rapid and grace-
   ful methods of proof, which reached a result with the least number of
   steps in the process, attaining thereby an artistic or literary character; and
   partly from the quality of his mind which tended to regard any mathematical theorem as a particular case of some more comprehensive one,
   so that we were led onward to constantly enlarging truths.[138]

Although Josiah Quincy, who was president from 1829 to 1845, supported widespread reform in electives, making a radical departure
from traditional college curriculum, his administration was other-

Benjamin Peirce with a class, ca. 1864. Peirce's son Benjamin Mills Peirce is standing on the right, next to his father. By permission of the Harvard University Archives, call # HUP Peirce, Benjamin AB 1829 [11].

wise marked by harsh discipline and reactionary teaching methods. Changes that had been initiated by such professors as George Ticknor in the number of classes taught by recitation were now abolished. Not only were more classes taught by recitation, Quincy also required professors to give grades on each student's recitation. Doing so greatly strengthened the adversarial feelings between student and teacher, making it nearly impossible for the professor to appear as a mentor or aid. Quincy believed that the ideal college course should consist of a thorough drilling. Under Quincy that's what the students got.[139]

Quincy was Harvard's most unpopular president since Leonard Hoar, who had served from 1672 to 1675.[140] Quincy's discipline was unyielding and harsh. These tactics resulted in the rebellion of 1834, one of Harvard's worst. The entire sophomore class was suspended for a year; in addition, seven freshmen, one junior, and seven seniors were dismissed from the college.[141]

In this atmosphere, Peirce appears as a demanding and often perplexing, but benevolent instructor. Although Peirce may have confused his students, they admired and respected him. There's no evidence of his having trouble with discipline with his classes, as did Asa Gray when he first taught at Harvard during the same period.[142] Peirce's students affectionately referred to him as "Benny." If he did not always make mathematics lucid, he did not demean students who were struggling unsuccessfully.[143]

Another of his students, George A. Flagg, remembered Professor Peirce's manner as being kind and genial with his small class, respecting their courage for attempting the rigors of his mathematical program, but his talk was often beyond his students' ability. " 'Do you follow me?' asked the Professor one day. No one could say Yes. 'I'm not surprised,' he said; 'I know of only three persons who could.' "[144]

The excitement Peirce generated in lecturing about mathematics extended to general audiences as well as to his students with mathematical talent. One of the first innovations made by Thomas Hill when he became president of Harvard in 1862, was to institute courses of lectures open not only to graduate and other advanced students, but to the educated public. They were called University Lectures. While these lectures were intended to be advanced, they were not to be technical. Peirce was one of the first to participate in the University Lectures and did so during five different years. Charles Eliot recalled that: "His University Lectures were many a time way over the heads of his audience, but his aspect, his manner, and his whole personality held and delighted them. An intelligent Cambridge Matron who had just come home from one of Professor Peirce's lectures was asked by her wondering family what she had got out of the lecture. 'I could not understand much that he said; but it was *splendid.* The only thing I now remember in the whole lecture is this—"Incline the mind to the angle of 45°, and periodicity becomes non-periodicity and the ideal becomes the real." ' "[145]

Perhaps the most flattering description of Peirce as a teacher, as a man, was given by Edward Everett Hale. Hale, author, abolitionist, reformer, and Unitarian minister, is perhaps best remembered today for his story, "The Man without a Country." Significantly, Hale was a student without mathematical interests. Hale entered Harvard hoping to receive instruction in the classics. In this he was disappointed, but not in Peirce:

> The experience . . . with Benjamin Peirce was that of real education. He was still young,—only twenty-eight years old. If we wanted to have anything explained to us, we might go to his table and sit down with him and have a perfectly friendly talk: of which the consequence was that we learned

something from a teacher. In the course of my life I have not had more than five such experiences with persons who took the name of teacher. . . . Here was Peirce, a leader of leaders, perfectly willing to take by the hand the most ignorant freshman. You felt confidence in him from the beginning, and knew he was your friend.

The common thing to say about Peirce was that the steps of his ladder were so far apart that, though he ascended it easily, other people fell through. Very likely this was true, but he kindled you with the enthusiasm which you needed. I have never forgotten the awful rebuke he gave to the class one day, when some fellow had undertaken to cheat at the blackboard. Peirce cut short the formal mathematics to give a lesson about truth. The mathematics were the voice of God; we were in that room because we wanted to find out what truth was; and here was a son of perdition who had brought a lie into that room. We went out from the recitation-room sure that we had been very near God when we listened to that oracle. That sort of revelation between teacher and pupil shows what is meant by "education."[146]

5

The Feeling of Mutual Goodwill

By the time he was thirty, Peirce was one of the leading mathematicians in the country.[1] His reputation was, in no small part, due to his activity with the nation's mathematical periodicals. Until the publication of the *Analyst* by Joel Hendricks in 1874, none of these journals were published for more than a few years.[2] All of them, including the *Analyst,* were principally problem-solving journals.[3] Contributors would submit problems which would be published in one issue; solutions would be submitted and published in a subsequent issue. These journals included a limited amount of other material, such as reviews and research articles, but the first true research mathematical journal in the United States, the *American Journal of Mathematics,* did not begin publication until 1878. Although ephemeral and often on a low mathematical level, these journals had value in disseminating mathematical knowledge. Despite the small number of subscribers to the journals, they had a surprisingly wide circulation. Peirce's former student, B. A. Gould, who had gone to Europe for graduate study, wrote, "In looking over the Observatory Library [Royal Observatory, Greenwich] here, I was exceedingly delighted by finding two numbers of the Math. Miscellany. They seemed like old friends, to one among so strange scenes & faces."[4]

Peirce continued to submit material to the *Mathematical Diary* after he started teaching at Harvard. At this time he became friends with Samuel Ward, who was also contributing to the *Diary* and was assisting James Ryan with editing the journal.[5] Ward, the son of a wealthy New York banker, graduated from the Round Hill School the spring before Peirce arrived there. In 1831, just after graduating from Columbia in two and a half years, Ward made a trip to Boston to visit Bowditch, who had agreed to help him in his studies. While there he met Peirce, the young Harvard tutor. Peirce and Ward corresponded about mathematical problems that appeared in the *Diary,* discussed details of the journal's production, and gossiped about other American mathematicians.[6]

Ward praised Peirce's work: "Your solution to question VIII is <u>beautiful</u> and looks very well in print, I should not say too much in asserting that it is by far the most interesting and astute paper that I have seen as yet in the *Diary*."[7] In another letter he mentions that Mr. Ryan "greatly admires your paper on perfect numbers."[8] This was indeed a work of distinction. Here, Peirce proved that no odd perfect number has fewer than four prime factors.[9] Perfect numbers are those whose divisors, not including the number itself, add up to the number, for example, $6 = 3 + 2 + 1$ and $28 = 14 + 7 + 4 + 2 + 1$. The Pythagoreans named these numbers around 600 BC and by the first century AD, the Greek mathematician Nicomachus listed four perfect numbers in his *Introductio Arithmeticae*.[10] Euclid proved in his *Elements* about 300 BC that if $2^n - 1$ is prime, then the number $2^{n-1}(2^n - 1)$ is perfect. Roughly two thousand years later Leonard Euler proved that all even perfect numbers are of this form, but the question remained open if there were odd perfect numbers.[11] Thus Peirce gave a partial answer to the question of the existence of odd perfect numbers. The question of whether there are odd perfect numbers may be the oldest open problem in mathematics.[12]

The above problem falls into the field of number theory. Problems in this field were popular with the largely amateur contributors to the *Mathematical Diary*. For example, one contributor submitted the problem: "To find four numbers, such that if the square of each be subtracted from their sum, the remainders shall all be squares."[13] Another asked: "To find two numbers whose sum shall be a cube, and the sum of their squares increased by three their sum shall be a square,"[14] and another problem solved by Peirce: "The fractions $\frac{2}{3}$ and $\frac{1}{3}$ are such that their difference is equal to the sum of their cubes: Are there other fractions having the same properties?"[15]

What makes Peirce's article on number theory notably better mathematics than the contributions of his contemporaries? For one thing, it is a harder problem. Its solution shows more mathematical ability than solutions to the other problems, whose solutions are similar to those of like problems. And, all of these problems merely ask for specific examples. They may be interesting and intellectually challenging, but they are not mathematically significant. Peirce's result deals with the basic properties of numbers.[16]

In 1832, James Ryan lost a sizable amount of money and was not able to continue publishing the *Diary*, which he had been doing at a financial loss.[17] Ward took over the editorship of the journal for one issue, and published an article that was offensive to some of its subscribers. Whether as a result of Ward's article or Ryan's financial reversals, this was the last issue of the *Diary* published.[18]

Soon afterwards, Ward visited West Point as a member of the examining board. He so impressed the people at the academy that they asked him to be on the faculty. Ward told his father that in order to do the job adequately he needed to go to Europe for extensive study. His father supported his ambition, and the following November Ward embarked for Europe to study mathematics. There he became the first American to earn a PhD in mathematics. Upon his return to the United States, however, Ward decided that he should be very rich before embarking upon an academic career.[19] After inheriting his father's fortune, he soon lost it trying to make a bigger one very rapidly. He then went to San Francisco, where he quickly made and lost a second fortune in land speculation. His mathematical interests having waned by this time, Ward found that his talents made him a good lobbyist, and he became known as the "King of the Lobby."

$$\Phi^n$$

In one of Ward's letters to Peirce, Ward described one of the new contributors to the *Diary:* "Mr. Gill, your fellow correspondent in questions 3 & 5, is the English mathematician of whom I spoke to you when last at Cambridge. He is a person short, thick & in look to my ideas exceedingly vulgar, but I am inclined to think from some specimens of his power in the English periodicals that he is extremely able in some respects. The Diophantine is his forte.[20]

Four years later, Charles Gill began editing the nation's next mathematical journal, the *Mathematical Miscellany.*[21] In a letter to Peirce, inquiring about his analytical geometry textbook, Gill mentioned that he intended to publish the *Miscellany* semiannually and hoped that Peirce would be a contributor.[22] Peirce responded:

> I should not have suffered your letter to lie a week unanswered if I had not hoped to find time to prepare something for your periodical. But all the time which I could spare from my college exercises has been necessarily devoted to the preparation of my Spherical Trigonometry, which I am hurrying forward as fast as I can in order to get it ready in time for my class. I shall however exert myself to send you something as soon as possible. I wish you success in your undertaking and if in any way I can be useful to you, do not hesitate to command my services. My own experience and slight acquaintance with the history of science have long convinced me that nothing contributes more to bring out mathematical talent than works of this kind. They serve as an arena, upon which we can readily come into collision with other minds and measure our wits with theirs; and we all feel that there is an irresistible fascination in the contest.[23]

Peirce did support Gill's journal by sending him solutions. Gill was pleased with Peirce's contributions to the *Miscellany:* "I received your solutions last night and they were as usual very welcome. After the stuff I have crammed down my throat—for in my unenviable capacity as editor—I have all kinds of crude corruptions thrust upon me which I am obliged to wade through—bring into order, and 'lick into shape.' You can have no idea of the absolute pleasure conveyed to me by the perusal of your solutions. They are like the oasis in the desert."[24]

Peirce supported the journal and Gill, even to the detriment of his personal interests. He wrote to Gill, "As to inserting my solutions or not, do, I beg you, have regard first of all to the interests of the *Miscellany,* and I promise you not to take offence at any apparent neglect."[25]

Peirce also supported Gill by writing a notice of the journal for the *American Journal of Science.*[26] This short article is interesting because in it Peirce draws a distinction between mathematicians and those who only pretend to be such. Peirce noted the value of actually solving mathematical problems, like those in the *Miscellany,* and then observed, "We are too prone to consider the mere reader of mathematics as a mathematician, whereas he does not much more deserve the name, than the reader of poetry deserves that of poet."[27]

The *Miscellany,* and other antebellum mathematics journals, had many contributors who were amateurs and philomaths. Much of Peirce's life was spent separating serious scientists from such novices in all scientific disciplines, but as Peirce's support for Gill shows, Peirce did not disdain scientists who were self-educated, as was Gill, the son of a Yorkshire cobbler, or who did not have the social and educational background he did, as long as they were competent and serious scientists. Peirce not only supported Gill in his efforts with the *Miscellany,* but, when the college Gill was teaching at closed, Peirce aided him in finding employment and endorsed his mathematical abilities.[28]

The highly competent Gill is in contradistinction to a self-styled expert in astronomy and mathematics like John Lee, who wrote to Peirce when he was a young professor at Harvard:

> The enclosed ticket will give you an opportunity of entering without charge, the apartment in <u>Boylston Hall,</u> Boston, No. 3, where I intend to deliver a lecture on next Thursday, commencing at 7½ P. M. the subject being, "the signs of the times." I know not whether I told you before, that I have for many years <u>disbelieved in toto</u> the <u>Newtonian astronomy;</u> my theory of longitude being merely planned (in the hope of reputation and

reward) <u>in conformity to the revised system.</u> I intend to make, on Thursday evening a <u>severe attack</u> on that system; also, at the close of the lecture, I shall make proposals for the publication of "<u>the universal analysis of equations</u>"—no part of which I have really communicated, except a demonstration which I gave you, in a detailed part of my summary, that when the coefficients or exponents of x in any equation cease to be <u>monomials,</u> such an equation may have more real roots than degrees. I will more candidly tell you that <u>one</u> reason for my having acted so singularly in my communications with you on that subject was:—I thought you had, in <u>conversation</u> with me, spoken more slightingly of the "longitude" than it deserved, for although the <u>diurnal</u> problem <u>might</u> be suggested to the mind by the observations <u>already</u> practiced, yet the <u>nocturnal</u> one is <u>altogether new</u> in its nature.

Moreover, I reflect upon the general tenor of your past intera[ctions] with me, with grati[tude] and [re]spect.[29]

Unfortunately no more is known about John Lee, but apparently this man had contacted Peirce several times. Very naturally Peirce would not want to waste his time with such a crackpot, and would want to distance himself and respectable science from men like Lee.[30]

$$\Phi^n$$

The *Mathematical Miscellany* never had an adequate number of subscribers,[31] and Gill was unable to continue publishing the journal. There was a small cadre of mathematicians who were enthusiastic about having a mathematical journal in the United States, and Peirce's mathematical prominence made him an obvious choice for the new editor. Faced with employment problems, Gill asked Peirce if he would take over the editorial duties of the *Miscellany* as early as the summer of 1836.[32] Although Gill was able to edit the journal for three more years, the journal's financial problems, as well as Gill's employment situation, resulted in the journal's demise. In 1842 Peirce took over the editorship of the journal, changing its name to the *Cambridge Miscellany*. Despite the change in name the journal was clearly a continuation of Gill's *Miscellany*. The first issue used much of the material from the old *Miscellany*, which Gill had passed along to Peirce.[33]

In order to do a satisfactory job of editing the *Cambridge Miscellany*, Peirce requested that the Harvard library subscribe to the principal European mathematical journals, including Liouville's *Journal des Mathematiques pures et appliqes, Le compte des rendues hebdomedaires des sciences de l'Academie des sciences,* Crelle's *Journal für die reine und ange-*

wante mathematik, and Schumacher's *Astronomische Nachrichten.* Peirce was willing to contribute approximately half of the cost of these journals if he could be assured that he would be allowed the first use of them, immediately upon their receipt.[34]

After publishing only one issue, Joseph Lovering, Peirce's colleague on the Harvard Faculty, joined Peirce in editing the journal. Although the *Cambridge Miscellany* was similar in format to Gill's journal, the co-editors introduced several innovations, including a substantial number of articles on mathematics, astronomy, physics, and meteorology. Most of these articles were by Peirce and Lovering, but there were other articles, including reprints of European work.[35]

Although these changes seemed to be well received,[36] the journal was short-lived, running through only four issues, all published in one year's time. It is likely that Peirce's failure with the journal was caused by the same problems that faced Gill—primarily the lack of a sufficiently large American mathematical community. But perhaps the busy Peirce did not have the time to continue it. He was involved in many other things at about this time. For example, in 1838, he wrote to Gill:

> I fear that it will be out of my power to go to New York this summer although I very much desire to go, as I have a great deal of work laid out—I have engaged to write a review of the *Méchanic Céleste* for the *North American* and to make a thorough index of the work for the family who will publish it—and then I have promised some of my friends (who I presume, will publish it for me) to write a translation of the fifth volume. The last sheets of the fourth volume were in the printer's hands at B's death and are now in type. He had done nothing to the fifth volume.
>
> Excuse my brevity as I am surrounded by young men waiting to get some explanations, to which office I devote my energies from 9 till 12 o'clock and hear my recitations in the afternoon from one till my dinner hour at $3\frac{1}{2}$ o'clock, by which time I am usually exhausted.[37]

By 1843, Peirce was also busy raising money for a new observatory at Harvard.

$$\Phi^n$$

The review of the *Méchanic céleste,* that Peirce alluded to above, was the first and longest article Peirce wrote for the *North American Review.*[38] A popular discussion of the relationship between mathematics and astronomy, the article was Peirce's first public tribute to his mentor, Bowditch. In the article Peirce struggled with the value of mathematics—a topic which was to concern him for much of his life. He began this article by observing: "Compared with the infinite variety and extent of physical phenomena, the domains of science of quantity would, at first sight, seem confined to very narrow limits. But when we consider, that mathematics treat of all forms and motions; when we find the sweetest tones and the brightest colors, the lightning and the rainbow, heat and cold, and the very winds and waves subject to the strictest laws of motion; when, in short, we learn, that the whole world is bound together upon mechanical principles; we must concede that the ocean, upon which the mathematician has launched his ship, is as unbounded as the material universe."[39]

Near the end of the article, Peirce confessed: "For, as humiliating as is the confession, it must be admitted, that practical astronomy is not dependent upon theory, and that the observer, with only a small degree of elementary mathematics, can deduce from his observations empirical laws sufficient for all his purposes. Judged, then, by the narrow standard of utility, which is generally adopted, and has been pronounced to be at the very foundation of the Baconian philosophy, transcendental mathematics can offer but little to redeem it from the reproach of being a mistletoe science, uselessly absorbing the highest energies of the most powerful minds. Even its noblest fruit seems to be blasted; its Newton was sensitive, fretful, and almost mean; and its La Place was an atheist."[40]

While Peirce pointed out that mathematics was essential to understanding physical science, he found the ultimate reason for studying mathematics to be spiritual. Mathematical knowledge makes man more divine: "Could any intellectual exercise be thought of, more nearly allied to that required at the creation? Is not the very labor which would absorb the powers of the archangel, to whom should be entrusted the sublime task of building a world?"[41]

$$\Phi^n$$

In the next several years, Peirce continued to form friendships, some of which were vital to him for the rest of his life. One of his lasting friendships was with the British mathematician, James J. Sylvester, the most outstanding addition to the American mathematical community in the early 1840s.[42] When Sylvester went to the University of Virginia in November 1841, he was already recognized as one of the most brilliant mathematicians in Britain. Only twenty-seven, he was a fellow of the Royal Society, and had been a colleague of Augustus De Morgan at what is now University College, London. Sylvester came to the United States, in part, to escape the strong anti-Semitism in Britain, but in this he met with severe disappointment at the University of Virginia. The university community welcomed him enthusiastically, but anti-Semitic articles in the Richmond newspapers had preceded him.[43] Articles against foreign professors soon followed them. After only a few months in Virginia, Sylvester had trouble with student discipline. One student waylaid Sylvester and attacked him with a club. Sylvester defended himself with a sword cane. Fortunately no harm was done to either party. But the university would not support Sylvester's request for expulsion of the student, and consequently Sylvester left the university and went to New York City, where he lived with his brother and sought other employment.[44]

On a trip to Cambridge, Sylvester met Peirce, and they became lasting friends. Sylvester wrote to Peirce, "I never remember more happy days than those I passed lately, under your roof, in the enjoyment of the pure pleasures which spring from the feeling of mutual good will and adaptation."[45] Sylvester hoped to stay in the United States and work with Peirce; he wrote:

> I think we could together make out a beautiful system of Mechanics much more satisfactory and more agreeable to common sense than any yet attempted at least so far as regards the deduction of laws from *first principles*.
> I would make almost any sacrifice consistent with independence to have the advantage of a long course of study with you.
> Without envy or jealousy by joining our forces in a constant and faithful cooperation in reading and reflection we might achieve wonders: Could any situation be made for me at your University? My expectations as to pay would be very limited as I know that your University cannot not [sic] afford large salaries.[46]

During Sylvester's stay in New York, he fell in love with a Miss Marston.[47] There were problems of religious differences, however, and his affections were not completely returned. Sylvester's romance gave him an additional motivation to stay in the United States long enough to win Miss Marston's affections. He desperately sought employment. He wrote Peirce:

> I came to the decision to give the affair up and arose the next day, as imagined cool and indifferent, like one awaked from a fever or fit of delirium. But fate had decreed otherwise; we met and the first glance like a lightning stroke of inspiration dispelled all doubts all ungenerous feeling and brought her back to my mind pure and noble as in the springtime of our acquaintance, . . . we met again yesterday[.] 1 joined her alone in a short walk and something like an explanation took place. For the first time she evinced warmth of feeling and of this I am fully satisfied that she never designedly fostered expectations which she knew were not to be fulfilled.

I can scarcely help flattering myself that I have inspired her with some regard; her <u>friends</u> are dead set against me and it is to their opposition I must attribute all that has appeared cold and ambiguous in her deportment. How far her regard extends I am still in doubt or whether it has ever transcended the bounds of friendly esteem.

She has heard of my determination about leaving for England and were I now without good or tenable reasons to draw back, that announcement would appear to her in the light of a <u>base</u> to elicit from her some manifestation of feeling that would otherwise have been withheld; on the other hand if I leave she is lost to me forever.

Now therefore in the name of love and <u>friendship</u> do if it be possible in the nature of things find me something with a <u>home,</u> (no matter <u>how trifling,</u> how unworthy as you might consider the <u>remuneration</u>) to keep me in the country and with you.

Think upon this; get your friends to canvass for me; I care not what the employment may be if it only afford a fair pretext for delaying my return. . . . June the first, the vessel sails.[48]

Unfortunately Sylvester was unable to find employment at Harvard, or elsewhere, and returned to England, the object of his affections not won. Neither he nor Miss Marston ever married.[49]

$$\Phi^n$$

Alexander Dallas Bache had planned on visiting Cambridge to meet Peirce in 1842, but unexpected business prevented him.[50] The men did not meet until the next year, when Peirce went to Philadelphia to attend a meeting of the American Philosophical Society, to which he had been recently elected.[51] Bache was to become Peirce's closest friend and staunchest scientific ally. In Bache, Peirce found a man with similar background and ambitions. Bache was a great-grandson of Benjamin Franklin, the most notable American scientist before Joseph Henry.[52] Bache's uncle, George M. Dallas, became vice president of the United States. His aunt's husband served as secretary of war.[53] His grandfather, his uncle by marriage, and his brother-in-law all served as secretary of the treasury. But also like Peirce, Bache was not independently wealthy. Both men were strongly committed to the growth of American science and dedicated themselves to this cause for the rest of their lives. Although Bache generally overshadowed Peirce, Peirce never had a conflict with him, and the two men seldom disagreed on the course of American science.

After graduating first in his class at West Point, still not nineteen, Bache spent three years working as an army engineer and teaching at West Point. At twenty-two he became professor of natural history at the University of Pennsylvania, and at thirty he was chosen as the president of the newly founded Girard College. Although called a college, Stephen Girard had, in fact, endowed a boarding school for white, fatherless boys between the ages of six and eighteen. To prepare for his administrative responsibilities, Bache went to Europe to study educational institutions there firsthand.[54] Bache was a congenial, vigorous man, who easily won people's confidence. Although committed to scientific excellence, he was a better administrator than scientist, his scientific accomplishments falling short of those of his friends Henry and Peirce.

Benjamin's spinster sister, Charlotte Elizabeth Peirce, whom everyone in the family called Elizabeth or Lizzie, described Bache: "Ben had an unexpected visit from Dr. Bache & his wife. . . . He looks like the pictures of Franklin. He is a man of great talent and capacity & works so quick that it is astonishing. One morning I was at Ben's when Dr. Bache received his morning's mail about 15 (not as many as he often has) long & important letters all of which he read and answered at once—all his answers had to be copied—so I copied some and Mrs. Bache some. He signed the copies after reading them & then folded and packed up the letters to be sent to the post just as if he was working by steam. I looked at his hands, supposing they must be somewhat the worse for wear, but they were not & are remarkably handsome, clean, nice gentlemanly but capable looking hands."[55]

$$\Phi^n$$

When Peirce met Bache, the Coast Survey was still under Ferdinand Rudolph Hassler, its first superintendent. Hassler was born in Switzerland to a wealthy and politically influential family. He was educated at the University of Bern, but also studied for short periods of time at the school of mines in Paris, the École Polytechnique, and Göttingen. In addition to his academic training, he obtained practical geodetic experience by working on several surveys in Switzerland. When Hassler was almost thirty-five, he left Switzerland to escape political and financial troubles caused by France's occupation of Switzerland. Hassler had supported an attempt to found a Swiss colony in the southern United States. When the project failed, he and his family were left financially destitute. He still had, however, an art collection, an excellent library of some three thousand volumes and a fine collection of scientific instruments. During his first years in America he was forced to sell some of these items to survive financially.[56]

When Hassler came to the United States in 1805, there were few men in the country with his scientific expertise. Members of the

Ferdinand Rudolf Hassler, first superintendent of the Coast Survey. Courtesy of the NOAA Photo Library.

American Philosophical Society had for many years desired to have a federally sponsored coastal survey, and Hassler's arrival in Philadelphia provided the catalyst for them to actively pursue the undertaking. Thomas Jefferson, president of the society as well as of the United States, recommended to Congress that the country's coast be surveyed, and Congress authorized the Survey on February 10, 1807. But it was not until four years later that Hassler was sent to Europe to procure instruments for the Survey. He stayed there until 1815, overseeing their construction.[57]

Actual work on the Survey was not begun until the fall of 1816.[58] The Coast Survey was responsible for mapping coastal islands and shoals, thus improving the safety of shipping, one of the most, if not the most, important sector of the economy at that time. Unfortunately things did not go smoothly. As soon as work had begun under Hassler, the army and navy began ceaseless lobbying to have the Survey under their control, rather than under the secretary of the treasury. Hassler, quite correctly, insisted on highly accurate maps based on astronomically determined latitude and longitude at crucial points along the coast. Critics thought Hassler's methods were too expensive and too slow. Proud of his European scientific training, the arrogant Hassler lacked the diplomatic and political skills to make congressional leaders understand the importance of using his exacting methods.[59]

The navy succeeded in gaining control of the agency in the spring of 1818, and Congress excluded all civilian employees from the Coast Survey, Hassler included.[60] However, by 1828, the secretary of the navy reported to Congress that the work done by the Survey since it was under the direction of army and navy officers had been largely incompetent.[61] Still, Hassler was not reappointed head of the Survey until 1832, and even then the army and navy persisted in their attempts to control the Survey. The navy again gained control of the Survey from 1834 to 1836, which again resulted in Hassler's expulsion. But from 1836 Hassler headed the Coast Survey until his death in 1843.[62]

In a country with little opportunity for instruction in applied science and engineering, the Coast Survey served as an important scientific training ground. Realizing the importance of this function of the Survey to the nation, Hassler collected a large scientific library, which was available to Coast Survey personnel to further their scientific knowledge.[63] Hassler had also hoped to expand the scope of the Survey to include a broad spectrum of scientific work, including geodetic pendulum experiments, geomagnetic studies, tidal research, and refraction experiments. But Hassler failed to achieve these broader goals.[64]

Φ^n

When the position of superintendent of the Coast Survey became available after Hassler's death, Peirce joined many other American scientists in strongly supporting Bache's bid for the position. He wrote to Bache, "I feel truly interested in your success, as I regard it to be intimately blended with the best interests of science in the country."[65] But Peirce did more than merely offer words of encouragement; he promised "to leave no stone unturned, which it is in my power to turn, to obtain for you the appointment."[66] With the assistance of his brother-in-law, Charles Henry Davis, he solicited the support of leading Bostonians, including John Quincy Adams, and secured the promise of support from Rufus Choate, the distinguished lawyer, who was then a senator from Massachusetts.[67] He contacted Judge Story, his mother's former suitor; John Pickering, the lawyer and philologist; and George Bancroft, who after leaving the Round Hill School had become an influential member of the Democratic Party.[68]

Once Bache was appointed, Peirce was not disappointed with his performance. In a letter containing observations that determined the latitude of the Harvard Observatory, Peirce told Bache that he was pleased with what he had done with the Coast Survey. "I was delighted with your work upon the Survey. We all knew your thorough mastery of science and your administrative genius, but you have taken us somewhat by surprise in placing yourself at once upon a level with Hassler himself in your skillful use of his own instruments. You have undertaken the Survey in a glorious spirit of nationality, you are pursuing it into its most minute details with unflinching thoroughness, and God grant that you may succeed in making them understand this at Washington."[69]

Five years later Peirce reaffirmed his high regard for the Coast Survey, and its importance to the progress of American science. He wrote, "leaving all my friendship for you out of the question, I regard your interests and those of science as identical; and I consider it the solemn duty of every scientist, who has a drop of patriotism in his veins, to battle with his whole might against any attack upon your administration of the Coast Survey."[70]

Bache did do spectacular things for the Coast Survey. Under Hassler it had been a small undertaking, generally lacking congressional or executive favor. Under Bache it grew into the largest employer of mathematicians, astronomers, and physicists in antebellum America, with a budget of half a million dollars a year.[71]

In the spring of 1845, Peirce and Bache discussed Peirce's joining the Coast Survey. Bache wrote to Peirce:

You are right . . . my object was to enlist you to receive some of our numerous calculations, but you have been so kind to me personally that I did
not know how to talk of pecuniary business. So one day I broached the
matter to the Lieut.—I asked if you should be offended at the idea of
being called upon to do an occasional calculation. He thought that you
ought not, if it was only to help [out]. . . .

Do you mean to indicate that you would take charge of computing the
predictions for me? I can lend you the aid of ready computers if you will
undertake the supervision of their work. The difficulty which I find is this.
It is impossible for me to examine all the work which is done. If entrusted
to those who have imperfect views of theory, lapses occur which are inju-
rious & mortifying besides.[72]

Peirce volunteered to superintend the computations for predicting occultations. Peirce, as well as Bache, saw this not only as an
opportunity to do scientific work, but as a school for American science. "There would be at least, two computers," wrote Peirce, ". . . and
I shall think it part of my duty to give these computers all the knowledge of theory, which I may find them capable of acquiring. You may,
then, find my study a convenient place for giving a little instruction,
and for testing the accuracy of some of your employees." Peirce
stressed that this was merely an offer of his services to use if Bache
found it convenient. He insisted that "in all our business transactions,
friendship shall not intrude itself, and I shall have no scruples in
regarding you as the head of the work with all the rights and privi-
leges there to appertaining."[73] Peirce had frequent spats with other
people, but none with Bache. They occasionally disagreed, but their
disagreements never affected their friendship.

In 1852, Peirce took over the longitudinal observations for the
Coast Survey, although Peirce at first declined the position. He
explained that he wouldn't have sufficient time unless he could be
relieved of most of his teaching duties, but doubted that the Coast Survey could pay him adequately to compensate for the loss of income.[74]
But an arrangement was worked out, and Peirce began overseeing the
longitudinal work. He continued to direct this division of the Coast
Survey until he was, himself, made head of the Survey.[75]

Peirce's association with the Coast Survey freed him of many of his
unpleasant teaching duties at Harvard. Although Peirce retained his
professorship at Harvard until his death, his teaching duties were
often light and restricted to a few select students. In 1852 he wrote to
Bache, telling him that if the Coast Survey could not pay him a sufficient salary, he would have to go back to the gruesome task of teaching sophomores.[76]

The Coast Survey under Bache was besieged by the same opponents as it had been in Hassler's days. In addition, the former Coast Survey employees had a difficult time accepting Bache. In particular, James Ferguson, who had hoped to succeed Hassler himself, not only resented Bache, but was also doing faulty work. In March 1846, two of Bache's assistants independently informed Bache that Ferguson's work on the Chesapeake Bay contained numerous errors. After inspecting the work, Bache suspected that some of the results had been fabricated. Bache confronted Ferguson, giving him an opportunity to explain the mistakes. Ferguson would neither admit wrongdoing nor resign.[77]

Wishing to avoid a public scandal that might damage both Ferguson and the Survey, Bache asked his brother-in-law, Robert Walker, the secretary of the treasury, to convene a committee of external investigators. Both Bache and Ferguson selected two men for the committee. Then each man selected one of the two names put forward by the other. These two men then selected the third member of the committee. Benjamin Peirce was selected from Bache's list, and the West Point-trained engineer, Andrew Talcott, from Ferguson's. The third member was Charles Hackley, also a West Point graduate and professor of mathematics and astronomy at Columbia.[78] The committee confirmed Bache's allegations, and Ferguson was dismissed.

Although there were other attacks upon Bache's administration of the Coast Survey, his lobbying, administrative prowess, and diplomatic skill allowed him not only to avoid Hassler's troubles, but also to expand the Survey into a broad governmental scientific agency. When the Survey was under attack, Peirce was always happy to lend a hand in supporting it.

When Peirce visited Washington on Coast Survey business, he found other things in Washington to interest him besides his work. In particular, he was impressed with Horatio Greenough's new statue of George Washington. Greenough, often considered America's first professional sculptor, was born in Boston and was a graduate of Harvard. His sculpture of Washington, a huge seated figure of the president, was the first major commission given by the United States government to an American.[79] Back in Cambridge, Mrs. Greenough had been anxious to hear Peirce's appraisal of her son's work. Peirce told his wife to tell her, "I have been to see Greenough's statue of Washington, and I can truly say that I was greatly disappointed. I was not, in the slightest degree, prepared for such a grand figure. At the first sight of it, as you look from the capital, it appears to be a figure of colossal size in the sky itself, a sublime figure of Iupiter Americanus.[80]

And as you approach, the impression is not at all abated. I am confident that all the old objections to it must have arisen from its being improperly placed."[81]

This statue of Washington was Greenough's first major work and also the first major commission given to an American sculptor. Few Americans agreed with Peirce's evaluation of the statue. It was done in a neoclassical style and was met with widespread ridicule. It was originally placed in the rotunda of the newly completed Capitol in 1842, some five years before Peirce saw it. But it was moved outdoors when it was discovered that the rotunda's floor could not support it. After being subjected to the elements for many years, it was eventually sheltered in the Smithsonian.[82]

Φn

As noted in a previous chapter, Peirce and Agassiz were also close friends. Although Peirce treasured his friendship with Agassiz,[83] he had his differences with him. He wrote to Bache that "he is not to be depended upon. But still he is true to science, and we must pardon his human frailty. I love him dearly but I have long ceased to rely upon his support in any emergency. How different from you."[84]

Nathaniel Shaler, one of Agassiz's students, reported that Peirce and Agassiz "were off and on the nearest of friends and the bitterest of foes."[85] At the time Shaler took his exams for his degree at the Lawrence Scientific School, Peirce and Agassiz were at odds because Peirce had accused his friend of lacking thoroughness in the instruction of his students. Agassiz tipped Shaler off about their tiff, because he expected Peirce to give Shaler a thorough grilling when it came his turn to examine him. He did.

Φn

In Peirce's eyes, men like Bache and Agassiz epitomized the best hope for American science: an undeviating commitment to sound scientific work and the political savvy to successfully promote it. Matthew Fontaine Maury, on the other hand, represented the worst aspects of American science: sloppiness and speculation. To Peirce and his close band of scientific friends, Maury became the scientific Antichrist.[86]

Born in Virginia, Maury was raised in poverty by a tyrannical, deeply religious father on a small farm in Tennessee. Maury's formal education had been limited to a bit of elementary instruction followed by seven years at a local academy. Maury's brother John went to sea as a midshipman. His tales of his sea voyages fascinated Maury, and, much against his father's wishes, he too obtained a position as a midshipman.[87] During his nine years of sea duty he schooled himself. He was interested in observing wind and weather, compass variations, currents, coastlines, and tides. He published his results in an article in *Sillimans's Journal*[88] in 1834. Upon being relegated to shore duty, he prepared a textbook on navigation for midshipmen, entitled *A New Theoretical and Practical Treatise on Navigation.* It was a watered down version of Bowditch's *New American Practical Navigator.* He also increased his scientific expertise by studying astronomy and geology.[89]

From 1836 to 1842, Maury was assigned to survey southern harbors. During this time he wrote articles arguing for various naval reforms. In particular, he urged more extensive technical and theoretical education for naval officers. Although his articles often irked his superiors, they enhanced his reputation and led to his appointment as superintendent of the navy's Depot of Charts and Instruments. Up to this point his relations with Peirce and Bache were cordial.[90]

Maury's predecessor as the superintendent of Charts and Instruments, Lieutenant James M. Gilliss, had studied astronomy in Paris. He had been assigned to make astronomical observations for the Wilkes expedition.[91] Despite inferior equipment, Gilliss did excellent work that won him the praise of professional astronomers. Furthermore, by 1842, Gilliss had convinced Congress to authorize a respectable naval observatory.[92]

Although it might seem that Gilliss had earned the right to be director of the observatory, in the eyes of Peirce and men like him, the founding of a naval observatory was a singular event, and for the sake of American science, the very best person available must be at its head. Peirce strongly recommended William Mitchell to his former employer at the Round Hill School, George Bancroft, who was now secretary of the navy. Peirce saw the Naval Observatory, not Harvard, as the center for American astronomy and thought, "It would be a sad and unfortunate tarnish to our National honor, for those magnificent instruments to remain unemployed and perish without contributing to knowledge, and one which would not, I think, be less likely to occur than under the present administration of the Navy." After praising Mitchell, Peirce added, "I cannot conclude my recommendation without alluding to the aid which Mr. Mitchell will undoubtedly receive, in his labours, from an accomplished mathematician and excellent observer in the person of his daughter."[93]

Unfortunately, Peirce and his confederates were not influential enough to have Mitchell appointed. When Maury took over the Depot of Charts and Instruments, Gilliss went to Europe to purchase instruments for the new observatory. Maury began amassing data from the collection of ships' logs stored at the depot on wind, weather, and currents. In 1844 Maury delivered two high-profile papers on the Gulf Stream and ocean currents at scientific meetings. The next summer the secretary of the navy, a fellow Virginian, required all midshipmen to read Maury's text on navigation. Despite Maury's rising reputation, Gilliss was astounded when he heard that Maury, not he, was to be the new superintendent of the Naval Observatory.[94]

Maury was more interested in oceanography than astronomy, so although he had grandiose plans for the Naval Observatory, they did not come to fruition, and little astronomical work was done at the Naval Observatory until Maury resigned just before the Civil War to offer his services to the confederacy.[95]

Maury did achieve marked success in oceanography. His work on currents made it possible for ships to reduce their sailing times. For example, using his charts, sea captains were able to reduce the sailing time from New York to Rio de Janeiro by thirty-five percent. And using his charts the *Flying Cloud* made a record run from New York to San Francisco in less than half the usual time. Despite Maury's practical success, his science was sloppy and speculative. Maury wrote his most important work, his *Physical Geography of the Sea,* in one spring, using his wife and his children as critics. The work reflected the haste in which it had been produced, as it was disorganized and repetitive. From the point of view of the professional scientist, it was highly deficient. It contained numerous errors, and advanced sweeping and preposterous statements as sound hypotheses. Yet the book was an immediate popular success. Maury also received extensive recognition in Europe.[96]

At first Maury was friendly with Peirce, inviting him to visit the Naval Observatory and inspect the new instruments.[97] Even after Maury sided with Leverrier in the controversy over the discovery of Neptune, Peirce was willing to pay him his due. He wrote his friend Mitchel: "His labors upon the ocean currents are very creditable to the science of the country. I am fully of the opinion that American science will not be respected unless votives cling together, forgetful of one another's defects and strive to love one another."[98]

But Peirce and his circle became increasingly irritated with Maury and his scientific ability. Peirce's friend, Bache, became annoyed with Maury when he hired Ferguson after Bache had dismissed him from the Coast Survey. Peirce and Bache were also irate when Maury dismissed their friend and excellent astronomer, Sears Cook Walker, from the Naval Observatory. Maury, no doubt, likewise resented Bache's then hiring Walker.[99] Bache was also angry with Maury for refusing to cooperate with the Coast Survey's project to observe the transit of Venus in 1849 and 1852, thereby seriously impeding the Coast Survey's expedition. Bache also proposed that the Coast Survey should direct the sounding survey of the Telegraphic Plateau. This was a project that Maury had already done work on and understandably felt was the provenance of the Observatory.[100] Maury also had a turf dispute with Peirce's friend Joseph Henry when Maury attempted throughout the 1850s to extend the Observatory's meteorological studies to the land, thus threatening the existing Meteorological Bureau of the Smithsonian, which was then the principal bureau for inland meteorological studies.[101]

Maury's science was also largely discredited. Peirce's good friend, John Le Conte, wrote a rebuttal of Maury's *Geography of the Sea,* which appeared in the *Southern Review.*[102] The most important response to Maury's science came from a self-taught Tennessee schoolmaster, William Ferrell. Ferrell not only scorned Maury's wild speculations, but replaced them with sound ones. Peirce's friend Benjamin Gould found Ferrell his first scientific employment with the *Nautical Almanac;* after nine years, he left the *Almanac* to work for Peirce in the Coast Survey.[103]

$$\Phi^n$$

The failure of the Naval Observatory to become the leading observatory in the country left a void that was eventually filled by an observatory at Harvard. Observational astronomy at Harvard went back to 1672, when Governor John Winthrop gave Harvard its first telescope. From that time on, the college possessed at least one such instrument, and faculty members Thomas Brattle and John Winthrop had made notable observations during the colonial period. But since the Revolutionary War astronomy had languished at Harvard, as it had in the entire country. In 1830, there was not a respectable observatory in the land.[104] The college had toyed with the idea of having an observatory for many years, but it was not until 1839 that serious steps were taken to obtain a college observatory.[105] By this time several other colleges had observatories, including Yale and Williams.[106] Many other observatories were soon to be built, most notably the Cincinnati Observatory through the efforts of the astronomer, Ormsby Mitchel.[107]

In an effort to rectify Harvard's deficiency, President Quincy attempted to lure William Cranch Bond away from his home in Dorchester. Bond, an instrument maker, lacked formal training in astronomy, but had taught himself well and was a competent observer.

He had built a private observatory in Dorchester where he followed his passion for astronomy whenever time allowed. Harvard had little with which to entice Bond, other than temporary quarters and the dream of a first-rate observatory. There was no salary with the position—Bond was to support himself with work for the government that he was currently doing. Understandably, Bond was reluctant to accept. Beyond the lack of financial support, Bond was apprehensive about the responsibility and notoriety of being in an observatory at a leading institution, such as Harvard.[108]

Nevertheless, Bond did come, bringing his instruments with him. Harvard began raising money at first to remodel the Dana house, which stood where the Lamont Library now stands, into an observatory. When it became apparent that this location would not meet Harvard's ambitions for an observatory, other property was acquired.[109] Peirce described it as being "the most favorable spot, which there is in the vicinity, for the location of an observatory, and a better one could be hardly desired."[110]

Peirce had high aspirations for the new Harvard Observatory, and was active in raising money for it. In July 1842, he reported to the corporation:

> A strong desire to obtain a large telescope, equatorially mounted, for the observatory has induced me to solicit the pecuniary aid of gentlemen for the purchase of such an instrument. My attempt has proved tolerably successful, 3500 dollars have been received, ... Great interest has been expressed in the success of the undertaking. I have also visited Philadelphia for the purpose of examining a telescope, which is connected with the Central High School, and is altogether superior to any other in the country.
>
> But a larger and more costly telescope [is] contemplated in obtaining the subscriptions and seems to me more [fit]ting the honor of the College—one, for instance, of which the expense should be about 5000 dollars. Independently of cost, I should think it advisable to adopt as our standard the celebrated Dorpat telescope which . . . is exceeded in performance only by one telescope . . . at Pulkova Russia.

Peirce said that such a telescope would cost between four and six thousand dollars, but he pointed out that: "As a stimulus, not wholly unworthy of our attention, [I] would state that I heard in Philadelphia that Prof. Mitchell of Ohio had raised 8000 dollars, in the course of three days, in the city of Cincinnati for the purchase of a telescope with other astronomical instruments and was promised as much more. He is gone to Europe, resolved to return with a telescope which should be unparalleled in the country."[111]

A month later, the college sent Bond's brother-in-law, Joseph Cranch, to Europe to consult with leading astronomers and investigate prices for telescopes. Until recently the best telescope in the world was in the Russian observatory at Dorpat, and Peirce hoped to obtain an instrument of equal or better quality. He instructed Cranch to go to Fraunhofer's in Munich: "The telescope must in every respect [be] equal or superior to the Dorpat telescope . . . the contract should be founded on an exact description of the Dorpat telescope and of the excellence of its performance . . . equal in all the following particulars—the perfection of its achromatism, its power of giving definite images of the larger stars, its power of separating double stars and of measuring their distance asunder, its power of exhibiting the colours of stars, the firmness of its mounting and the accuracy of its clock work; the eye pieces should be of the highest degree of excellence and the delicacy and accuracy of the micrometers should be capable of standing the severest tests."[112]

By early fall, Peirce was happy to report that a small telescope, valued at $300, had been donated to the university.[113] But the much more ambitious task of raising funds for a large telescope remained.

Peirce's money-raising efforts were given a boost with the coming of the comet of 1843. This was a comet of exceptional brilliance, even at midday, with a long tail and rapid flight. Such a comet generated great interest among the public. Ever since Edward Halley had accurately predicted the comet of 1758, the appearance of these heavenly objects had lost much of the ominous aspect they had held in the Middle Ages.[114] In 1843 some of this apprehension remained with the public, especially with the concomitant appearance of such a spectacular comet and the prediction of the end of the world that year by the Millerites.[115]

On March 22, 1843, Peirce gave a lecture at Odeon Theater in Boston to approximately one thousand people, hoping to raise money for a telescope for the observatory. He reassured his audience that their fears about the comet were groundless. Then Peirce pressed home his main point, and "referred to the imperfection of the instruments used at Cambridge, and the necessarily imperfect character of the calculations founded upon the observations taken with those instruments. . . . Philadelphia and Yale enjoyed great advantages over Harvard in this respect, for their instruments were larger and of a superior construction."[116]

Peirce used Arago's observation that calamities occurred frequently with or without the occurrence of a comet. "But if a comet, he jested, seemed to bring extreme heat in Europe, severe cold in America, earthquakes and influenza in the West Indies, fires and neuralgia everywhere, and to be 'the prophet of the end of all things to all of us,' in Massachusetts, 'the generous spirits of Boston consenting,' it would bring a telescope."[117] No doubt Peirce's appeals to civic responsibility and the fear of being outdone by Philadelphia and Cincinnati were much more effective means of raising money than lingering fears about comets. Although some religious groups believed that the great comet was a sign of the end of the world, it is unlikely that many of the donors to the observatory did.[118]

Peirce followed up this plea with one made to a meeting of the American Academy of Arts and Sciences a week later. This meeting had been prompted by the public's dissatisfaction upon learning that the Harvard Observatory had no telescope of sufficient quality to study the comet.[119] While agreeing that a new telescope was badly needed, Peirce highly praised the observations of the comet that Bond had made with inadequate instruments. Again, he cannily pointed out that the Philadelphia High School was better equipped to make observations of the comet and had been able to do so.

Over the next several years many generous contributions were made for an observatory and for a large telescope. By the time the lens for the telescope was purchased, the college hoped to at least match the fifteen-inch refractor at the Imperial Russian Observatory at Pulkovo, near St. Petersburg, this telescope now having eclipsed the 9½-inch telescope at Dorpat.[120] The Harvard Observatory was able to acquire a lens of the same size as the one at Pulkovo, and of slightly better quality. Thus Harvard acquired, what was at that time, the best telescope in the world.

Peirce was on hand when the lens arrived from Europe in November 1846. Bond recorded, "We have been waiting for the object glass all week. . . . It arrived on Friday being provided with a private car for its exclusive accommodation, it arrived in the forenoon looking very much like a bale of old rags. We began unpacking with Prof. Peirce between 3 & 4."[121] The mounting for the telescope didn't arrive for another six months. Peirce enthusiastically described the new instrument in a letter to his friend, the astronomer, Elias Loomis: "Our telescope is mounted and will be in working order in about two months. All its workmanship is the very best German style, superior to anything which Mr. Bond has ever before seen from Germany. . . . We have had only one decent night for testing its performance but that one night was sufficient to prove it to be as remarkable in all the other qualities of a good telescope as it is in its dimensions. Mr. Bond and myself have always had, upon this point the most serious misgivings and have looked forward with great anxiety and even fear, to the period of testing; we know that our own demands in this sphere are above the mark usually required . . . of the operator."[122] Peirce was convinced that Harvard now had the best telescope in the world.

He was also excited about the less spectacular instruments in the observatory. Two years before, anticipating a visit from Loomis, Peirce had written, "Our little equatorial is just mounted in readiness for you, provided you should like to amuse yourself with observations—That is provided we give you any time to indulge in such celestial dissipation."[123]

The large telescope captured the public imagination and dollars much better than smaller more mundane, but equally valuable, instruments. When Peirce pressed for a transit circle, explaining that this instrument was necessary to make the large telescope useful, the college's treasurer, Samuel Eliot, replied, complaining about the already considerable expense of the observatory. He pointed out, quite reasonably, that the college's resources were finite, the expenses for the observatory were only partially met by direct subscriptions, and that a large portion of the money for the observatory must come from the general funds of the college. "Who will accuse us of sparing? And who will not accuse us of taking the bread of elementary knowledge from the mouths of the pupils, in the shape of disposable funds, to fill them with the luxury of science? You say the Transit Circle is necessary to make the telescope useful; but equally so are the buildings." Eliot's tone was certainly not unsympathetic or antiscientific. He was merely trying to be a good steward of Harvard's resources. He had "looked over all the original subscription papers," and found "that all but about $850 was subscribed to papers or books, that said nothing of any other instruments than the great telescope, & in all the receipts I have given, I have only acknowledged so much 'towards the purchase of a first class Telescope.'" He goes on to say, "Do try again & see if you cannot find somebody to make up $1000 for a transit circle, which you know so well how to represent as indispensable, & for which there are not funds enough."[124]

For many years, Peirce enjoyed a close and warm relationship with Bond. He lauded Bond's work and made sure that Bond received proper recognition for it.[125] Bond was offered the directorship of the Naval Observatory, but eventually declined it because of his strong attachment to his work in Cambridge.[126] Peirce, however, seized this circumstance to obtain a salary for Bond at Harvard. He wrote to Bache: "We are preparing to make a movement to get some more aid for our observatory from our Boston friends. For this purpose, we shall make full use of the offer which Mr. Bond received, to go to Washington. But this is between ourselves for the present. We wish to secure a salary for the director of the observatory. Bond and I have talked a great deal about what we would do, if we were to go to the Washington Observatory together; we talked about it, until we were actually half in earnest, and we broke all ties which bind us to Cambridge as if we were lovers of sixteen anxious to elope together. Which are the greater fools, men or children?"[127]

Two things combined to make the Harvard Observatory so successful: the astronomical observers, William Bond and his son, George, and the generosity of Boston's wealthy, who liberally funded the observatory. Compared to other American observatories and even to other departments of Harvard University, the observatory was well endowed. The Bonds were also given great freedom in their work in the observatory. Aside from occasional visits by groups of students, casual visits to the observatory by the public were prohibited. In addition, Bache supported the observatory by giving it work from the Coast Survey. All these factors combined to make it *the* national observatory in antebellum America.[128]

Φ^n

Although Peirce and his friends had not been able to block Maury from becoming head of the Naval Observatory, they met with considerably more success in the founding of the *American Ephemeris and Nautical Almanac*. The *Almanac* was established in 1849 under the Naval Observatory budget; but it was headquartered in Cambridge, far away from Matthew Maury. Furthermore, the head of the *Almanac* would report directly to the secretary of the navy. Uneasy about Maury's increasing popularity, the secretary of the navy was happy to adopt this plan, but the scheme came from Bache and his protégé in the Coast Survey, Charles Henry Davis, the new head of the *Nautical Almanac*.

Davis entered Harvard in 1821, at about the age of fourteen, but dropped out of school two years later to become a midshipman in the navy.[129] He spent most of the next seventeen years at sea, devoting as much of his time as possible to studying hydrography and astronomy.[130] During one brief hiatus from sea duty in 1835, he studied mathematics with Peirce. He and Peirce became close friends and corresponded regularly after this.[131] He then returned to Harvard to complete his degree, again studying mathematics under Peirce.[132] After finishing his degree at Harvard in 1841, Davis married Harriette Blake Mills, Sarah Peirce's younger sister. Thus he and Peirce became brothers-in-law.

For a short time Davis worked at the naval rendezvous in Boston, before becoming an assistant with the Coast Survey in April 1842.[133]

In December 1843, Bache became superintendent of the Coast Survey; Davis and he subsequently became good friends. While working with the Coast Survey, Davis did outstanding scientific work in hydrography. His first work was the investigation of the velocity and direction of the tides in New York harbor and on Long Island Sound. He also studied the current of the Gulf Stream and the tides and currents of the Nantucket shoals. This work led to a general study of the laws of tidal action, as well as to several important navigational improvements.[134]

Cambridge had many advantages as the home of the *Nautical Almanac,* in addition to being far away from Washington and Maury. Printing could be done cheaply there, and the best scientific libraries in the country were nearby. The Harvard Observatory, the best in the country, was also there. But probably more important to Davis, he had family and friends there,[135] including Peirce, who was put in charge of the theoretical work.

Simon Newcomb, one of the men who worked on the *Nautical Almanac,* described its purpose: "The term 'Nautical Almanac' is an unfortunate misnomer for what it is, properly speaking, the 'Astronomical Ephemeris.' It is quite a large volume, from which the world draws all its knowledge of times and seasons, the motions of the heavenly bodies, the past and future positions of the stars and planets, eclipses, and celestial phenomena generally which admit of prediction. It is the basis on which the family almanac is to rest. It also contains special data needed to enable the astronomer and navigator to determine their position on land or sea."[136]

An impressive group of men were gathered to work under Davis and Peirce in the *Nautical Almanac* office. They included the established astronomer, Sears Cook Walker, and a host of aspiring mathematicians and astronomers, including Simon Newcomb, George W. Hill, Joseph Winlock, Chauncey Wright, John D. Runkle, William Ferrel, Truman Safford, and James Edward Oliver.[137] Davis ran a congenial ship in Cambridge, which proved highly productive. Duties were light, being about five hours a day. Many of these men took advantage of the light load and the proximity of Harvard's newly founded Lawrence Scientific School to further their education.

Peirce and Charles Henry Davis were fast friends as well as brothers-in-law. When Davis returned to sea duty years later, Peirce's sister, Elizabeth, wrote that Henry Davis "is one of the most amiable and excellent of men—we shall miss him very much, especially Ben, who is more intimate with him than any one else."[138] Peirce's son Charles noted that "Davis . . . would pace up and down with my father by the hour."[139] In 1864, the Davises built a house on Quincy Street very

Charles Henry Davis. Courtesy of the NOAA Photo Library.

near the Peirce house. Sarah's mother sometimes lived with one of her daughters and sometimes with the other. The two households dwelt almost as one.[140]

There were times, however, when the relationship between Peirce and Davis was stormy. Peirce complained to Bache that the two men had a serious difference of opinion as to the quality of Asa Gray's scientific work. The friction between the two men was aggravated by Davis's assumption that Gould was influencing Peirce on this matter. Peirce wrote, "I have avoided this subject with Davis, and fearing lest it should disturb the harmony of Cambridge Society, I close my lips to everybody, even my wife."[141]

Nevertheless, Davis persisted in discussing the matter, igniting Peirce's fiery temper. Davis "said so many things, which were so bitterly severe upon my course that I opened my lips upon Gray and we separated in anger. The breach is destined to widen. I shall leave Harvard. I shall go to N. Y. . . . But nevertheless I will not turn back. My duty to my day and generation demands sacrifice of myself and I will make it. Oh that this cup might pass from me. I shall have nobody to thank me, not even my mother and my wife. But duty, as I see it, points but one way, and I will follow it even to the breaking of my heart."[142]

Bache urged Peirce to smooth things over. Peirce reported back, "As soon as I received your note I went to see Davis as usual. At first things went coldly, but at last you will be happy to tell that your loving spirit did the good work and things go as well as ever. God bless you!"[143]

6

A Prince of the Humbugs

CHARLES F. WINSLOW, A PHYSICIAN FROM TROY, NEW YORK, HAD MADE what he felt to be monumental scientific discoveries. Whereas Isaac Newton had shown that gravity was an attractive force in the universe, Winslow had discovered that repulsion was a solar and cosmic force. He likened himself to Isaac Newton, and was confident that his receiving like recognition was only a matter of time.[1] The principal hurdle to his obtaining the recognition, indeed the immortality, he so richly deserved was, in fact, Benjamin Peirce, backed by a cabal of American scientists, bent on running American science for their own purposes. Winslow wrote: "That clique at Cambridge including my friend! Hill,[2] are all bound together by that more solemn & responsible than (probably) all Masonic oaths 'You tickle me & I will tickle you'. So being self-created supreme directors of scientific affairs in this country no man must think without their permission & when he does think & originate & write, they or one of them it seems must oppress the real fathership of the thought & bolster himself higher on the pinnacle of fame to the complete forgetfulness & exclusion of the humble & modest soul on whom the discovery first dawned."[3] In another letter, he added: "I have the strongest reasons to believe, that a class communion exists between the leading gentlemen of the Smithsonian & the leading gentlemen of Cambridge. At Washington they have their men to whom to refer matters & investigations beyond their own reach."[4]

Winslow was right—about the existence of a scientific clique, that is—not about his own scientific prowess. The members of this elitist group at first called themselves the Florentine Academy, but soon adopted the name Lazzaroni, or scientific beggars, Lazzaroni originally being Naples' lower class. They were a loosely associated group of men whose primary goal was to rid American science of quacks, charlatans, and old fogies. Ultimately, they were determined to make America the world leader in science. To achieve these goals, the Lazzaroni worked to organize and professionalize American science.

Before Peirce's birth, the gentleman amateur dominated American science. By the time of his death, most American scientists were highly trained specialists, whose sole, or principal, source of income was from their scientific work. The emergence of this community of professional scientists was the most important development in American science during Peirce's lifetime.[5] And Peirce and his Lazzaroni associates were at the very center of this process.

A professional scientist to the Lazzaroni, however, was not necessarily one that was academically trained. Although the younger members of the Lazzaroni, such as Benjamin Apthorp Gould and Oliver Wolcott Gibbs, were trained in Europe, and Gould boasted a German PhD, Joseph Henry was largely self-taught. Peirce, too, lacked European training. Lazzaroni members were generally supportive of self-taught scientists, and even men who followed science as an avocation, as long as they took science seriously and did scientific work carefully and competently. Peirce, for example, was highly supportive of such self-taught astronomers as William Bond, William Mitchell, Lewis Rutherford, and Edward C. Herrick.

The group that comprised the Lazzaroni got its name around 1850,[6] but some of them had already been working together to further excellence in American science at least five years before this. After the organization was named, its membership continued to change. In many ways it was a social organization. One of the Lazzaroni stated that its purpose was to "have one outrageously good dinner together."[7] Peirce described one of these dinners to his wife: "At five o'clock we went to Frazer's to dinner, and the dinner was worthy of the Lazzaroni, but my powers of description are quite unequal to making you comprehend it. The wit was not indeed of any very marked and decided character, but there was an uninterrupted flow of conversation and discussion from the beginning to the end, and when any one undertook to tell a story, he was eternally interrupted by pundits on this life and the other, so that the point of the story was scarcely equal to that of the play in getting to it."[8]

The members of the Lazzaroni loved to call each other by pet names. Bache, the center of the Lazzaroni, was called the Chief, the Chief of the Coast Survey, the Chief of the Lazzaroni, and to the members of the group, the Chief of American science. Peirce was called Functionary, and signed his name in letters to members of the Lazzaroni as a phi with a superscript n.[9] Henry was called Smithson, and Agassiz was called Fossilary.[10]

Based on the voluminous correspondence between the members of the Lazzaroni, the closest friendship between any two of its members was that between Peirce and Bache.[11] Peirce opened his letters to Bache with such salutations as "Most Darling Chief," and "My dear, true, noble-souled friend."[12] Peirce often wrote to Bache using words generally reserved for a family member or lover. For example he wrote, " Most lovely of your sex! Come to my arms! Let us embrace! A fond and loving embrace!"[13] Bache, although less demonstrative than Peirce, usually addressed him as "Darling Functionary." But he also expressed strong affection for Peirce; for example, Bache, expecting to see Peirce soon, closed his letter with, "I long to embrace you my darling Functionary."[14]

Despite many strong friendships, the Lazzaroni were not a unified group and often had tiffs with one another. The core Lazzaroni members were Bache, Henry, Peirce, and Agassiz, but even these four were sometimes at odds about Lazzaroni activities. Agassiz stayed clear of the Dudley Observatory fracas, and in the same controversy, Henry declined to support Gould, causing a permanent rift between Henry and Bache. Henry was also purposely excluded from one of the Lazzaroni's most important successes, the founding of the National Academy of Scientists. Furthermore, the membership of the group changed over the twenty odd years of its existence. Thus there was little unanimity in the group. To be sure, the Lazzaroni was a clique, but it was not a cabal.[15]

$$\Phi^n$$

Although the members of the Lazzaroni sometimes had differences of opinion, they were united by their dedication to the progress of American science. To them, being a scientist was a noble calling, and raising the level of American science was a sacred trust. Addressing the opening of the American Association for the Advancement of Science in Cleveland as its president, Peirce spoke of science as a religious experience. He told his fellow scientists:

> We have again come together at our appointed rendezvous, to make each other glad with the tidings of truth which we bring from the heavens and the earth, and to reanimate our fainting zeal by the story of the successful search for the philosopher's stone, the true *elixir vitae,* the fruit of the tree of knowledge, and the footprint of Him of whom the earth is the footstool. . . .

> Gentlemen, we are not convened for a light duty,—our self-imposed task is not an amusing child's play; and we have not accepted the liberally offered hospitalities of this beautiful city for the enjoyment of a social festival. We come to give and receive instruction and inspiration. . . .

Alexander Dallas Bache. Courtesy of the NOAA Photo Library.

Gentlemen, let us stand here reverently. This is holy ground. Let us not presume to make these walls resound with the bickerings of angry contention for superior distinction, and the foul complaints of mortified vanity. Let us not raise the money-changer's cry of mine and thine, lest the purifier come, and, taking the royal jewel into his own possession, thrust us out into the ditch, and turn our fame into infamy.[16]

But in Peirce's mind, the members of the AAAS had an additional allegiance to that of science; they had an allegiance to their country and their country's science as well. While reminding the association that science was universal and belonging to no country or clime, Peirce called for an American science that not only equaled, but surpassed that of Europe. To Peirce, this was "not asking too much."[17]

A year later when Peirce addressed the same association as its retiring president, he said "to recall our duty to the country to which we owe our allegiance. We must despise the base servility to foreign superiority, which affects to look down from the heights of the cosmopolite upon the duty of patriotism, and scorn it as an abomination."[18]

$$\Phi^n$$

Organizing American science and building scientific institutions was vital to Peirce and the Lazzaroni. One such institution was the Smithsonian. Peirce and the Lazzaroni worked diligently to determine its direction and character; doing so, in their eyes, was important to the development of American science.

The half million dollars that came to the United States from the will of the Englishman, James Smithson, in 1835 "to found . . . an establishment for the increase and diffusion of knowledge among men,"[19] provided an exciting opportunity for American science. But Congress couldn't decide what to do with it. For one thing, Congress wasn't sure whether a proud republic should accept money from foreigners. One congressman feared that every person wanting to memorialize himself would make a donation to the United States. But Congress accepted the bequest relatively quickly in 1836. Deciding what to do with it took much longer. Among the possibilities were a national university, an observatory, a great national library, and a natural history museum. By 1846, Congress was sufficiently embarrassed by its delay that it did establish the Smithsonian Institution, but it avoided any detailed description of what the Smithsonian should be, largely allowing a board of regents to decide the direction the institution should take.

Alexander Dallas Bache was not only the sole scientist on the board, but by far its most influential member. Bache succeeded in having Joseph Henry appointed secretary of the new institution, but it was a doubly difficult task. Bache not only had to persuade the board to appoint Henry, he had to convince Henry to accept the position. Bache wrote to Henry: "All is as you wish we offer you $3500 & a

house. I can make the arrangement you desire in regard to temporary connection to fall back upon your Professorship if you do not
like us. That should be done quietly. . . . Science triumphs in you my
dear friend & come you *must*. Redeem Washington. Save this great
National Institution from the hands of charlatans. Glorious result. . . .
Come you *must* for your country's sake. What if toils increase & vexa-
tions come. Is a man bound to nothing for his country, his age. You
have a name which must go down to History the great founder of a
great Institution. The first Secretary of *the* American Institute."[20]

Unlike Bache and Peirce, Henry's parents were poor, so poor, in
fact, that when Henry was seven, he was farmed out to an uncle.
Largely self-educated, he had never graduated from college. Nevertheless, Henry was one of the nation's leading scientists and its leading physicist. His scientific work had made him respected abroad as
well as in the United States. He made fundamental discoveries in
electromagnetism, and the henry, the unit of inductance, is the only
such unit named after an American. His discoveries in electricity and
magnetism were important to the invention of the telegraph, tele-
phone, and electric motor. Henry took out no patents, seeking no
reward for his scientific discoveries other than the extension of
human knowledge and the recognition such discoveries brought.[21]

Henry was reluctant to leave his professorship at Princeton. But his
devotion to American science and his Presbyterian religion of self-
sacrifice prevailed, and he reluctantly accepted the post. Henry's
vision for the Smithsonian was vastly different from the two most popular ones, a library and a natural history museum. Henry wished to
use the Smithsonian funds to encourage basic research, a course
warmly supported by Bache and Peirce. That Peirce and Bache were
both members of a three-member executive committee of the Smithsonian, made things easier for Henry and his goals for the Smithsonian.[22] But he still was forced to compromise. Henry opposed having a building for the Smithsonian, let alone a building that housed
a natural history museum and a library. He eventually did accept a
building, a library, and a natural history museum, but he continued
to resist the growth of the latter two, and dismissed his assistant in
charge of the museum, when he circumvented Henry's authority in
attempting to expand the museum.[23]

The greatest challenge to Henry's plan to use some of the money
to support basic research came from one of the regents, Rufus
Choate. Choate, a distinguished trial lawyer, had served in both the
House and Senate as a representative from Massachusetts. He envisioned the Smithsonian as a great national library, and in 1847, he

had been successful in having half of the endowment income allocated for library purchases. Choate had an able ally in the Smithsonian librarian, Charles Jewett. The full allocation to the library was deferred until the building was completed. Then in 1854, Henry proposed canceling the 1847 agreement, causing a power struggle between him and Choate and Jewett.[24]

Peirce aided Henry to lobby against a library during the 1854 ruckus. Peirce was concerned about the Smithsonian, and opposed both an expanded library and museum, since they would only be of value to Washington, and not the whole nation.[25] In July 1854, he dined with Choate at a dinner party in Boston, and "at last I told Choate that I hoped he would fail in his efforts, which daring speech seemed greatly to arouse and astonish the rest of the table. We then talked together in a low tone and I expressed myself repeatedly and decidedly opposed to the plan of the library. . . . I tried to make him feel the importance of the present labours of the Smithsonian. Among other things I said that by proper appliance to aid them in their investigations, I was persuaded that some of the greatest men of Europe might be induced to accept professorships in America. This was a new thing to him."[26]

Peirce also worked to enlist others to maintain the research orientation of the Smithsonian.[27] He wrote to Nathaniel Prentiss Banks, a congressman from Massachusetts, arguing against turning the Smithsonian into a library: "I beg you, also, if you have an American heart, to let it beat freely in this cause which, which it has gained the applause of all the scholars of Europe, is most truly and emphatically an American cause. The plan of operations is wholly new and exclusively American in its origin. It fosters American intellect and has given the world the opportunity to witness what triumphs American genius can achieve in literature and science—as brilliant as it had hither to performed in arts and arms."[28]

Henry was able to halt the expansion of the Smithsonian's library. When Jewett attempted to have the library separated from the rest of the Smithsonian, Henry dismissed him.[29] Eventually Henry was able to divest the Smithsonian of the library by transferring it to the Library of Congress. Although Henry's budget in the Smithsonian was small and much of it was diverted away from supporting basic research, the Smithsonian was the first institution in the country to support research. Furthermore, the money from the endowment was free from being cut by congressional whims. By giving grants to support research and by publishing the findings of that research, Henry gave a significant impetus to American science.

Φ^n

Another key step in organizing American science was the founding of the American Association for the Advancement of Science. The beginnings of the association go back to 1838, when Edward Hitchcock, of Amherst College, proposed a national organization of geologists. Even then, he had a more inclusive association of American scientists in mind.[30] The organization quickly expanded membership to include naturalists, and in 1847 the time was ripe to expand the membership to all scientists. The meeting that year took place in Boston, and Agassiz, Joseph Henry, and Peirce were in attendance. Although just seven years after the first meeting of the American Association of Geologists, slightly more than half of the papers given were on topics other than geology and paleontology. Agassiz's presence added a special distinction to the meeting, as did Joseph Henry's report on the newly founded Smithsonian. Catching Agassiz's enthusiasm for a national organization of scientists, the association voted almost unanimously to change from the American Association of Geologists and Naturalists to the American Association for the Advancement of Science.[31]

American scientists had not always been so enthusiastic about a peripatetic national organization for scientists. Although many had attended meetings of the British Association for the Advancement of Science, and saw its advantages, some thought the country was not ready for a similar organization. Earlier in 1838, Joseph Henry was especially wary of such an organization. Since many men who professed scientific interests were dilatants and amateurs, he feared that men without real scientific expertise might control the association, and it would become a national embarrassment.[32]

Henry's fears were not unfounded. The Washington-based National Institute for the Promotion of Science was in competition with the American Association of Geologists. It exemplified the amateurism and political involvement that Henry and Peirce feared in such an organization. When the specimens of the Wilkes expedition[33] came under its control, its secretary, a scientific hack, removed labels from the specimens and gave many of them to friends and tourists.[34]

Nevertheless, ten years later the mood was more optimistic, if guarded.[35] Peirce's friend, Bache, who more than any other one man dominated the course of American science during this period, did not attend the 1847 meeting, but had been campaigning, nevertheless, for such an organization. He hoped that such an organization would help recruit able men into a scientific career, foster communication among scientists, help channel private and government dollars into science, and most importantly govern the American scientific community.

Peirce, Agassiz, and Henry Rogers were given the task of writing a new constitution for the AAAS. Rogers, who referred to the constitution as his, apparently did the actual writing, but Agassiz probably dominated the organization of this more general association, which was patterned after that of the British Association for the Advancement of Science.[36] The constitution certainly served the Lazzaronis' goals. Furthermore, the Lazzaroni dominated the early years of the AAAS leadership. Lazzaroni members Henry, Bache, Agassiz, Peirce, and Dana, were presidents of the organization from 1848 to 1855.[37]

$$\Phi^n$$

Peirce and his Lazzaroni friends were happy to expose unscientific charlatans as well as scientific ones. On June 25, 1857, Peirce attended a séance at the Albion building in Boston. The séance was held in response to an offer of five hundred dollars to anyone who could successfully demonstrate the validity of spiritual phenomenon. The Bostonian, Dr. H. F. Gardener, took up the challenge and was joined by a number of other spiritualists, including the two well-known mediums, Miss Kate Fox and her sister, Mrs. Brown. Peirce was there as chairman of a committee to ascertain the validity of the proceedings. He was joined by his two Lazzaroni friends, Louis Agassiz and B. A. Gould, as well as his Harvard colleague, Professor Horsford, the holder of the Rumford Chair. Attempts to demonstrate communication with spirits went on for three days with no positive results.[38]

Peirce was an old hand at investigating spiritualism. Several years before he had performed an experiment on a young woman claiming susceptibility to spiritual phenomenon. Baron von Reichenbach had claimed that a universal force, or "Od," pervaded all of nature and manifested itself in some people who were sensitive to it, usually women. The force was evident in magnets, heat, crystals, light, and some chemical processes. The young woman in Peirce's experiment "was almost thrown into convulsions when brought near a powerful steel magnet. Unfortunately for the supporters of the new force, the same effect was produced when the lady was submitted to the influence of a piece of wood formed and painted like a magnet."[39]

Peirce's scientific friends generally sided with him against spiritualism. His southern friend, John Le Conte, wrote to Peirce, "We are very much gratified to see that you have been exposing that impudent piece of rascality, 'Spiritualism.' Wherever there is a '<u>Rap</u>' there is a '<u>Rascal!!</u>' "[40]

$$\Phi^n$$

An incident at the AAAS meeting in Montreal in the summer of 1857 began Peirce's most notable controversy involving amateur scientists. The confrontation reflects many of the problems that professional scientists faced in building American science. Peirce's sister, Elizabeth, relates the details:

> She [Sarah] and Ben returned from Montreal last Saturday. She says that she was very glad she went however[Sarah had been ill], if only on Ben's account, for if a gentleman had no wife, or his wife was not with him, he was packed into a room with ten or 16 other gentlemen. The Hotels were all so crowded. Sarah was able to go to two balls, . . . Sarah was the first lady handed in to supper & she was conducted by the Lord Bishop—In general the members of the association did not receive nearly so much attention as at the other places where they have met.
>
> Ben, as I suppose you saw in the papers, was accused by a Mr. Winslow of having stolen some of his (Winslow's) discoveries. Of course Winslow told a lie without a shade of truth & everybody of any sense knows he did. It is very strange what induces Winslow to make such an assertion so injurious to himself. When Ben first made the discovery, he wrote to the Rev. Dr. Hill telling him of it & Dr. Hill wrote about it to Winslow who in his answer, expressed great delight in Ben's discovery. Dr. Hill still has Winslow's letter.[41] At one of the meetings of the Association, Winslow made some sneering remark concerning Dr. Hare, a very learned & excellent man of Philadelphia, 80 years old, but unfortunately a believer in the spiritual manifestations in reference to which Winslow said something about putting him in the Insane Hospital. Ben was very indignant at an attack on a man so amiable, wise, & highly respected as Dr. Hare. So when he spoke next he said "that he had been accused of appropriating to himself another person's discoveries while if he were indeed the originator of the vagaries which truly belonged to that <u>person</u> he (Ben) should be entitled to a place in that Asylum Institution, to which one of their most beloved and venerated members would have been so unjustly consigned."[42]

In 1853 Winslow gave Peirce a prepublication copy of his book, *Cosmography, or Philosophical Views of the Universe*.[43] In this book, Winslow put forth his new theory of the cosmos. His basic theory was that large masses of matter, such as the sun and planets, acted much like atoms. Although they attracted bodies at a distance, they repelled them if they were brought sufficiently near. Thus, for each central body there is an inner sphere of repulsion, extending to only a limited distance, and without it a sphere of attraction. Not only did Winslow's theory have no basis or connection with physical law, it was filled with internal inconsistencies. The sphere of repulsion of the sun, Winslow claimed, was very close to the body's surface, the planets being in its sphere of attraction. Yet Winslow claimed that the sun's sphere of repulsion extended all the way to the stars, which are vastly farther from the sun than the planets.[44]

Winslow used his theory to account for earthquakes and volcanoes. In the winter when the earth draws nearer the sun, the sun's repulsive force squeezes the earth into a smaller size, thus squirting out liquid beneath the earth's crust.[45]

After the publication of his book, Winslow sought to extend his theories. He took his scientific work seriously, even traveling to the Sandwich Islands to gather further evidence. Winslow was so excited about his findings that he cut his trip short, came home, and presented his results at the 1856 meeting of the AAAS in Albany. Although Winslow claimed his paper was enthusiastically received, the newspapers paid little attention to it. The copy of the paper that Winslow presented to Joseph Lovering, the secretary of the AAAS, was not published in the proceedings of the AAAS, nor was its title even listed. Winslow had to repeat a request for the return of his paper, and even then was never given the last section that dealt with continental shape.[46]

Then Winslow noticed an announcement in the Christian Examiner of a discovery by Benjamin Peirce. The essence of Peirce's discovery was: "If we elevate a terrestrial globe until the Arctic and Antarctic Circles are tangent to the wooden horizon, and then cause the globe slowly to revolve, we shall find that a *majority* of the lines of elevation in the earth's crust, i. e. lines and mountain ranges, will, either as they rise or as they go down, coincide in passing with the wooden horizon."[47] Peirce also hypothesized that the sun was responsible for the formation of the planets, but from the sun's heat, not its gravitational force, as Winslow had hypothesized.

The *Christian Examiner* heaped lavish praise on Peirce's discovery. It felt that the principal value of Peirce's discovery was to religion, as it showed "by geological facts, that the obliquity of the elliptic has been essentially unchanged since the dawn of creation, and that solar heat was the agent to carry out the command of the second day, to let the dry land appear."[48]

Winslow was rankled by the praise Peirce had received from the *Examiner* and wrote to the *Boston Traveler,* accusing him of plagiarism. An article in the evening edition of the *Traveler* reiterated Winslow's accusations, and praised him as "a gentleman of the highest respectability, and unquestionable scientific attainments." Although it added that Winslow's "egotism is amusing."[49]

Peirce's underlings at the *Nautical Almanac* quickly wrote an unsolicited refutation of Winslow's claims.[50] In doing so, they not only defended Peirce, but the emerging body of scientific professionals, who wished to garner international respect for American science:

> Neither the book to which Dr. Winslow refers, nor, judging from his own description, the paper which he presented to the American Association, contains any ideas which either Prof. Peirce or any one else who values his scientific reputation would purloin—supposing him to be in the habit of doing such things. . . .

> The *Traveller* [sic] pronounces the author of "Cosmography" a gentleman of "unquestionable scientific acquirements." Such may be the case; but if it is, his acquirements are in some branch widely different from the subject on which he wrote. Omitting the two introductory paragraphs, there is scarcely a principle stated in the whole book which will bear the test of criticism by a competent person; and, with the same exception, I challenge the production of three successive paragraphs in the first part of the book (I am not sure that I might not include the whole work), which do not betray a total misapprehension of the principles, methods and results of modern astronomy, and which any eminent astronomer in the world would not pronounce to be directly opposed to what we *know* of the constitution of the universe and the laws of Nature. It is true that there is not lack of self-glorification, flowing language and learned phraseology; but we would not suppose the *Traveller* considers these qualities as indicative of the scientific attainments of the author. . . .

> Many persons suppose it to be the principal occupations of men of science to propound new and curious theories, and to speculate upon mysteries which they cannot comprehend. That a mere *theory* can be proven true beyond a doubt, they are not willing to admit. There is a class of would-be scientific men in this country, who, having imbibed these notions, and being possessed of considerable imaginative and speculative powers very naturally consider themselves inferior to none in their powers of speculating and propounding, and thereupon originate some new theory of heat, of light, or the solar system. Their knowledge of the phenomena is probably limited and inaccurate, as is also their knowledge of the foundations on which principles rest, and the degree of certainty which they can attain. . . .

> When such persons bring their speculations, unsupported before a scientific body, what is that body to do? Spend the valuable time of a hundred men in listening to a discussion which has no more interest to them than the narration of the vagaries of a dream, and then publish the paper in their transactions? . . .

A very strong reason for excluding papers which do not possess scientific accuracy is that the proceedings of the Association are spread over most of the civilized world, and will very naturally be taken as an index to the state of science in America. What opinion then would foreign scientific men form of us if they saw in the transactions of the largest scientific body in this country, a jumble of the work of Newton by a person who did not understand mathematics?[51]

Peirce was deeply touched when Winlock, Wright, and Newcomb, his subordinates at the *Nautical Almanac*, came to his defense. Peirce wrote to John Le Conte:

I confess my dear professor the venom of the newspaper attacks is infinitely overbalanced by the readiness of the defense; and I thank my stars for this opportunity to feel that so many hearts are capable of beating in human breasts with grateful affection. I cannot deny that I have been kind to these men—but I never [expected] any return for them, for my kindness were the simple acts of justice to evident merits and which it was my duty to perform, and which I performed as a duty, and assured them when I performed them, that they were acts of duty. Now this generous and uncalled for return makes me almost weep like a stray-hearted woman. Thank heaven that such hearts are found in the bodies of mathematicians. They are the finest pride of our glorious science, and makes one hope that the mystic key of geometry will sometimes open the chambers of diverse affection.[52]

$$\Phi^n$$

Winslow went to the Montreal meeting of the AAAS in August 1858, seeking redress for the injustices perpetrated upon him by Peirce and the scientific establishment. He was astounded that the same scientific establishment that had refused to publish, or even mention, his paper, had allowed Peirce to give what seemed to him a similar paper at Montreal.[53]

Peirce deftly defended himself at Montreal. The *Boston Daily Courier,* reporting on the activities of the AAAS at Montreal, felt he had easily dispatched Winslow's accusations: "There is a well-known work in English theological literature, called Leslie's Short Method with the Deists. We commend to our readers Prof. Peirce's short method with Dr. Winslow, as will be found recorded in our letter from Montreal. The Professor uses few words; but they do their business thoroughly. He strikes but one blow, but that is as effectual as a blow from an elephant's trunk; the victim never knows what it was that hit him."[54]

The *Courier* recorded Peirce's actual words: "In reference to the charge with which certain papers have recently been laden, of my having appropriated the labors of another person, I must be permitted to state that there is no allusion to the facts which I have brought before the Association, either in his published book or in his memoir, which has also been printed at full length, and which every one may read and judge for himself, and which no one can now steal. With regard to his speculations, they are so opposed to my own that if he is right I am altogether wrong, and if I am right he is wrong; and I have no desire to claim his wild conglomeration of vagaries, which seem to me worthy only of the institution to which he would have consigned one of the most venerated and beloved of our members."[55]

Φ^n

John Warner, a man with mathematical interests from Pottstown, Pennsylvania, had seen a notice of Winslow's accusation of Peirce's plagiarism in the *Rochester Union and Advertiser*,[56] and found Winslow's story to be much like his own. He too had been robbed of his scientific work by none other than Benjamin Peirce.

Warner had asked a friend, P. W. Sheafer, to deliver a paper he had written "treating the subjection of organic forms to mathematical law"[57] to Louis Agassiz. Sheafer delivered the paper to Agassiz in November 1853. At this time, Warner was not aware that Europeans had done research in this area and was hoping that Agassiz would confirm the direction he had taken and give him encouragement.[58] Subsequently Warner discovered, quite independently of Agassiz, that work in this area had indeed been done by European scientists, and felt justified in the direction of his pursuits and assumed that he was the first American to pursue them.

When Warner did not hear from Agassiz, he attempted, through Sheafer, to procure his paper, despite the fact that his original objective in communicating with the distinguished zoologist no longer existed. Then three years later, the paper still not returned, he became impatient with the delays.[59] At this time, having the paper returned became an obsession with Warner. He wrote to Agassiz,[60] and, being unsuccessful in this avenue of endeavor, again recruited the aid of his friend Sheafer to procure the paper. Sheafer, in turn, wrote to a W. W. Whitcomb, in Boston, requesting his aid in recovering the paper. Whitcomb visited Agassiz at Harvard and was cordially received. Agassiz recalled Shaefer's and Warner's letters, but did not know what had happened to the paper. He explained that since it

dealt with a mathematical topic, he had likely passed it on to someone with expertise in that field, probably Benjamin Peirce. If so,
Peirce may have submitted it to the American Association for the
Advancement of Science to be considered for its meeting. Many such
papers were rejected. If this was the fate of Warner's paper, Agassiz
wasn't sure what became of such rejected papers—whether they were
filed or not. At any rate, Agassiz promised to inquire after the paper.
Before Whitcomb left, Agassiz showed him a stack of correspondence
that he had received in the last week, indicating that it was impossible to deal with all of it.[61]

Several weeks later, Whitcomb visited Peirce. At first Peirce had no
recollection of the paper, "but afterwards having a faint recollection
of having something of the kind—said he had a great lot of papers to
enter at the meeting—still thought that if this one was among them
he should have a more distinct idea of it. His impressions were that
he never entered it."[62] Peirce sent Whitcomb to Lovering, Peirce's
colleague at Harvard and secretary of the AAAS. Lovering's records
confirmed that the paper was not among those presented and
entered. Whitcomb noted that Agassiz had not mentioned the matter to Lovering, as he had promised. It was Lovering's opinion that
Agassiz had simply mislaid the paper three years earlier.

This explanation did not satisfy Warner, who still wished to have his
paper returned or at least an apology from Agassiz or Peirce for having lost it. By this time he had already published a pamphlet on
organic morphology,[63] and he implied in the letter to Agassiz that his
researches had progressed considerably since his having written him.[64]
Thus the paper could have little value to Warner, even if he had not
kept a copy of the original.

Beyond being aggravated at the loss of his paper, Warner also suspected Peirce of plagiarism, since Peirce gave a talk at the 1855 Providence meeting of the AAAS on organic morphology.[65] Warner also
suspected that Peirce had served as a referee for his paper when he
submitted it earlier to Joseph Henry at the Smithsonian, so Peirce
may have seen the paper before he gave Agassiz the paper for the
AAAS. Henry sent the work to an unknown referee, and he did on
occasion use Peirce as a referee for mathematical work.[66] Henry
replied to Warner in December 1853, saying that the referee recommended against publishing the paper, although he praised Warner's
talent. He also suggested that after Warner had studied more mathematics he would find a more fruitful branch of the subject in which
to do research.

Exactly when Peirce saw Warner's work, if he ever saw it at all, is
quite immaterial. Peirce's colleague, Agassiz, was familiar with organic

morphology as early as 1837, well before coming to this country.[67] Although organic morphology was never one of Peirce's principal scientific interests, he did some work in it, as early as 1849.[68] Peirce also wrote about analytical morphology to his friend John Le Conte in 1852.[69] Significantly, Peirce applied organic morphology to botany, and his impetus for studying it certainly came from Agassiz, not Warner. Furthermore, what little work Peirce did in this field bore only the most superficial resemblance to Warner's.[70]

When Warner read the article in the *Rochester Union*, he wrote to Winslow hoping to secure his "cooperation in the recovery of an original memoir upon Organic Morphology."[71] Warner complained, "A proclivity to lose original manuscripts upon scientific subjects is to be deplored both in individuals and in scientific bodies."[72] Winslow responded promptly to Warner's letter, and they corresponded frequently over a period of four years.[73] Winslow wrote frequent long letters; they are often over ten pages; one is twenty-seven pages. All of them are filled with vitriol towards Peirce and his associates.

Winslow explained to Warner that his only recourse had been through the press, otherwise he would have been ignored: "In sending my letter to Lovering to the *Trav.*[74] I sent the remarks respecting Pierce which were personal enough to make him very angry at Montreal & to provoke his reply there in his defense, he said there was not one line in my Book or Memoir (This showed he was familiar with the contents of my paper.) about the coastlines being arcs tangent to the polar circle etc.—& charging my vagaries to be fit for a lunatic hospital etc.—This is the first time I had opportunity to speak to him directly—for he now announced his discovery himself etc. & it is only now that I have been able to get my reply out by paying for it."[75] Publishing his accusations against Peirce was expensive, but Winslow had a higher purpose than merely his own recognition. He explained to Warner that the thirty dollars it cost "is a good deal of money to pay these times but the 'Good of Mankind' demands the sacrifice."[76]

Warner turned to the press for justice as well, writing a letter to the *Boston Daily Journal*.[77] He also wrote to Horace Greeley, the noted journalist and politician. Greeley edited the *New York Tribune*, which he had founded. In writing to Greeley, Warner was hoping to get wider coverage in the press.[78] Greeley responded in a kind letter; he felt, however, that Warner's case would neither convince nor interest the general public and urged Warner to publish his work in scientific journals. Warner wrote Greeley, saying that he would drop the matter, except for Winslow's encouragement.[79]

$$\Phi^n$$

Although a physician, Winslow received no scholastic training after he was fourteen.[80] Winslow's lack of formal education does not necessarily separate him from serious scientists of the period, but his baseless speculations do. Although Winslow deeply resented being called a speculator by the AAAS leadership,[81] he cheerfully confided to Warner, "You see I use no caution with my brains.—dash on till I reach for every abode of the great I am. I have no fear while I reason on facts & hold truth in my grasp. I would shake my fist at all the Corporated Institutions & their pasteboard professors from Cambridge to Washington."[82]

Winslow also freely admitted his ignorance of mathematics. "It is true I am inpatient of the higher calculus, but few persons making researches have ever been more patient or speculated less. Mathematics is not required in the consideration of my doctrines."[83] Furthermore, Winslow expressed his dubious opinion of mathematics: "Mathematics is only the handmaid to the discovery of physical law & not physical law itself; & its admirers show great narrowness of comprehension & physical learning to assert authoritatively & with cold imperviousness that in as much as it proves the truth of gravitation[,] it a priori proves the existence of any other force to be impossible & an absurdity."[84] Winslow was proposing new and highly advanced ideas about new forces of repulsion that were not understandable in mathematical terms. He, like Galileo, was far ahead of his time.

Winslow's ignorance of mathematics did not deter him from handing out awards, however. He had high hopes for Warner, as Warner no doubt did for himself, and told Warner that he hoped he could find a situation where he could develop to his full potential. He told Warner such a situation would "give you a chance to exhibit what I believe you possess[:] the most original & highest mathematical genius in this country."[85]

But even Warner, despite his mathematical aspirations, had limited mathematical knowledge. His paper on organic morphology is well done and lacks the baseless speculation of Winslow's work, but the level of mathematics used in the paper is relatively modest. Warner, in fact, admits he doesn't know much mathematics: "I never pursued mathematics very far, although it may seem different here [Pottsville, Pennsylvania] where but little attention is paid to it."[86] In addition, Warner realized that even his knowledge of the narrow field of organic morphology was limited.[87]

Φn

At first Winslow reluctantly gave Peirce credit for his abilities as a scientist. He wrote that Peirce "is a vampire in an intellectual way." He "is a scamp without any doubt although very acute in geometry."[88] Even this grudging credit soon disappeared. Winslow's hatred of Peirce blinded him completely to Peirce's ability. He said, "I should like to see him produce something wholly original. I do not believe he has yet done one thing to make the least impression on the progress of science in this glorious age we live in. He is a sham—a pretender & a prince of the humbugs & I believe time will determine him to be so in his morality at least if not in part of his mathematical claims."[89]

Φn

The Warner-Winslow correspondence makes it clear that although they regarded Peirce as an unprincipled plagiarist, their deeper frustration was not with him, but with the American scientific establishment. Warner and Winslow were distressed because they were being excluded from this community; leading American scientists were telling them that they were, in fact, not scientists. Warner's pathetic pleas for recognition, if only in the form of an apology for losing his paper, are cries to be treated with the respect a scientist deserves.

Winslow resented the emerging scientific aristocracy; he complained that the Cambridge scientists have "mistaken themselves for Gods."[90] He added that: "Such men as Pierce wish to be Lords of Creation, exclusive to the last degree & they are worse than men stealers & Negro overseers. They use the original thoughts and labors of other men to aggrandize themselves, as a Southern Planter grinds the sweat & labor out of his Negroes, to buy himself bread[,] & clothes himself in sins & broad cloth."[91]

Φn

Peirce was, of course, annoyed by the attacks of Warner and Winslow. He wrote to his friends the Le Contes, telling them of his frustration with the attacks and his resolve not to respond to them in the papers.[92] No doubt the attacks at least temporarily tarnished his reputation among the public.

Even so, it was a minor aggravation; for example, Peirce did not bother to mention the affair in his correspondence with Bache.[93] Serious scientists, as well as non-scientific friends, easily dismissed the attacks, and they became in fact, something to joke about. This is born out in a description of a Harvard commencement by Winslow himself: "I was at Commencement at Cambridge a few days ago. One of the graduates in pronouncing a dissertation made some allusions in a facetious way to plagiarism in organic morphology & at the same time turned in a complementary manner to Pierce who sat with Agassiz & a hundred other dignitaries on the Platform. Pierce was much flattered & amused & it was intended to be bitingly ironical toward any who should charge the great American analyst . . . with plagiarism about curves & numbers & inclinations of the divine mind, no doubt. The amusement and merriment escaped my own attention; but my son who was present said he observed & heard it plainly."[94]

The problems reflected in the Warner-Winslow affair reflect larger problems with the leadership of the AAAS and the American scientific community. Although it was reasonable for the Lazzaroni to discourage scientific quacks, the distinction between competent scientists and fraudulent ones was not always clear cut. Many reputable scientists disliked the Lazzaroni's handling of AAAS procedures. Among them were David Wells, who was trained at the Lawrence School, and the oceanographer, Matthew F. Maury.

Yet whatever their feelings about the AAAS, Warner's friends discouraged him from continuing his assault on Peirce in public. Most felt the newspapers were not a suitable arena for scientific disagreements.[95] The astronomer Daniel Kirkwood proposed that Warner's researches were not appropriated by Peirce, but "suggested to him the idea of his investigations on the subject."[96] J. W. Yardley cautiously advised Warner to give a presentation at the Springfield meeting of the AAAS, but not make a direct attack against Peirce until he stood on firm ground.[97] Another friend, the geologist J. Peter Lesley, also advised Warner to publish his work.[98] Then, after discussing the matter with John L. Le Conte, the Philadelphia entomologist,[99] Lesley urged Warner to drop the whole matter, pointing out that he had no case against him, the similarities in their work being superficial. Furthermore, Le Conte and he agreed that "your friend's [Winslow's] appeal to the public through the daily press is no organ for a scientific discussion involving questions of the greatest difficulty such as priority & identity of thought."[100] Winslow alone urged Warner not to give up the fight. Although when faced with the advice of his friends and the crusade's futility, he eventually did.

Winslow never gave up hope that he would some day receive the recognition he felt he deserved. In one of his later letters to Warner, he explained:

It matters not to me how the two forces work together as long as they are proved to both exist. Newton is immortal by his discovery of gravitation. The discovery of that & its place in nature, have their pages in the general history of the inductive sciences. Repulsion as a solar and cosmic force must hereafter have its pages of history too. Newton's discovery was dormant much longer than mine.—& he had his Leibnitz to contend against who was a fairer opponent than Prof. Pierce has been to me. But my Book is written. Four years afterwards Peirce sneered at it. In one year more, after the appearance of Donati's comet,[101] he announced his unqualified belief in its doctrines; & in his elation, hastily & erroneously published conclusions which hereafter will be discredited to his mathematical reputation.

These are facts. All this must become history. Keep this letter—& use it after I am dead.[102]

7

As the Tree Grows Towards Heaven

In APRIL 1850, PEIRCE WROTE HIS FRIEND BACHE, APOLOGIZING FOR NOT giving him and the Coast Survey his usual strong support.[1] Peirce had slacked off partly because he didn't feel the Coast Survey was in any real trouble and partly because "Dr. Webster has driven everything else out of heads in Boston."[2] John W. Webster, professor of chemistry at Harvard, was appreciated more for providing the seniors with a lavish spread at graduation than for his teaching or accomplishments as a chemist. Entertaining friends lavishly as well, Webster lived far beyond his means. In 1847, to cover some of his debts, using his mineral collection as collateral, he borrowed over two thousand dollars from George Parkman, a wealthy Boston merchant and real estate speculator. Although Parkman was trained in Europe as a psychiatrist, after inheriting money and property, he spent most of his time collecting rents and making private loans. Despite his generosity to the Harvard Medical School, Parkman was generally regarded with disfavor among Bostonians due to his business dealings and eccentricities.

A year later Webster obtained another loan from Parkman's brother-in-law, using the same cabinet of minerals to secure the loan. When Parkman learned of this chicanery he argued with Webster, but agreed to discuss the matter further with him at Webster's laboratory. Parkman disappeared on the day of the appointment. Several days later, most of Parkman's dismembered body was found under the privy of Webster's laboratory. Webster was arrested, tried, convicted, and executed for the crime.[3]

Peirce, convinced of Webster's guilt, was surprised that Bache had doubts:[4]

> We are quite astonished in Boston that there should be any hesitation as to the justice of the verdict in Webster's case. His staunchest friend here admits it, and believes the murder to have been premeditated. First of all the murdered man's property is found in Webster's possession, property which might have been in Dr. Parkman's pocket book at the time of the

murder, so that the actual possession of the pocket book or even of Dr.'s coat and pantaloons would not have been a more damning evidence of guilt. And then the remains were found under Dr. Webster's lock and key—and part of the body was found destroyed in Dr. Webster's furnace. It was destroyed in the furnace in which it was found, for it adhered to the sides and had to be broken off. And then there is Littlefield's[5] testimony in which the defense was unable to find any flaw whatever, but on the contrary confirmed it in the most singular manner by the evidence of Dr. Webster's own family.

But I will not argue the case for I do not believe that if you have seen a full report of the case, you can have any doubt of his guilt. I do not believe the story of the medical student. It has presented itself in too great a variety of contradictory shapes, and I am rather inclined to believe that its sole foundation is another story which I believe be authentic. A student was reported to have said that he believed Dr. Webster to have killed Dr. Parkman, and an officer was sent to enquire of him, and he said, . . . "that Dr. Webster's last lectures had been so tolerably stupid that he thought they must have been the death of Dr. Parkman." Dr. Parkman was so harsh and cruel a man with his debtors, that his murder seems almost to have been a retribution of Providence designed to teach us an appalling lesson. After the identification of his remains, which was so complete and satisfactory, that Sohier,[6] the junior council, was asked what he thought of it. "Well," said he "it must be admitted that the body resembles Dr. Parkman in one respect; it has no bowels." At another time Sohier's attention was drawn to the fact that Dr. Webster was getting more dejected as time proceeded and he observed "Yes, he is beginning to get the hang of it." At the close of the trial, Sohier said to the senior council, "The attorney general has destroyed our facts, the chief justice our law, and our client has destroyed our character. There is nothing left us but religion; come Merrick, let us go home and experience religion." Webster has disgusted every body by his levity and there is a very faint hope that he may prove to be insane. His whole mind seems to be running upon his food, and upon the delicacies with which he may pamper his appetite; and it is very remarkable that he has never protested his innocence of Dr. Parkman's death. In the course of the trial, Prof Treadwell[7] told him that he should see his family in Cambridge, and ask if he had any messages to send. "Yes," said he, "ask them to send me some custards." On the morning of the sentence, as he was leaving the jail for the courtroom, he asked the jailor for a piece of paper on which he wrote in his most eager and excited manner an order for Parker to send him some cream cakes for dinner. The first of these two stories I have from Treadwell himself and the second from Dr. Codman[8] who was told it by the jailer himself; and the jailer said that of all the men, whom he had ever had the charge of, no one was so strange and unaccountable in his conduct as Dr. Webster. He said that the Dr. slept so soundly, that it was with great difficulty he could awaken him for his breakfast; but perhaps this over sleeping in the morning may have been the consequence of his having kept awake half the night.[9]

Despite the macabre beginning, the 1850s started out well for Peirce. Even Webster's execution opened the way for John P. Cooke, who worked with great diligence to improve chemistry at Harvard.[10] In 1850, at forty-one years of age, Peirce had accomplished much. He had a wife he loved, five children, and a professorship at the oldest university in the land. The telescope had been installed for several years in the Harvard Observatory that he had worked so hard to build, and he had been honored both at home and abroad.

Although things looked good, Peirce felt uneasy. He wrote, "If I am not mistaken, there are clouds hovering on the horizon—and I think I have seen a flash or two in some averted faces, which indicates mischief. But sufficient unto the day is the evil thereof."[11] Unfortunately Peirce was right; as he watched the nation progress toward civil war, something he felt and feared strongly, trouble came to his own life.

$$\Phi^n$$

On August 18, 1851, Peirce set off for the meeting of the American Association for the Advancement of Science in Albany, leaving his wife in Cambridge. He took his eleven-year-old son, Charles, with him as far as Springfield, where he left the boy to catch a train to Northampton to visit his mother's family. Peirce continued on to Albany, where he met his other Lazzaroni friends and their wives.[12] In a letter to his wife, Sarah, he said that their hosts in Albany proposed "on Wednesday next, to take the Association to Troy in carriages, when the corporation of Troy will show us their wonders, and then fill us with the good things of this carnal world and send us back to Albany. . . ." Peirce closed the letter in typical fashion, "Good bye, my darling Sarah, my life, my happiness, my all that is sweet and good and true. Goodbye and may blessings of providence surround you, night and day."[13]

When Sarah wrote Benjamin, she, too, expressed her love for him. "This is my fifth day of weary widowhood. . . . Dull beyond all description in spite of the children's gaiety and noise. . . . But if I had all the world beside without you I should feel alone." Yet Sarah also wondered, "How many attractive young ladies Prof. Peirce has discovered in Albany."[14]

Peirce did meet one attractive lady, Josephine Le Conte. He not only took to Josephine, but to her husband, John, as well. He became the closest of friends with both of them. During the next decade he would pour out his heart to them in his letters in a way that he did not even do with Bache. Only war would separate them.[15]

John Le Conte. By permission of the Bancroft Library.

John Le Conte was an outstanding scientist. He had eclectic interests and did fine work in many fields. Trained as a physician, he made heavy use of statistics in studying cancer, and conducted experiments with the nervous systems of alligators. His specialty, however, was physics, a mathematical science. It was a field Peirce identified with and regarded as solid.

Not only was Le Conte a scientist himself, he came from a family of distinguished scientists. Probably no other family of the time contributed as much to American science. Le Conte's brother, Joseph, was a noted geologist, who studied under Agassiz at the Lawrence Scientific School, but unlike his teacher, wrote extensively supporting the theory of evolution.[16] Their father, Louis, had strong interests in chemistry and botany; in addition, he had one of the best botanical gardens in the United States on his Georgia plantation.[17] Although Louis had no interest in publishing his scientific investigations, he instilled a love of science in his sons. Louis's brother, Major John Eatton Le Conte, and his son, John L. Le Conte, were two of the country's best entomologists.[18] With the exception of John Le Conte's father who died in 1838,[19] Peirce became good friends with all of these men.

Le Conte was an aristocrat. His father had owned a plantation in Georgia that had two hundred slaves on it.[20] More than being a southern aristocrat, the Le Contes proudly traced their ancestry back to Guillaume Le Conte, a Huguenot nobleman, who left France seeking toleration for his religious views.[21] But Le Conte wasn't an aristocrat simply because of his ancestry and his being a wealthy southern landowner. He was an aristocrat in spirit and in belief.[22] He wrote to Peirce in 1860, blaming the bad political situation in the country on universal suffrage: "The boasted <u>universal suffrage</u> is the <u>fountain</u> and <u>origin</u> of all of our political difficulties; which are rendered more incurable by that other <u>boasted privilege</u> the <u>freedom of the press.</u> We, of the South are more fortunate than you are, in so far as the great mass of our laboring class is excluded from voting. I am fully satisfied, that the <u>normal condition</u> of society is, that <u>intellect</u> and <u>wealth</u> should exercise a <u>controlling influence.</u> This is a <u>fundamental law</u> of sociology, and any attempt to contravene it, must prove as impotent and as imbecile, as our vain efforts to thwart the laws of the physical universe."[23]

Like most southerners, John hated abolitionists, but more than abolitionists he hated charlatan scientists.[24] This made him a perfect candidate for the Lazzaroni. He and his brother John quickly became close friends not only with Peirce, but with Bache, Henry, and the other members of Peirce's scientific circle. Although their geographical separation made them unavailable for many of the Lazzaroni activities, John did write a scathing review of *The Physical Geography of the Sea*, written by that archenemy of the Lazzaroni, Matthew F. Maury.[25]

Reserved and formal, John did not make friends easily.[26] Not so his wife. She was exceptionally charming and as gregarious as her husband was aloof. "Mrs. Le Conte was a woman of wonderful personal magnetism, queenly in bearing, and of extraordinary beauty."[27] "Josephine . . . reveled in the attention her comeliness attracted. 'She had constant adorers from all classes,' recalled Joseph's daughter Emma, and the callow college boys fluttered around her like moths 'while John's acquaintances bowed at her shrine.' "[28] Elizabeth, Benjamin's sister, described her:

> I wish you could see Mrs. Le Conte. She is a <u>celebrated</u> beauty & in my opinion she deserves her celebrity, for she is very handsome. She is tall & rather large—but has a beautiful figure, magnificent dark hair—a fine complexion & the prettiest feet and hands you ever saw. Her hands are very attractive and busy hands—they sew, cut & are always doing something. When they have nothing else to do, they flourish about, sort of talking. . . .

> Mrs. Le Conte dresses in the height of fashion & when she came into our parlour she almost filled it up in back—She did quite occupy one of our sofas. She wore black satin exceedingly full & with a long train, which followed after her, dragging along. Her bonnet was a pink silk covered with black lace, a beautiful lace veil & large red roses inside her bonnet.[29]

Yet Josephine Le Conte was far more than attractive and fashionably dressed. She was a woman of rare intelligence, wit, and force of character.[30] One evidence of the strength of her personality is the position she held in the Lazzaroni circle. The wives of the Lazzaroni members had nicknames as well as the men. But except for Josephine, they were simply modifications of their husband's name. Bache was the Chief; his wife was the Chiefess. Peirce was Functionary; Sarah was Functionarya. But Josephine retained her own identity among the scientific elite of her time. She was the Queen. The Queen of the Lazzaroni; the Queen of science.

Everyone in the Peirce family was quite taken by her. Bejnamin often asked for her opinion and advice.[31] Sarah valued her friendship.[32] Peirce's son Benjamin had a crush on her; Peirce's daughter, Helen, idolized her.[33] Peirce wrote:

> You cannot think how often the children speak of you and your haunts. When ever they see a pretty woman—they compare her to Mrs. Le Conte, and always to her disadvantage for they have very good taste. Yes, my dear friend. We all love you in Cambridge, and the doll which bears your name is quite a pet. . . .[34]

> You would feel pleased, I think, if you knew how freely your memory was preserved in this house. The children are almost every day alluding to you[,] and Benjie is as much your devoted page as ever he was. Mrs. Peirce always speaks of you with great affection and interest.[35]

Josephine Le Conte. By permission of the Bancroft Library.

The January after the meeting, Peirce received gifts from his new friends in Georgia.

What shall I say to these exquisite presents? to this charming handicraft? to these palpable proofs that you are, as I have reported you, a true-hearted friend? What can I say? In a single mathematical formula, I could express the motions of a universe, all the past, all the future, and even all the possible motions of all possible worlds—but not a single emotion of the human heart. . . . Poor mathematician! Is he not to be pitied? However great he may be, however capable of arranging perturbations by which nobody is disturbed, and of predicting the future which nobody will live to see—when you bring him into the presence of one of those beautiful and bewitching descendants of Mrs. Eve and witness his nervous stammering and stuttering and how the brilliancy of his intellect is instantly eclipsed and occulted into dark and silent midnight by the brightness of the Eve—Do you not, then, pity him? Do you not, then, feel that his life is a sort of splendid misery? . . . There—my wife will interrupt me—For many years a proud Peirce she is now purse-proud. She commands me to thank you, in her name, most earnestly and sincerely and kindly and warmly—and a warmer heart than hers never yet loved—for your most beautiful present.[36]

$$\Phi^n$$

Peirce had met other friends at AAAS meetings. Earlier in the year Peirce first met Joseph Winlock at the AAAS meeting in Cincinnati. Peirce was impressed with the young man, and hired him to work on the *Nautical Almanac*.[37] Years later he would recommend him over his former student and fellow Lazzaroni member, Benjamin Apthorp Gould, for the directorship of the Harvard Observatory.

At the Albany meeting, Bache made a proposal that would become one of the Lazzaroni's most important goals. In his address as retiring president, Bache called for the founding of a national organization of scientists that would have government support and be given the official task of guiding the government in scientific matters. As head of the Coast Survey, Bache had firsthand experience about the problems that arose if laymen were left to make scientific decisions. He said, "first a few observations on the ordinary modes of promoting science; in connection with which, I would throw out for your consideration some reasons which induce me to believe that *an institution of science, supplementary to existing ones, is much needed in our country, to guide public action in reference to scientific matters.*"[38] Bache went on to say: "Suppose an institute of which the members belong in turn to each of our widely scattered States, working at their places of residence, and reporting their results; meeting only at particular times, and for special purposes; engaged in researches self-directed, or desired by the body, called for by Congress or by the Executive, who furnish means for the inquires. . . . The public treasury would be saved many times the support of such a council, by the sound advice which it would give in regard to the various projects which are constantly forced upon their notice, and in regard to which they are now compelled to decide without the knowledge which alone can ensure a wise conclusion."[39]

Φⁿ

But beyond the intellectual stimulus of giving and hearing scientific papers, and the joy of social pleasantries, exciting news was afoot in Albany. There was talk of building a great university there, a national university that would compare favorably with the best European institutions.

The modern research university was an outcome of the Napoleonic wars. In 1806, when Napoleon defeated the Prussians at the Battle of Jena, the Treaty of Tilsit left Prussia a third-rate power. Prussia retained about half its former territory, had an army that was severely restricted in size, and was crippled by heavy indemnity payments to France. Faced with this dismal situation, King Frederick William III stressed education as a way of revitalizing his kingdom. Part of this educational initiative was the modern research university, which stressed research more than teaching. It was here that modern academic freedom was born. German universities became the best in the world. Consequently, during the discussion of the prospective university in Albany, enthusiastic advocates suggested modeling it after German universities, especially the University of Berlin, although some proponents also used France's École Polytechnic as a model.[40]

Americans had dreamed of a fine university since the beginning of the Revolution.[41] A nation founded on moral principles should have superior institutions to perpetuate those principles, and the lack of a university was a serious flaw in the American fabric. Benjamin Rush, the Philadelphia physician, signer of the Declaration of Independence, and a founder of Dickinson College, made the first formal proposal for a national university emphasizing graduate education and research in 1787. George Washington mentioned the possibility of a national university in his first annual address to Congress in 1790.[42] Four years later he wrote, "That a National University in *this* country is a thing to be desired, has always been my decided opinion; and the appropriation of ground and funds for it in the Federal City, have long been contemplated and talked of. . . ."[43] Washington regretted that Americans should be sent to foreign countries to be educated,[44] so much so that he bequeathed money to start the endowment of a national university.[45]

Jefferson struggled to found the University of Virginia for almost a half a century. He purchased the land, planned the grounds and buildings, supervised the construction, hired the professors, and ran the university.[46] He had hoped to found a university equaling European ones, but the school failed to meet his expectations and brought him disappointment and frustration.[47] Despite Jefferson's disillusionment with the University of Virginia, he requested that his founding the university be one of the three things noted on his gravestone. The other two were his authoring the Declaration of Independence and his writing the Virginia Statutes for Religious Freedom.[48]

Well before Peirce had become a student at Harvard, Thornton Kirkland had hoped to transform Harvard into a great university. Kirkland stated in an article in the *North American Review*[49] that Harvard University, which he served as president, would soon become a full-fledged university. He need only wait for the return of four young men who had gone to Europe to study. Once they were on his faculty, the institution would be a fine university. The four of them returned to Harvard between 1819 and 1822, but of the four only George Ticknor stayed any length of time, and he was constantly frustrated by the reception his proposed reforms met. Far from producing a great university, Kirkland's administration ended in disaster, and he was forced to resign in 1827.[50]

Other attempts, such as one at New York University,[51] were made to provide the equivalent of European higher education, but they also failed. Thus by midcentury the United States still had no institutions that approached the German university. Most of its colleges were small and taught Latin, Greek, mathematics, morals, and a smattering of other subjects. Small libraries also limited research opportunities, especially for professors.[52] In 1846 Yale had founded the Sheffield Scientific School. A year later the Lawrence Scientific School was founded at Harvard. These were both steps towards a German university, but still far from it. Some prominent American scientists had discussed a technical institute in New York City the year before the Albany meeting in 1850, but nothing had materialized before brighter prospects loomed in Albany.[53]

Albany was a leading American city at this time. It was not only the capital of the Empire State, but an important commercial and trading center whose relative importance among American cities was much greater then than it is today. Albany already boasted a school of medicine.[54] A new law school would commence instruction with seventy-two students in the fall of 1851.[55] Furthermore, Albany was the home of the New York Geological Survey, one of the best in the country. The Geological Survey, led by James Hall, already served as a de facto center for graduate work in geology.[56]

Hall was dedicated to the pursuit of science and the money to support it. He was successful in obtaining funds from a reluctant New

York state legislature by a combination of promises and threats. Hall, for example, had no trouble persuading Albany's business and educational leaders to liberally support a series of lectures by Louis Agassiz, shortly after he came to this country. Hall's reputation as a scientist reached beyond the confines of the country; he was, for example, one of the few scientists sought out by Sir Charles Lyell, the Scottish geologist, during his tours of the United States.[57]

Hall was the key man in organizing his fellow citizens to persuade the legislature to incorporate the university in April 1851, but the *Albany Argus* had announced the university as early as February.[58] Among the citizens of Albany that were working for the formation of the University were Thomas Olcott, president of Albany's leading bank as well as an officer of the New York Central Railroad, and James Armsby, a prominent physician and civic worker. It was through Hall's efforts that the August meeting of the American Association for the Advancement of Science was held in Albany in order to recruit the best scientists in the country for the proposed school's faculty.[59]

Yet there was more enthusiasm for the university than substance. For one thing, to circumvent the difficult problem of raising sufficient money to support the new university, its proponents proposed that it be supported, not by beneficent endowment, but by fees paid by the students directly to the professors. The supporters felt sure that the faculty at the institution would easily attract students and be self-supporting. In doing this, they were following a plan first proposed by Brown's president, Francis Wayland, almost a decade earlier. But Wayland's motivation for the proposal was to assure a quality faculty, not compensate for lack of endowment.[60] The lack of sufficient funding was perhaps the principal reason for the scheme's failure.

Peirce's friends were enthusiastic about a German-type university, and they were willing to make sacrifices for it, but they were cautious. Agassiz, for example, wrote to Hall, "You know how my whole soul is bent upon the project of an American University and what sacrifices I am ready to make if I can contribute to bring it about in any way."[61] Later Hall wrote, "we might have in Albany a university equal to European universities and become in time equal to the celebrated University of Berlin."[62] Yet Agassiz pointed out that his resources were limited and he needed to support his family. He had been offered two thousand dollars a year for four consecutive years to deliver a series of lectures to the medical school in South Carolina. He wanted assurances from the New York legislature before giving this up.[63]

Bache failed to mention the new university in his retiring speech as president of the AAAS that summer in Albany, but he alluded to the limitations of existing colleges by pointing out that teaching science was far different than producing science. He also observed that with regard to teaching, "It is not necessary to found institutions especially for its encouragement; nor should the power to diffuse science in successful courses of lectures be considered as a substitute for exertion in its advancement."[64] Bache, of course, was enthusiastic about a research university. He later wrote: "There is a great and growing demand in our country for something higher than college instruction; and one great University, if fairly set in motion, would thrive. To make it, the Professors must leave positions in which they are now comfortably established. For the sake of being together, I know that the leading scientific men of the country, with few exceptions, who are bound to carry out plans which they have in hand, would leave their present homes. They could not do this for an uncertainty. There are men enough to make one very brilliant institution by their high qualities and learning."[65]

Both Agassiz and Bache objected strongly to the fee system, and called for endowment and regular salaries.[66] The impetuous Peirce, championed the concept. He wrote to James Hall, "To assist in establishing the first University yet founded on a thoroughly scientific basis, a University in which the self sustaining principle is to be manfully recognized, and one which cannot fail to be sustained because it will supply the honest and earnest demand of a highly cultivated community, is a task, from which, however onerous it may be, no man will shirk, whose actions are inspired by a truly patriotic spirit and a generous sympathy with his fellow citizens."[67]

Yet Peirce was not so enthusiastic that he did not realize that the university could not survive without startup money. He carefully detailed a plan that required one hundred thousand dollars in endowment to support nine professors over the next five years.[68] Peirce also made it clear that going to the new university would be a personal sacrifice. "I am now so pleasantly situated that I have no personal temptation for a change in my condition, and nothing could induce me to disturb my present happy course of life, but a profound sense of my duty to the cause of American Science."[69] Peirce also listed a number of things that had to occur if the university was to succeed. Every effort had to be made to attract the nine professors he listed. "It is of the highest importance that Agassiz and his collections should be transferred to Albany; he should be provided with the means of increasing his Cabinet, and relieved from the necessity of lecturing about the country."[70]

In order to attract the best scientists in the country to the university, Peirce suggested that the professors should be part-time, giving lectures for only three months. That way faculty could maintain positions at other institutions.

Peirce was also egalitarian with regard to the university's students:

> I am, first of all, persuaded that no great enterprise in education can be permanently successful which is not in the strictest harmony with the ranks of the country. . . . Search out carefully what the Community actually requires; and conscientiously satisfy the demand. The soundest inferences upon these matters can be described from the investigation of those institutions, whose praises are sounded throughout the land. It is in the common schools, that the fundamental ideas are to be sought, which should guide us in the organization of the new university. It must be arranged with reference to them, designed with engrafting upon their stock, and planned upon the model to which they are tending, as their eliminating vertex. It must be free from every false Element of Aristocracy, and equally accessible to all students of any condition of life, who have been gifted by their Maker, with the power of profiting by its opportunities.[71]

A month later, Peirce heard from John Norton, the Yale agricultural chemist. Norton was extremely enthusiastic and hopeful; he was the first scientist not residing in Albany who had actually committed himself to give a course of lectures at the new University. He was currently traveling back and forth between Albany and New Haven, attempting to cover his responsibilities at both institutions.[72] He now wrote, "Dr. Armsby writes me from Albany that your letter and Prof. Mitchell's speech have produced a wonderful effect there; the feeling is perfectly enthusiastic. The governor he thinks would like to propose some plan in his message; it seems to me that something might be furnished him which would tell strongly in our favor."[73] A few days later, his enthusiasm untarnished, Norton wrote suggesting that a general plan should be made public, so people would not criticize details.[74]

Hopes for a great university in Albany ran high during the winter that followed the AAAS meeting. The trustees promised to provide facilities to support outstanding research and sought legislative support. Many scientists went to Albany to lobby the state legislature, which was then in session, for support for the university.[75] Agassiz regretted not being able to be there, but Peirce was there, full of enthusiasm, as was Bache, who stressed that the university must have state funding.[76] Peirce wrote to John Le Conte: "I came to Albany on Monday to take part in an effort to enlist the empire state in the establishment of a truly <u>national university.</u> On the first day I tried in jest, the phrase of 'preestablished harmony.' But almost every movement which we have made, since our arrival has been so extraordinary a manifestation either of preestablished harmony or of the action of a higher power, of which we were the unconscious instruments that I have almost been afraid to look myself in the face fearful lest I might find that it was not I."[77]

One man who matched Peirce's passion for the new institution was Samuel B. Ruggles. Ruggles, a successful lawyer, was heavily involved in land development and railroading, being a founding director and comptroller of the New York & Erie Railway Company. He increasingly devoted his time to civic affairs, and successfully championed the development of Manhattan Island, rail and canal transportation, and graduate education.[78] A close friend of Peirce and the Lazzaroni circle, he remained one of the strongest and most persistent advocates for a research university throughout the coming decade.[79] Ruggles made a passionate appeal for state support. He envisioned a new university that would "exert its influence not only within this State but far beyond it," and would "unite and combine in one mass a body of learned men, far exceeding in number and strength anything that has been presented to the American world."[80] Throughout the struggle in the next decade, Ruggles was a staunch and enthusiastic proponent of higher education. Peirce wrote to Bache, "The times afflict me sadly and my heart is heavy laden. Do you not like Ruggles' letters? They are lights amid the gloom."[81]

But enthusiasm blinded both the trustees and the scientists to rifts in the harmony. There was a seminal misunderstanding between the two groups. The scientists sought a university that would give them time and facilities to do basic research, and the concomitant recognition this research would bring. Local supporters talked more and more of the importance of the new University to commerce and the professions.[82] Although the local press thought the meetings went well, Bache had a different impression.[83] When asking Peirce if he was going to Albany the next winter, he noted, "What a botch they made of their last meetings."[84]

Spring came, and the legislature had failed to fund the new university. At this point it should have been apparent that the dream of such an institution at Albany was dead. It was such a beautiful dream, however, that those who had it were reluctant to relinquish it, and it died slowly. So, although Peirce still had hopes of a great university in Albany, his euphoria was beginning to wane. He wrote home to his wife, "I think I see indications of breakers ahead and have no kind of belief that the University is going to succeed without a struggle."[85]

In the winter of 1852—53, Peirce and other backers of the Albany university met there again, hoping to get legislative support, but the participants lacked the zeal of the previous year, and were unsuccessful.[86] Peirce wrote to the Le Contes: "The New York University scheme moves very slow, so slow that you cannot see it move. Yet as the shadow does advance in the wall, and as the tree grows towards heaven at an imperceptibly slow rate, so I think the cause is gradually gaining and my ultimate hopes of success are great. I have indeed, just received a Valentine from my friend Mr. Ruggles in which he invokes my immediate presence in the great emporium to advise him in regard to some steps which are to be undertaken at once, and I answered him, 'In eight days or in nine, Oh ecstasy divine! Our arms will intertwine, my Valentine?['] Which means that in about a week I shall go to New York."[87]

In August the ever-optimistic Ruggles informed Peirce that New York had levied a tax on canal traffic, which would supply the state with sufficient funds to support a university. "Rely on it, the next session of the legislature, will be *the golden moment*."[88] But unfortunately Ruggles was wrong; the support never came.

There were many reasons for the university's failure. The proposed university did not intend to compete with existing institutions of higher education, but to offer graduate courses for which Americans now had to travel to Europe.[89] Even so, Union College and Rensselaer Polytechnic Institute were less than enthusiastic about the venture. Union, in particular, was disturbed, since it had begun providing graduate work itself.[90] The New York Agriculture Society did not want agriculture to be taught at the new institution, feeling that the subject was of such importance that it deserved an institution dedicated to agriculture alone. The society maintained this position, despite its high regard for Norton and his heroic sacrifices in providing instruction at the new university in Albany.[91] Attempting to teach in both New Haven and Albany took its toll on Norton's health. He contracted tuberculosis and collapsed from exhaustion before finishing his first course at Albany. He died shortly afterwards.[92]

Another reason for the failure at Albany was that other people in other places wanted a university. They were all competing for the nation's limited resources to establish such an institution. In the winter of 1851–52 Bishop Alonzo Potter suggested to Bache that the University of Pennsylvania, of which Potter was a trustee, might be transformed into a true university.[93] When support for the Albany venture failed to come from the New York legislature, Bache approached trustees of the University of Pennsylvania about turning that institution into a European-level university, hoping that it would "gather the harvest of students" that the university at Albany would have attracted.[94]

Despite his enthusiasm for a university at Albany, Peirce himself had no particular partiality for that location. His goal, and that of his

Lazzaroni friends, was to establish a great American university, and Peirce was quick to support efforts to launch one in other locations. Peirce had not given up on Albany, and would not for years, but he was quick to respond to Bache's suggestion. "Your promotion appears to have many advantages over any other, and is well worthy of many oysters and much hock," he wrote to Bache. Peirce suggested a meeting of the Florentine Academy, in which Ruggles should be included to discuss the matter.[95]

There was talk of Bache being provost of the university, and Peirce, Agassiz, and Henry being on the faculty. The trustees questioned the demand for advanced education in America, suspecting that few young men would forego the opportunity to begin earning a living, and the plan died.[96]

Benjamin Gould, Peirce's friend and former student, wrote to him in the spring of 1852, suggesting that if a national university was not to be in Albany, why not in New York City, or in Massachusetts. In particular there was hope that with timely action Columbia might be raised to university status. Gould pointed out, "The other point is that the Albany scheme has failed, for the present in the N. Y. legislature, & there seems so much good sense in the old proverb of striking while the iron is hot."[97] Following Gould's suggestion, the Lazzaroni shifted their efforts toward New York again. Although President Charles King was committed to changing Columbia into a university, their staunchest ally in this endeavor was the Columbia trustee, Samuel Ruggles.[98]

Ruggles convinced the trustees to sell their expensive downtown property and move to cheaper uptown property, thus freeing money for the university's growth and development.[99] Peirce discussed the situation with Ruggles and George T. Strong, also a trustee and Ruggles son-in-law. He suggested that professors be required to publish regularly to prevent their "degrading into drones, like Renwick."[100] When, in fact, Renwick did resign, opening up a professorship in chemistry, the Lazzaroni sought to procure the position for fellow Lazzaroni member, Wolcott Gibbs, as an essential step in Columbia's rebirth.[101]

Peirce supported Gibbs's candidacy, despite inquiries for support from other friends. John Le Conte wrote asking him if Gibbs's election wasn't as sure as he'd understood: "At my suggestion, friend Bache made some inquiry in relation to the vacant chair in Columbia College; from which I inferred that it was a settled fact that W. Gibbs would get it. But from some remarks dropped by Gould (who came on to this City with us), I conclude that there is considerable doubt about the matter. . . . Have you any information in relation to this subject? Would it be worthwhile for me to make an effort to get the situation? I do not wish to make attempts, unless there is a strong probability of success. Write me candidly your opinion. I will place the whole matter in <u>your</u> hands: you have proved yourself to be our best and truest friend."[102]

Peirce wrote to Josephine: "Tell the Professor that I was so unwilling to recommend another man to Columbia College that I waited a fortnight before I would write to New York, and not then till I received a letter stating that there was a very great danger of Gibbs failing of the appointment and, if I did not write by return of mail it would be too late. There can be no doubt that this was the only practicable course."[103]

And later to John:

> I was quite surprised to hear that Wolcott Gibbs was a candidate for the Professorship at Columbia College. But since he is so, we can but presume the simple course of abstaining from urging your claims. For all our strength lies in his especial friends and, so far from introducing any division among them, we must aid them all in our power for there is reason to fear that even with our help, they may not succeed. I hope, however, that good will come of it, for I am quite sure that they will ere long be compelled to divide the professorship, and that Gibbs will assume the Chemistry and leave the Physics free to the efforts of your friends. I have not yet taken any part in the matter, but after considerable deliberation and with much reluctance I have decided to write President King in favor of Gibbs. By going heartily for him now, we may command the aid of his friends bye and bye. I understand Gibbs to undertake this office, in the hope of making a renewed effort for the university; but I do not anticipate success, although I shall do all in my power to promote it. Science flourishes at Function Grove and Clover Den and there is a big hen in the hen house all ready to hatch.[104]

After graduating from Columbia, Gibbs studied at the College of Physicians and Surgeons in New York. He earned an MD, not with the intention of practicing medicine, but because this course was one of the few avenues open to someone who wished to extend his knowledge of chemistry in the United States. Gibbs then studied in Germany and France. He was one of the best, if not the best, chemist in the country.[105]

Ruggles and Strong pressed hard for Gibbs. To the largely Episcopalian Columbia board of trustees, the primary objection to Gibbs was that he was a Unitarian and worked on Sundays.[106] Ruggles himself, despite his advocating Gibbs, was a devout Episcopalian. Six of the trustees were Episcopalian ministers, all of whom opposed Gibbs.[107] Peirce was sensitive to the public's perception of scientists as infidels. He wrote to Bache, "Have you heard of the meeting of naturalists at Göttingen last September and of the query there propounded. Is the mind a phenomenon of brain or of soul? All but one insisted that it was all brain, and nothing but brain!! Is not this a dreadful blow to science? For a time it will, I fear, give the 'know nothings' the real Simon pure <u>know nothings</u> a mighty advantage. If the anti-scientists in Col. Coll. [Columbia College] get hold of it, what will they not crow?"[108]

Although Gibbs's religion was the overriding issue, the Lazzaroni hurt their own case by pressing too hard. Ruggles mentioned that one of the objections to Gibbs was that "his appointment has been unduly and disrespectfully urged by his friends."[109] Gouverneur M. Ogden, the trustee who led the fight against Gibbs, perceptively observed that "I have a suspicion very strong that the most favorable recommendations [of Gibbs] . . . , extravagant as they are, have proceeded from interested motives, from men who expect places in a grand scheme for education expected to be carried into execution by us; of which his elevation is to be the entering wedge."[110]

This, of course, was the goal of the Lazzaroni. Ruggles wrote to Peirce on the same day, "We <u>have almost elected</u> Gibbs & yet may lose him." But, Ruggles continued, "It is not the question of electing <u>Wolcott Gibbs</u> or not—but whether <u>Science</u> is to be confined, influenced, & controlled by the Church. . . . The War will be long and bitter—but a University will spring out of it."[111] The Lazzaroni strongly believed that religious and academic freedom were necessary for scientific independence. Ogden, on the other hand, felt the laymen who supported science should have a voice in its direction[112]

After failing to get Gibbs the professorship, the Lazzaroni ceased to meddle in Columbia's affairs,[113] but irreparable damage had been done to the Lazzaroni's influence at Columbia. A few years later, a position in mathematics became available for which Peirce recommended his friend, William Chauvenet, one of the best mathematicians in the country.[114] Chauvenet failed to obtain the post, however; it went to Charles Davies, a mathematician of limited ability. Davies wrote extremely popular textbooks, but even these lacked innovation. When New York papers gloated about the fact that a recommendation from Cambridge had not carried the day, Davies wrote to Peirce, assuring him that he had no part in the triumphal air of the article, and expressing his gratitude to Peirce for befriending him in the past.[115]

In 1855, after attending a meeting of the AAAS in Providence, Peirce and Bache went to New York City to attend a meeting of the American Association for the Advancement of Education, of which

Bache was president. Addressing the association, Bache said that "a great University . . . is, I am persuaded, the want of our country."[116] Bache made it clear that what was wanted was faculty who were active researchers who "would thus inspire the pupils, and professors would have not only hearers but followers. Such an institution requires a large endowment, not to be expended in costly buildings, but in museums, laboratories, collections of nature and art, and in sustaining liberally a corps of professors worthy of the institution and of the country."[117] Bache pointed out that it was not necessary for America to produce such a faculty, only to bring them together. "There are in all branches of science enough men in our own country of the highest class of mind to adorn such an institution, and to make it the equal of the best establishments of the old world."[118]

Peirce responded to Bache's talk at the meeting, saying that he was deeply disappointed that a great university had not been founded. It was something for which he was willing to make a personal sacrifice. "There was a time, when we were engaged in our efforts at Albany, when I should have been willing to embark in such an institution, when against the entreaties and almost the tears of my family and friends, I should have been willing, for the sake of the cause of education in the country to have abandoned existing connections with another place of learning to join that institution."[119]

Peirce went on to stress that "the state will be benefited by educating every man to the highest point that he can be; and it will be the best investment it can make of its funds to invest them in intellect developed to its utmost capacity."[120] He also stressed that even genius needed encouragement. "I know it is a popular doctrine that genius will find its own way; but I doubt whether genius will necessarily be developed of itself. We have another popular doctrine which is much nearer to the truth, which is, that opportunity makes the man. We can not have a great man unless he has great ability, but, neither can we have a great man who had not an opportunity worthy to develope [sic] him."[121]

Henry P. Tappan, the president of the University of Michigan and an important educational reformer, also addressed the meeting on the "Progress of Educational Development in Europe." Tappan's remarks made a profound impression on Peirce. He said, "This learned and profound discussion of the progress of the University seems to be of the greatest importance to the understanding of what the University ought to be, and what ought to be the relations of our colleges to education. I confess that for the first time, I have had a perfectly clear understanding of this whole subject."[122]

Peirce was not only frustrated that a first-class university had not materialized, he was unhappy with conditions at Harvard at this time. Many of the reforms that had been made at Harvard in the thirties and forties had now been abandoned. Peirce wrote to Bache: "I am sorry to say dearest chief, that our college has fallen from 55 to 40 in the matter of education. I find it goes against the grain to continue in an institution, with which I am in such a want of harmony—not that there is any—the least want of kindness between me and my employers or colleagues on the contrary, our present president is the best the college has ever had—but his ideas of education are but just worldly wise, without any prospective enlargement of comprehension—and he is willing to let the others do what they will—even to the total abandonment of the elective system."[123]

With hopes for a university dashed at Columbia and the University of Pennsylvania, and highly unlikely in Albany, Tappan, Peirce, Bache, and the rest of the Lazzaroni joined forces to lobby for a new university in New York City.[124] It was a good matchup. Tappan's broader perspective nicely complemented the Lazzaroni's technical expertise. This time success seemed likely. Peter Cooper was expected to heavily endow education.[125] In addition, the Astor Library was already in New York, well supported and directed by the German-trained scholar and Peirce's former employer, Joseph Cogswell.

Cooper had made a fortune as an inventor and manufacturer. Among his many inventions was "Tom Thumb," the first successful steam engine built in America. Cooper had had only a year of formal schooling, and despite his enormous success, he recognized that he was working under a handicap, and wished to make the road easier for other laboring-class people.[126] Tappan sent Cooper a plan for a university drawn up by his son-in-law, the astronomer Francis Brünnow. Brünnow's principal innovation was to found a combined university and academy of sciences and arts of the city of New York. This arrangement would foster a close association between the faculty and other educated members of the community, and Tappan thought it was an organizational improvement over the European universities.[127]

Tappan lobbied hard to raise funds for the university. He sought public funds as well as money from William Astor, Cooper, and other private donors. By April 1856 he was extremely optimistic when he wrote to Ruggles, again a key player in the scheme, "There is such an intense earnestness on this subject now on the part of the best men in the country, that we have every thing to hope for. . . . It would argue such weakness—detestable weakness to fail now, that I fear we should lose our sense of manhood if we did fail. No, my good friends, we must not fail—we must do it. Shoulder to shoulder we will make it go."[128]

Peirce was equally optimistic when he wrote to Tappan, "Let us pull together in this grand enterprise—Now is the time—the harvest is ripe, and we must be up & doing."[129] Tappan replied to Peirce that money should be forthcoming from both Cooper and Mayor Fernando Wood of New York. The big problem was to convince Cooper to endow a research university, as Cooper envisioned a school that would teach engineering and art.[130] Tappan wrote again to Peirce saying he had promised the mayor the support of leading literary and scientific men. He urged that such men draw up and sign a paper assuring the mayor of their support and publish it in New York papers.[131] Apparently this was done, as the mayor spoke to the aldermen about the unanimous support for a university and academy from the nation's leading intellectuals.[132]

Yet the Tappan-Lazzaroni coalition failed. Evidently there was a split between Tappan and the Lazzaroni over Brünnow's place in the proposed university. Tappan later recollected that Gould and Peirce had joined to oppose him, perhaps not wanting competition from another astronomer. In addition, no money appeared from the city. Fernando Wood, a man of tarnished integrity, may have never intended to provide any. At any rate Tappan's correspondence with New York ceased, and he put his full efforts into building up the University of Michigan, a task with which he had great success.[133]

Although Tappan gave up on a university in New York, Peirce and the Lazzaroni did not. Peirce developed a plan for a national research university much along the lines of his proposals for the university at Albany, some five years before. In addition to stating the need for a fine library, observatory, and laboratories, he again proposed having professors being in residence for only a small portion of the year, so that the best men in the country could be attracted without their having to leave their present positions. This way "all the powerful minds of the country can be concentrated upon this institution, not even excepting the presidents of the colleges, the historians and poets, the retired statesmen, the secretary of the Smithsonian Institution, or the superintendent of the Coast Survey."[134]

Although Peirce labeled his document "confidential," this restriction was only to the press.[135] It was printed and probably saw wide circulation. Not everyone liked Peirce's plan. For example, George J. Adler, a philologist, who was professor of German at New York University, wrote to Peirce saying that he preferred "the natural development and gradual elevation" of existing institutions, rather than a single centralized university.[136] Armsby, with whom Peirce had been involved in Albany, liked Peirce's proposal for a university. "I have received your . . . excellent plan for a University. It is simple, beautiful, grand, and is nearly perfect, as is necessary or desirable for a "working plan." All that is necessary to begin the movement is to find some <u>one person</u> who will give $100,000 to endow seven professorships. . . . I have conferred with Chancellor Finns of the N. Y. University who is delighted with your plan and feels confident he can secure at least 10 professorships in the city of New York."[137]

Moreover Peirce's friend Bache hailed it as "the Lazzaroni Light."[138] The plan influenced an address he gave before the American Institute of New York City in the fall of 1856. Bache suggested an institution that would combine Tappan's program of combining a university and an academy with Peirce's ideas on part-time faculty.[139] Encouraged by the response to his talk, he wrote to Peirce, "The idea is fermenting and will go."[140]

Bache, Gould, and Peirce met with Cooper in Washington, D.C. during January and February 1857. Peirce wrote that Cooper was unhappy about "the neglect of his Institute and may perhaps be induced to give it such a direction as the men of science would much prefer."[141] Peirce traveled back to New York with Gould and Cooper. While in New York, Peirce saw his old friends Cogswell and Bancroft, as well as Ruggles. Peirce stayed on a few days to attempt to raise money for a university. He talked with Astor, and Cooper invited him to tea at his house.[142] "Mr. Cooper was quite agreeable and his house is a superb one. He seems to me to have no enlarged idea of what he ought to do and although he is modest and ready to hear and receive all which may be said to him, he is by no means easy to fix to anything."[143] Astor, though pleasant, was equally noncommittal. Discussions continued with various parties until the summer, when the whole enterprise fizzled out.[144]

Cooper, unlike Abbott Lawrence, was not dissuaded from founding a school that offered practical education, the Cooper Union. While not a graduate university, Cooper's school was a wonderful institution, offering instruction in engineering, science, and art, free of tuition.[145]

Although they had languished, the dreams of a national university at Albany had not disappeared, and Peirce was still involved:

The gentlemen of this city have almost persuaded me to devote the next year after finishing my volume to the getting up of the University. They told me, but this is not for the public, that they had waited on Mr. Peabody with regard to the Dudley Observatory, and he had told them that to whatever institution he might give, his intention was to give the whole so that it should not depend upon the generosity of others and that his donations to the university would be counted by millions. . . .

Mrs. P. thinks that I ought not to waste my time upon such a subject; but I do not feel that it would be wasting my time to accomplish the establishment of a true university upon a grand scale—for I should not ask for less than a million dollars, and none of this money should be buried in bricks and mortar nor any of it in collections of any kind. It should be exclusively devoted to salaries, and I think it possible to arrange the funds in such a way that they should double in less than fifty years without [endangering] the progress of the institution. Here then would be a true university built that it should expand with the country to all time.[146]

Unfortunately, again, nothing came of Peirce's dreams. Despite the growing intellectual and economic state of the country, there were probably too few resources in the United States before the Civil War to provide for the kind of university that Peirce and his colleagues envisioned. In order to have had such a school everything would have had to have worked out flawlessly: public and private financial support, and the dedication of the right faculty. Too many people with diverse concepts of what the new university was to be would have had to agree. Peirce's insistence that nine specific men should be professors at the Albany University is telling. Things seldom work out perfectly; they didn't in this case.

$$\Phi^n$$

Yale, not Harvard or a new national university, was the first American institution to award a PhD. The university awarded the first three American PhD degrees in 1861. Two were nonscientific, but one was in physics.[147] The second PhD in science, or at least engineering, was awarded to Josiah Willard Gibbs in 1863. Gibbs went on to become not only one of America's but the world's greatest scientists.[148] The first university dedicated to graduate work in the United States was Johns Hopkins, founded in 1876. Peirce's friend, James J. Sylvester, would return to America to head the graduate program there in mathematics.

Harvard lagged somewhat behind the college in New Haven in granting PhDs. It was not until 1873 that the first one was awarded to William Byerly, who wrote his dissertation under Benjamin Peirce.[149]

8

No Idea of an Observatory

THE SUMMER OF 1855 BEGAN TRAGICALLY FOR PEIRCE. HE WROTE TO Bache: "A few weeks ago, and I had never seen one die. But it is little more than a month that I closed the eyes of my wife's brother, and last Saturday, my own brother died in my arms after years of excruciating suffering. Death, my dear friend, is the best angel God has ever sent to man, next to his own Son. God be blessed for death, resurrection and the gospel. And now, my dear friend, you are dearer to me than ever. More than ever do I recognize your truth and purity of heart."[1]

By August, Peirce had shaken off the sadness of the passing of his two family members and was excited about visiting Bache in Bangor, Maine.[2] Although Peirce's primary motivation for going was social, he warned Bache that he intended to do some serious work while he was there: "To be frank with you, I hope that you will have some work to do while I am with you, . . . because I intend to mix some work with my play; for play without work is very dull—it is mere child's play. For at least four or five hours of each forenoon therefore, I intend to encase myself in a shell harder than steel, . . . harder than pharaoh's hardened heart for that would have nothing against the smiles of your daughter (pity that she was not with the Israelites in Egypt; she would have done what all the other plagues failed to do.) I am going to surround myself with the impenetrable veil of higher geometry and bid defiance to the shafts of beauty and the artillery of wit."[3]

Peirce saw Bache and his other scientific friends again several weeks later, when he went to the AAAS meeting in Providence. Unfortunately Sarah was ill and did not attend the meeting.[4] He wrote to her about the entertainment provided for the scientists and their wives: "I must tell you that the party the evening before was at the house of a nice rich Quaker gentleman, the beaux ideal of bonhomie and who Thee'd and Thou'd with a distinctness of enunciation which was peculiarly charming. But his wife 'what shall I say of her?' Everybody was in love with her—of all sexes. Her face glowed with the

165

ardor of hospitality like the face of an angel, and though quite past the middle age, and subjected to the severe ordeal of her plain Quaker garb, she was so beautiful that I fell in love with her, and I grieved that you were not there to see her. . . . She would satisfy the utmost fastidiousness of your highest poetic requirements."[5]

At another assembly of the association, presumably after a banquet, Peirce was full of wit. After cracking jokes for some time, he said:

> Mr. Chairman, you ought to have been satisfied with calling upon me once—But however the honor, which the standing committee has entrusted to me is one of which my heart is full, of which I am full, of which we all are full. It is this elegant entertainment [a low wave of laughter began at the speaker and swelled to a storm as it burst upon the distant boundaries of the tent.] But when it was given to me this morning I retired to my study and sat down with my slate and pencil to work out by formula some spontaneous wit with which I might enliven my remarks. (laugher) But it was all in vain. The mystic theory of mirth transcended all the powers of modern analysis, and the lives of laughter follow such a zigzag electric course, with such constant solutions of continuity that they have absolutely no radius of curvature and engulf no geometric laws of progress. (The last words were scarcely heard amid the roars of laughter.) Mr. Chairman, the other sciences . . . have been more successful. The zoologist sees in the caricature of man written on the monkey and the ape, that even these may be a quiet joke in heaven; and our brothers Rogers would tell you that the good old mother earth was wrinkled into mountains in its earthquake's fits of laughter. Rogers got up at once, and said "no[,] fits of rage!!" Very well, I replied, "That is another theory. There used to be in geology the conflicting theories of fire and water; and now there are those of rage and laughter—it is always the same thing." (low laughter) But since we geometers are altogether out of place on a festive occasion, even as astronomers we are more congenial to the unique bird of wisdom, and it is our turn to go to sleep when the lark begins to sing its song of quiet. Our very language is unfitted, by its weight in profoundness, to such an occasion as this. For we know by the name of Aries that which you call mutton, by the name of Taurus that which you call beef, by the name of Gemini that which you call a pair of chickens, by the name of cancer that which you call lobster salad, and so with all the dinners ending with Pisces (immense laughter) and our very characters show the confusion of our ideas upon all these phenomena of this occasion, for we do that which the worst old bachelor would not approve, we confound by characters scarcely differing from each other the virgin and the scorpion.[6]

$$\Phi^n$$

James Armsby, the prominent Albany physician, was also at the meeting in Providence, following up on the plans for a university at

Albany that had made the AAAS meeting so exciting four years ago. In 1851, Armsby and other civic-minded citizens had planned an observatory as an important element of the university. Although the dream of a great university at Albany had faded, the plans for the observatory had gone ahead. Money had been raised and a building erected.[7]

Back in 1851, the instigators of the observatory had called upon Ormsby Mitchel, the head of the Cincinnati Observatory, to generate interest in an observatory at Albany. Mitchel had mesmerized audiences in Cincinnati with popular lectures on the wonders of the heavens, and raised money there for an astronomical observatory.[8] He then went to Europe to buy instruments for the observatory and receive two months of training under G. B. Airy, the Astronomer Royal of England. The Cincinnati Observatory had no endowment for a director, a position Mitchel assumed, but he supported himself largely by delighting audiences throughout the country with popular lectures on astronomy. Mitchel also founded an astronomical journal, the *Sidereal Messenger.*[9]

In Albany, Mitchel encouraged an observatory, assuring the promoters that "twenty-five thousand dollars, for building and instruments, would be sufficient to realize at the beginning, my views of the matter, and to lay the ground work upon which immediate *action* and consequent success, could be based."[10] At the 1851 meeting of the AAAS, Mitchel selected a site for the new observatory, and Stephen Van Rensselaer, who owned the property, donated seven acres of land for the observatory.[11]

Despite enthusiasm from several fronts, Peirce, Bache, and Henry discouraged the idea at a meeting with supporters of the observatory, while they were at the AAAS meeting in August 1851.[12] In their minds, the country had far more pressing scientific needs. The refractor at Harvard was as good as any in the world, and there were many other fine telescopes in the United States.[13]

Ignoring their advice, the Albany supporters went full steam ahead with the project. Mitchel brought his lectures on astronomy to Albany during October 1851. After his lecture, he addressed a meeting of two hundred interested citizens on the topic of an observatory in Albany. James Hall wrote Agassiz about the tremendous enthusiasm Mitchel had created.[14] By March 1852, twenty-five thousand dollars had been raised for the observatory, thirteen of which had been donated by Blandina Dudley, the wealthy widow of a United States senator.[15]

The trustees had assumed that Mitchel would become the director of the observatory, a notion Mitchel did not discourage. However, financial difficulties had forced Mitchel to resign from the Cincinnati Observatory in 1852 and go to work for the Ohio and Mississippi Railroad. Despite his railroad duties, Mitchel designed a building for the observatory, and the construction was completed before the end of 1854.[16]

Since much of the twenty-five thousand dollars pledged for the observatory had been done so on the understanding that Mitchel would be in charge, it was imperative that he remain on as nominal director. Mitchel, however, could not fulfill the responsibilities of the director, and the trustees wanted someone who would take the position as his assistant until all the money was in, at which time the assistant would become the permanent director. They, in consultation with Mitchel and the astronomer Sears Cook Walker, chose Benjamin Apthorp Gould.[17]

Gould, an extraordinarily well-trained and promising young scientist, was not only an obvious choice, but contrasted sharply with Mitchel, who was essentially self-taught. A child prodigy, Gould had been Peirce's student at Harvard. He then went to Europe where he studied under the mathematician, physicist, and astronomer Karl Friedrich Gauss, and the astronomer Friedrich W. A. Argelander, earning a PhD at Göttingen. Before returning home, Gould visited all of the important observatories of Europe, often staying for several months at a single observatory. During all of his stay in Europe, Gould easily made friends of both Europe's leading scientists and of other Americans seeking further education in Europe.[18] Unfortunately, when he returned to his own country, this ability eluded him.

When he returned to the United States in 1848, he found slim pickings for employment. For two years he supported himself by teaching mathematics, French, and German in Cambridge. He published over twenty articles on astronomy before he joined the Coast Survey in 1851. There Gould replaced the terminally ill Sears Cook Walker, continuing Walker's work on the telegraphic determination of longitude.[19]

Gould was not only brilliant and well-educated but was dedicated to the advancement of American science. Soon after returning to the United States, Gould wrote to his friend and mentor, Baron von Humboldt: "I dedicate my whole efforts, not to the attainment of any reputation for myself, but to serving to the utmost of my ability the science of my country—or, rather as my friend Mr. Agassiz tells me I must say, science in my country."[20]

These were not hollow words. As soon as Gould returned to Cambridge he founded, conducted, and personally financed the *Astronomical Journal*, the first American research journal in astronomy.[21]

Markedly different from Mitchel's *Sidereal Messenger,* a popular journal largely for the amateur astronomer, the *Astronomical Journal* was strictly for the professional. Patterning it after the prestigious *Astronomische Nachrichten,* Gould maintained uncompromising standards for scientific excellence. Such a journal was necessary to the progress of American astronomy and received high praise from European astronomers. Peirce was a frequent contributor to the journal.[22]

Yet editing the journal was done at great personal sacrifice. Peirce wrote Bache, "Gould is very gloomy about the Journal and says that he does not hope to be able to continue at it any longer. I am very sorry and feel that it will be a great loss to American science. Cannot something be done at Washington about a work of this kind, which is of much national importance?"[23]

Gould's dedication to American science wasn't limited to his editing the *Astronomical Journal.* In 1851 when only twenty-seven years of age, Gould was offered a professorship at Göttingen, with the promise from Gauss that he would become director of the observatory there.[24] The prestigious position offered social position and security, both things lacking in the United States. Although a significant honor and attainment for any scientist, it was an extraordinary one for an American; only one man, the geologist Henry D. Rogers, held a professorship in a European university before the Civil War.[25] Gould's friends urged him to accept the position. Peirce wrote to Bache, "We have advised him, by all means to accept this offer."[26] Gould did accept the offer, but quickly changed his mind. He wanted to stay in the United States and build American science. He was that dedicated.

Despite his dedication, the ambitious Gould was understandably less than excited about such an uncertain position as the one at the Dudley Observatory. He was also reluctant to leave his friends and family in Cambridge as well as the intellectual milieu at Harvard. Gould declined the position. The trustees repeated their offer to Gould several times, the last one being in April 1855.[27]

$$\Phi^n$$

So four years after launching the observatory, still without instruments and a director, Armsby went to Providence. There he met once again with Peirce and Bache; this time they were more receptive to the idea of an observatory in Albany. There were several reasons why. For one thing, at this same meeting, Peirce give a paper on a new method of determining longitude, a problem of great practical importance. To make these determinations, a heliometer—a telescope capable of great accuracy—was necessary. At the end of his presentation, Peirce noted the necessity of a heliometer to make these precise measurements and the lack of one in America.[28]

Peirce, Bache, and Armsby discussed the possibility of installing a heliometer in the Dudley Observatory. This was strictly a research instrument. The public would not come and gaze through it on Tuesday evenings if the sky was clear. It was also a state-of-the-art instrument; not only were there no heliometers in America, there were only two in Europe.[29] Bache suggested that in return for allowing the Coast Survey to use this instrument, he "would supply a Transit instrument and observers, from his corps of United States employees, free of expense to the Institution."[30]

While still in Providence, Bache requested written assessments of the value of a heliometer from both Peirce and Christian H. F. Peters, who was employed by the Coast Survey. Peters wrote to Bache extolling the value of a heliometer. "Though, after the recommendation of a man as Prof. Peirce, I may appear audacious by instilling upon your mind the idea of establishing a heliometer in this country, you will allow me a few words, as an instrument of such a kind is of the greatest utility, both for the astronomical sciences and for geography."[31]

A few days later, Peirce, now in New York for the meeting of the American Association for the Advancement of Education, also wrote to Bache about the advantages of a heliometer, going over much of the same ground as he did in his presentation in Providence:

> The investigation of the Pleiades Occultations is embarrassed with a serious difficulty. . . . For this purpose a heliometer of the largest dimensions seems to be indispensable and I know of none, with which this task could be undertaken, except with the heliometer of Konigsberg, . . . or . . . Oxford. I am aware that such an instrument must be placed in a fixed observatory, and that the Coast Survey is not permitted to hold such an establishment. But I am informed that it may not be impossible to obtain such an instrument for the observatory at Albany; and that the liberality of the citizens of that place of which we have been witness more than once may be successfully invoked to contribute this grand boon to American astronomy, as soon as they are informed of its importance. You are well aware that this instrument was not only capable of being used in all observations in which any other great equatorial can be used; but that it can also undertake the solution of the higher problems of astronomy, of which there is no other substituting form of research.[32]

While still in New York, Peirce and Bache heard from Armsby that Mrs. Dudley had agreed to donate additional money for the new heliometer.[33] Now enthusiastic about the project, Peirce suggested

naming the new telescope "the Dudley Heliometer." He also suggested having a scientific advisory council for the observatory. He
offered himself, Bache, Davis, and Gould. Armsby was happy to
accept, although Henry was quickly substituted for Davis, as Davis and
Gould did not get along.[34]

There were two reasons, then, for Bache's and Peirce's change of
heart about the Albany observatory. In the first place, an observatory
equipped with a heliometer would not be just another American
observatory, but a fine research observatory that could compete with
the best European observatories.[35] It would significantly raise the
quality of American astronomical research.

Not only would the proposed observatory be better equipped than
any in America, it would be under Bache's control. From the beginning, Hassler had planned on having two observatories as part of the
Coast Survey,[36] but Congress prohibited having a permanent observatory under the Coast Survey. Astronomical work had to be farmed
out, much of it, in fact, to the Harvard Observatory. But Bache was
frustrated with the Bonds's lack of mathematical skill and their ties
to British methods when German research methods had replaced
them. The Bonds were practical, not theoretical astronomers.[37] In
particular, the Bonds were not interested in determining Solar Parallax, one of Peirce's principal interests.[38] A heliometer was particularly suited to studying Solar Parallax.

Having the actual control of such an observatory was immensely
appealing to a man like Bache. And Peirce's suggestion that a scientific council guide the observatory's scientific work would put the
Dudley Observatory in the firm control of the Coast Survey. Although
this arrangement circumvented the congressional restriction, Bache
was walking on thin ice. If it appeared that the Dudley was to be under
the control of the Coast Survey, Bache might garner Congressional
censure.

Although the Dudley now had a new direction, additional funding,
and a scientific council, it still lacked a director. Gould filled the void,
not as its official director, but by volunteering to oversee things. He
also declined an offer of generous compensation.[39] His first task was
to go to Europe to obtain instruments for the new observatory. Gould
devised an extensive list of equipment for the observatory that
included several other types of telescopes in addition to the heliometer, including an equatorial telescope and meridian circle.[40]

Armsby wrote Gould, "Our citizens are delighted with our new
arrangement and look forward to the commencement of observations with deep and hourly interest."[41] Indeed, like Bache and Peirce,
Armsby, Olcott, and other civic-minded citizens of Albany were

The Dudley Observatory, ca. 1854. By permission of the Dudley Observatory Archives.

excited about the prospect of having an observatory of unexcelled scientific excellence. The luster that such an institution would add to their city greatly appealed to them.[42] Armsby's letter to Gould showed the kind of institution he wanted. He asked regarding one of the instruments intended for the Dudley, "Is there a Prime Vertical Instrument in this country and are there many in Europe? Cannot this instrument also be the best in the world? My own hopes and trust in the future are as sanguine as your faith and aspirations are boundless."[43] Armsby added, "We appreciate most highly the generous offers of aid from Prof. Bache, and trust that neither Prof. B. nor his distinguished associates will ever have cause to regret their connection with our institution."[44]

Peirce, despite his reluctance to give popular lectures, went to Albany to give a lecture to promote the Dudley Observatory.[45] Yet despite his general enthusiasm, Peirce had a fundamental mistrust of Gould that led to a foreboding that things might not go well. He wrote to Bache, "Yes Gould's golden opportunity is now given him, and if he fails now, all is lost."[46]

Gould returned to the United States in late December 1855. While in Europe, he had ordered several telescopes and other instruments for the new observatory, but not a heliometer. He hoped to obtain this instrument from Charles A. Spencer of Canastota, New York, a maker of optical instruments. Spencer had just completed the equatorial telescope for the Hamilton College observatory, the largest made in the United States up to that time.[47]

Gould continued his role as the unofficial and unpaid director of the Dudley Observatory, but he maintained his residence in Boston, where he enjoyed the intellectual climate of Cambridge and the association with his family. He also continued to work for the Coast Survey. Thus the time he could devote to the new observatory was limited, and he spent very little time in Albany, never being there for more than one or two days at a time. Gould found Ormsby's building inadequate; it had to be largely rebuilt. Yet despite his remaining in Cambridge, Gould insisted on making all the decisions on the renovations. The workers often remained idle, but at full pay, while waiting for instructions to be sent from Cambridge. Work progressed slowly.

In May 1856, Gould met Bache in New York to discuss Gould's assuming the official directorship of the observatory. They decided against it.[48] The biggest drawback was the absence of any endowment for the salary of the director. The Cincinnati Observatory had been erected without adequate endowment and as a consequence had not been able to fulfill its purpose. Even the Harvard Observatory was seriously under-endowed. Understandably, Gould did not wish to follow in Mitchel's footsteps as the head of an observatory that could pay him no salary.[49]

As an alternative to Gould's being the official resident director of the observatory, Gould suggested that Bache appoint Christian H. F. Peters to the observatory on a permanent basis as Gould's assistant to oversee the work on installation of the instruments and other work on the observatory. This was the same Peters from whom Bache had asked for an assessment of a heliometer.[50] Peters had met Gould when Gould was studying in Europe and later contacted Gould about astronomical employment in the United States.[51] After immigrating, Peters, upon Gould's recommendation, was employed by the Coast Survey in Cambridge under Peirce.[52] A highly competent astronomer, Peters had studied under Gauss and Johann F. Encke, and earned a German PhD.[53] Bache, who did not have a high opinion of Peters, followed Gould's advice reluctantly, relying on Gould's assurances of Peters's ability. Although arriving in Albany as a foreigner, he soon won the respect and friendship of leading Albany civic leaders, including Thomas Olcott, and scientists, such as the geologist James Hall. The trustees were excited about having a competent astronomer working full time to install the telescopes at the observatory.

The problem of adequate endowment remained. In a statement that smacked of extortion, Bache, Peirce, and Gould stated that the "observatory can not be creditably conducted for less than ten thousand dollars of annual outlay." They suggested that the instruments remain idle until sufficient funds were raised to support their use. But they assured the trustees that "if the greatness of your giving can rise to this occasion as it has to all our previous suggestions with such unflinching magnanimity, we promise you our earnest and hearty cooperation, and stake our reputations that the scientific success shall fill up the measure of your hopes and anticipations."[54] An estimated hundred and fifty thousand dollars in endowment was needed to sustain the observatory at the envisioned level. A substantial portion of that amount was quickly raised; Mrs. Dudley donated fifty thousand dollars herself, but the total was still far from the goal.[55]

The problem of an endowment was perhaps the central sticking point between the scientific council and the trustees. In mid-nineteenth-century America it was comparatively easy to raise money for visible scientific equipment and buildings, like telescopes and observatories, but next to impossible to raise money to support a scientist to work with these facilities. Yet Peirce, Henry, and Bache were well aware of the importance of endowment to the furthering of scientific work.

Shutting the observatory down made no sense to the trustees, so despite the recommendation of the scientific council, work did proceed at the observatory. The relationship between Gould and the trustees deteriorated, however. In their eyes many of his demands seemed unreasonable and costly. They were also delaying the observatory's being put in working order. Some of the trustees, especially Armsby, took initiative to get things done, but the trustees' lack of scientific expertise proved to be as costly as Gould's absence. Furthermore, Gould found the trustees' actions to be intrusive and interfering. Another point of contention between Gould and the trustees was the esthetic appearance of the observatory. Gould had no interest in the appearance of the building, but the observatory was in a prominent scenic spot in Albany, and the trustees were concerned with its appearance.

The following incident illustrates the problems that arose between Gould and the trustees. When stone piers, upon which the telescopes were to be mounted, were delivered, they had small cracks in them. Gould's assistant, Peters, informed Armsby of the problem immediately. Armsby's main goal was to get the observatory up and running as quickly as possible, so he told Peters to have the piers installed, and not to bother Gould with the problem. After all, the fissures were very small. Peters, however, did postpone installation and wrote to Gould. In the interim, while letters were being exchanged, the trustees paid for the piers. Gould, of course, refused to accept the stones and was incensed with Armsby for attempting to conceal the damage.[56] Although Armsby was at fault for accepting faulty material, the problem would have been avoided had Gould been on hand when the piers arrived.

Armsby annoyed Gould on many other occasions. Gould later described him as "busy, untiring, gossiping, and wondrously meddlesome."[57] For their part, the trustees found Gould to be arrogant and rude. In addition, Gould was exacting in his demands. Worst of all, Gould, pointing out that he was not the director but only an unpaid advisor, would take no responsibility for the condition of the observatory. That Gould would not accept responsibility for his decisions, not only irritated the trustees, but made them question his character.[58]

In addition to the lack of endowment and friction between Gould and the trustees, Gould was suffering from mental illness. He wrote a friend, "The truth is my dear boy, I have been very ill. Not at all so as to keep me within doors, not sick in the belly or in the members, but in that worst of all places the brain."[59] Even when things were going relatively well, he was troubled with bouts of depression, and his enthusiasm for the Dudley waned.

Φn

Gould had hoped that most of the instruments would be installed by late August 1856, when the AAAS would return to Albany for its annual meeting. The trustees had planned the inauguration of the observatory to coincide with this meeting.[60] But by the time of the inauguration, the observatory was in disarray. Major portions of the building had been torn down and not reconstructed. None of the instruments were installed. Nevertheless, the inauguration went well. None other than Edward Everett, the foremost orator of his day, gave the principal speech. Everett had been a congressman, governor of Massachusetts, ambassador to Great Britain, and president of Harvard.[61]

Furthermore, at the time of the inauguration, relations between the trustees and the scientific council remained warm. Peirce wrote to Armsby, "Never, my most excellent friend, shall we forget the delightful visit which we had in Albany—nor the heartiness of your great hospitality."[62]

After the ceremony, delays continued to occur, however, and progress on the observatory was painstakingly slow. The trustees became frustrated; they were well aware that work on a similar observatory in Michigan was moving along much more rapidly. Even Bache was concerned about the lack of progress. He wrote to Peirce, "Whether our Albany friends can carry it or not is doubtful, but go it will, dear Φ^n [Peirce]. I hope we may live to see it, but who knows?"[63]

$$\Phi^n$$

A rift that developed between Peters and Gould eventually had serious consequences. The first source of bad feelings occurred when Peters, using a small comet seeker, the only operational telescope in the observatory, discovered a new comet. He named it after the trustee and prominent Albany banker, Thomas Olcott. Although doing so was not against astronomical tradition, Gould was personally opposed to the practice, preferring to simply number the burgeoning number of new comets and asteroids being discovered.[64] Since Gould had publicly denounced this practice, he was embarrassed that an underling would follow it. Unfortunately the trustees viewed Gould's open criticism as evidence that he not only resented Olcott's recognition but envied Peters's discovery.[65]

A more serious split between Peters and the Lazzaroni resulted from conditions resulting from the financial crisis and economic downturn of 1857. This downturn brought an end to a dozen years of growth and prosperity in the United States. The panic of 1857 began with a decline in investment from European countries, which resulted in lower prices in many stocks and bonds, reducing the assets of American banks holding these securities. In addition, some factories temporarily shut down due to large inventories. These circumstances led to a climate of nervous apprehension. When a midwestern bank had to suspend payment because an employee had embezzled funds, depositors made a widespread run on the nation's banks. By the fall of 1857, all but a few of the banks had suspended specie payments. Additional factories shut down, business failures multiplied, railroads went bankrupt, and crop prices plunged. Unemployment was rampant, and for the first time in a decade, the federal treasury ran a deficit.[66]

The economic downturn made raising money for the Dudley more difficult, and there were no funds available to mount the telescopes. A transit and a meridian circle remained in their packing cases, with only the small comet seeker in operation.[67] Not only was the observatory adversely affected by the recession, it also impinged upon the Coast Survey. Bache was under pressure from Congress to cut costs. With no decent telescope in place in Albany, having Peters there became increasingly hard to justify, and Bache, whose opinion of Peters remained low, reduced his position from "observer" to "computer" and reassigned him to work in Cambridge under Peirce.[68]

Understandably unhappy with being demoted and removed from Albany to become Peirce's lackey in Cambridge, Peters resigned from the Coast Survey.[69] In Gould's eyes, however, Peters was being disloyal in taking this action. Gould had not only befriended Peters, but vouched for him to Bache. When Olcott suggested that Peters might stay on at the observatory independently of the Coast Survey, Gould objected, claiming that Peters's disloyalty precluded him from the position.[70] Gould later wrote: "Loyalty to the service in which I [Gould] was engaged, precluded me from rewarding disaffection and insubordination; discretion forbade me from exhibiting to other assistants an example of faithlessness successful, and the authority of our own official chief insulted. Despite continued importunities, I refused to appoint him [Peters] assistant, if other Coast Survey employees were to be brought in contact with him."[71]

Peters's dismissal marked a serious deterioration in the relationship between the scientific council and the trustees. It was a serious mistake. Many people in and out of Albany felt that Peters had been badly treated.[72] Moreover, the trustees and others in Albany liked having a competent astronomer on hand working towards getting the observatory going. They liked Peters.[73] The *New York Times* noted, "But it was not until the trustees had been induced by Dr. Gould to dismiss Dr. Peters, a most valuable and accomplished astronomer, that his real hostility to the Observatory was suspected."[74]

Lacking a satisfactory director for the Dudley, the scientific council decided once again that the institution should be closed down until such a time as someone suitable could be found. To the trustees, however, such an action was incredulous. An observatory with precious instruments still in their packing crates sitting idly and unused was ridiculous. In addition, they were extremely pleased with Peters, whom they wished to hire as permanent director.[75]

The trustees were also anxious to have the observatory be an operating, viable institution. When Mitchel had been unable to assume the directorship, the observatory building had remained vacant from 1852 until 1855. Having the observatory dormant again would be an embarrassment. Far better in the trustees' minds to have it running at half-throttle than not at all. In addition, Bache and Gould had procured contracts with the railroads for the observatory to furnish them with exact time. This would provide significant income and demonstrate the practical value of the observatory. But all this would be lost if the observatory were to remain dormant.[76]

The scientific council was surprised at the trustees' reaction. From Bache's point of view, it was the only sensible action, and the trustees had generally followed the decisions of the scientific council in the past.[77]

The prospect of having Peters head the observatory was abhorrent to Bache. His already low opinion of Peters was reduced to loathing when Peters resigned from the Coast Survey. He wrote to Peirce, "Why it is a great burglary darling Φ^n [Peirce]. A national observatory with Peters as chief. Can't be Φ^n [Peirce] dear."[78] Alarmed at reports from the observatory, Bache sent Peirce to Albany to "see what is what and who is who."[79]

Once in Albany, Peirce found Peters to be the primary cause of the trouble. It was becoming increasingly clear that Gould would have to go to Albany and take the reins himself. With a loyal assistant in Albany, Gould may have delayed going to Albany for a year or two, but without one, there was little other option.[80]

Support for Peters remained strong in Albany. When Gould arrived in Albany in mid-December, Olcott urged that Peters be hired at the observatory to work in Gould's absence, or that Gould come to Albany permanently and hire Peters as his assistant. Gould flatly refused, stating that if Peters stayed at the observatory, support from the Coast Survey would end. Olcott countered that support was so strong for Peters in Albany that the only alternative to not retaining him would be for Gould to come to Albany and assume the full responsibilities as director.[81]

At this point, Peirce became exasperated with Gould. And, although the exact source of Peirce's displeasure is unknown, Peirce also disagreed strongly with Bache's course with the observatory: "I cannot endorse your view. I differ from you in toto. Nevertheless I shall go with you but in entire silence. As to Gould, I feel that he has thoroughly dishonored me in this affair, and I shall hereafter decline under any circumstances whatever to be his advisor. Let this hereafter be a closed affair between us. It would be folly for us to attempt to discuss it. But I must now say, once and for all, that Gould has been at the bottom of all the difficulties of my life—and I cannot permit him to disturb the remainder of my days."

Nevertheless, Peirce had no intention of breaking with Bache over the incident. He closed his letter: "With regard to yourself, dear Chief, you on the contrary are associated with all that is happiest in

Daniel Huntington. Benjamin Peirce (1809–1880), 1857. Oil on canvas; actual: 75.8 × 63.4 cm (29 13/16 × 24 15/16 in.). Harvard University Portrait Collection, Gift of George A. Plimpton to Harvard University, 1924, H254. Photo: Imaging Department © President and Fellows of Harvard College.

the events of my life since I have known you, and I shall always stick to you as the dearest of friends and the truest of friends."[82]

At this point, Bache was tempted to sever the scientific council's connection with the observatory and move on to more promising things, but the hope of having an observatory that would compare favorably with any in Europe under his control was too much to give up.[83] Bache and Peirce agreed that Gould must go to Albany and assume the directorship of the observatory, despite Gould's reluctance to take the job and his deteriorating mental health.[84] Bache convinced Gould to move to Albany and take charge of the observatory. Gould met with the trustees after Christmas. As a result of this meeting, Gould believed that Peters's connection with the observatory would be severed, but a few days later he learned that Peters had been given a permanent position at the observatory.[85]

In response to this turn of events, Peirce, Bache, and Henry all rushed to Albany. Their aim was to keep the observatory in the hands of the Coast Survey, install Gould as director, and oust Peters. The relationship between Gould and Armsby was particularly bad; Bache and Olcott attempted to smooth things over between the two men. "Finally it was arranged that in the evening there should be a grand reconciliation." Gould felt that "it was the hardest thing to go through with . . . that could be imagined."[86] Nevertheless Gould and Armsby met, shook hands, and agreed to bury the hatchet. Armsby assured Gould that he never bore him the slightest ill will. In an attempt to firm up the reconciliation, Gould invited Armsby to tea. "He sent a saucy reply that nothing would be gained by any meeting until I [Gould] had done him justice before the Trustees."[87] Gould felt that he had already done him more than justice before that body.

Despite the bad feelings, Bache, Peirce, and Henry worked to patch things up. Henry, a native of Albany, was especially effective in smoothing over the bad feelings between the trustees and the Lazzaroni. Finally the protracted squabble was settled. Olcott, who had strongly supported Peters, found him a position as director of Hamilton College's new observatory, where Peters had a distinguished career.[88]

Peirce described the events in Albany:

> The affairs at Albany, again were at one time in imminent danger of coming before the public in the form of an angry and disgraceful controversy and we should probably have felt the effects of the rupture at our next meeting in Baltimore. But thank Heaven we have conquered, and Albany, with which I have so many delightful associations, is not lost to us. In my last visit, I staid (sic) at the home of Mr. J. V. L. Pruyn. . . . Mrs. Pruyn with all the bravery of a brave woman, took our part with real persuasion and did us great good in routing our foes. At one time, the game seemed to be lost, and I told the Pruyns how sad it made me to think of associating such a wicked defeat with a place in which I had commenced one of the truest friendships of my life. One of the redeeming traits of this fight is that it has left me no serious scars, no real mortification of true men, and even the conquered party have yielded from the cause of justice and not from the want of power.[89]

From the Lazzaroni's point of view, the day was saved; all was well with Gould and the observatory. Peirce wrote to Bache, "Yes the plot was in vain, and so finish all plots against the Lazzaroni."[90]

But the Lazzaroni were not as invincible as it appeared. By the time Gould actually assumed command, the relations between him and the trustees had already seriously deteriorated. The antipathy between Gould, and Armsby and Olcott had grown into hatred. The words of a partisan supporter of Gould and the scientific council given later were probably close to the truth: "He claimed that before Dr. Gould came to this city it was decided by the aforesaid majority that he should be removed, and after he came here they concluded to starve him out."[91]

$$\Phi^n$$

The scientific council planned on meeting in February in conjunction with a meeting of the Lazzaroni in Philadelphia. Peirce, who had been having financial problems caused by the government's refusal to pay him, initially declined to go, but sent a list of toasts to Frazer.[92] Fortunately Peirce managed to go at the last minute, for he found it to be the best Lazzaroni dinner so far.[93] The Monday after the dinner Peirce wrote, "We had a big and exciting meeting of the Scientific Council of the D. O."[94]

Things did not go as well in Albany as they had in Philadelphia. Gould's lack of enthusiasm for being in Albany likely showed to the trustees. There were more delays and a serious lack of ready funds. Gould's methods were slow and costly. The trustees became increasingly frustrated. By spring, their relationship had seriously deteriorated. Gould was on particularly bad terms with Armsby and Olcott.[95]

Peirce was once again pessimistic: "B. A. G. will probably be driven from Albany, with his council in contempt. It is improbable that we shall succeed in recovering a position there."[96] Bache, too, was concerned about developments there and about Gould's ability to handle them. He was particularly concerned about Gould's health. "Perhaps Gould cannot stand it. Perhaps the council cannot."[97]

One source of prolonged irritation between Gould and the trustees arose over the trustees having access to the observatory. Gould felt that "individual Trustees have no more legal right to interrupt observatories than the Trustees of a Hospital have to visit the sick beds."[98] The scientific council agreed that Gould needed to control access to the observatory in order to protect the delicate instruments.[99] The trustees, on the other hand, felt they had a right to enter the building. Peirce described conditions to his friends, the Le Contes:

> The Scientific Council of the Dudley Observatory are getting rapidly into inextricable difficulties. The Chief . . . is resolved to sustain Gould . . .

through thick and thin—and I can assure you there is some thick covered over with a very slight layer of thin. Armsby undertakes to visit the observatory as a trustee. Gould sees him coming just as he is going to <u>town for the only time for a week,</u> and leaves the assistants in charge. The assistants ignore Armsby and his trusteeship—come in and go out of the observatory—almost in his very eyes—and have no keys to let the awful and indignant trustee in the trustee rooms—the assistants are provokingly sympathetic. The trustee thinks it the most extraordinary and indiscernible coincidence that the director is always absent when he comes and at no other times. The assistants think the same. The trustee wonders how some ladies have just been able to visit the observatory if the assistants have no keys—the assistants are equally amazed at the unprecedented success of the ladies. The trustee goes to his mates with a sad countenance. His mates resolve that the trustees shall have keys to their own rooms, and that such finished gentlemen shall not be done by such unfinished "boys." An elegant fight. But the trustees have the legal power. They are getting irritated. The intention to irritate them is too plain to be mistaken! We shall be turned off and our redress is with the town pump of the good city of Albany, whither we can go and wash ourselves to our hearts content. And the noble deutschmans will give a party in our honour. Oh! My dear Queen! I feel that I need a washing in the royal bath of the true honor; and I wished that I were free from this Dudley Observatory and as to Albany—I cannot see with composure the obscuration of the delightful reminiscences with which this city is hallowed in my memory.[100]

$$\Phi^n$$

Another incident that worsened the relationship between the scientific council and the trustees involved a computing engine Gould had purchased during his trip to Europe. The Scheutz calculating engine would aid the computational work that would have to be made at the observatory. A forerunner of the modern computer, it was an unusual and exciting machine. At the end of May, Gould gave a talk to the Albany Institute in which he explained this calculator. His assistant, Batchelder, who had obtained his position on Peirce's recommendation, gave the demonstration. Newspaper accounts of the presentation annoyed Gould and Bache in two ways; they identified Batchelder as an employee of the Coast Survey stationed at the Dudley Observatory, and they gave Batchelder credit for his part in the presentation, slighting Gould. Both Gould and Bache felt Armsby was responsible. "Gould was so excited that he rushes to N. Y. where Batchelder was to request him to say so like a man that he had merely done the mathematical part & was under Dr. Gould's direction." Since Batchelder was reluctant to write to the paper to clarify matters, Bache asked Peirce to urge him to do so. He wrote, "A man who is not loyal in small things will not be in great and I trust none such willingly."[101]

Peirce replied that he would ask Batchelder to write a letter, but he warned "that it is a very injudicious and unwise step to be drawn into a newspaper fight by such a silly squabble and that it will injure us all in the opinion of most judicious men. . . . Armsby's course is a dirty one—but you cannot remove dirt by erasing it with dirt of another kind."[102]

Φ^n

By this time, the trustees had had enough of Gould. Several days later the trustees issued a statement to the scientific council stating that there was a lack of harmony between them and Gould and that some other arrangement must be made.[103] The scientific council, both to forestall Gould's dismissal and assume authority they, in fact, lacked, replied that they regretted the trustees' action, and asked them for more information, so they might make a responsible decision.[104]

Yet the scientific council's posturing had some moral basis. One of the sources of trouble was that the relationship between the two bodies, the trustees and the scientific council, was not agreed upon and mutually understood.[105] The political setup made things ripe for misunderstanding. Men like Armsby and Olcott were civic-minded men, who had labored continually and without compensation to bring a creditable institution to their community and raise the level of American science; moreover, they were legally responsible for the welfare of the observatory. Yet the scientific council, Gould, and his assistants from the Coast Survey, worked with no compensation from the observatory itself.

Beyond their own desire to have a national observatory under the Lazarroni's control, Peirce, Bache, and Henry felt a responsibility to the people that had generously donated money to the observatory in large measure because the scientific council had associated itself with the institution. They had a responsibility to the observatory as well as the trustees. The scientific council was not too far from the mark, when they said: "It should not be forgotten that the Council was not merely the creature of the trustees, but was sanctioned 'at meeting of the friends of the Observatory,' and that large donations were secured through the use of their names. That, in fact, the institution was dead when the Council agreed to supervise it, and that it was revived by the pledge of their scientific reputations that it should be made a 'national,' a 'first class' Observatory."[106]

The trustees, were not intimidated. They accused the scientific council of an "unwarrantable assumption of power," and promptly severed their connection with the observatory.[107] This did not bring an end to the controversy between the two parties. And, as the *New York Times* observed, the controversy was "becoming decidedly personal."[108] Not only was the controversy personal, but nasty. When Olcott accused "Dr. Gould of gross mismanagement of the funds, . . ."[109] the scientific council countered by demanding an accounting of money given the Dudley, implying Olcott's negligence.[110] Olcott blamed the delays in mounting the telescopes on the "unwillingness of Dr. Gould to test his skill as a practical astronomer."[111]

Shortly before the trustees called for Gould's resignation, Mrs. Dudley wrote the scientific council, assuring them of her support and calling for the resignation of the trustees:

> I have watched your management of the scientific affairs of the Dudley Observatory, and the gratification with which I have seen its development under your skillful care, in spite of the embarrassments with which I am well aware you have been impeded.
>
> As stated in my letter at the time of the inauguration of the Observatory, my donation towards the required endowment was made upon the guarantee furnished by your names and reputations.[112]

In addition to feeling a responsibility toward the donors, and the hope of having a scientific institution that would surpass any in the United States under their control, the Lazzaroni were, at this point, reluctant to concede for yet another reason. They were highly desirous that scientists, not lay boards, should make important decisions affecting scientists. The scientific council had the potential of meeting this requirement and they were reluctant to abandon it.[113] Bache requested Peirce and Henry to come to Albany to lobby for Gould, hoping to achieve the same success they had the previous January. Peirce confided to his friend, LeConte:

> I was, during much of the week, in constant expectation of a call from the Chief to meet him at New York—but he finally decided to have us meet at Albany next Tuesday—so that on Tuesday Morning we shall go. It will be a hard fight and the victory will assuredly be with our opponents—but nevertheless we must die game and try . . . to bring off Gould without dishonor from the embarrassing position in which he is placed. For my own part, I think that his course has not been characterized by any excess of wisdom and good judgment, and he seems to me rather to have sought than avoided occasions of difficulty. All the legal power is in the hands of the trustees, and their passions seem now to be so much excited that I

believe they will scarcely hesitate to use their power and leave us for their own councils, to which Maury will very possibly be invited to assist.[114]

Gould not only alienated himself from the trustees and Peirce, but from many prominent American astronomers. He did this despite his contributions to American astronomy, including his publishing the *Astronomical Journal.* Henry observed that "unfortunately he [Gould] has more personal enemies than any person with whom I am acquainted."[115]

In particular, Gould had alienated George Bond at the Harvard Observatory. An article in the *New York Times* reported that Olcott said Gould's relationships with prominent American astronomers were so bad as to "preclude the hope of concert and cooperation between them."[116] Gould wrote to Bond asking for a letter refuting the *Times'* statement. Bond replied, offering no letter, and obviously piqued that Gould had not chosen the Harvard Observatory in the recent determination of longitude for the Dudley Observatory. Gould wrote again clarifying his request.[117] Bond then wrote that Gould's not using Cambridge for the longitude determination was "only one among too many instances, showing the existence of unfortunate personal relations . . . between the two observatories. If Mr. Olcott refers to the state of feeling of which such facts are the strongest evidence, it does not appear to me that his statement is very wide of the truth."[118]

$$\Phi^n$$

Bache's next step was to pursue legal action and to write a pamphlet supporting Gould. During the early summer of 1858, Bache and Henry wrote the *Defense of Dr. Gould.* [119] Peirce wasn't available to meet with them and help, but he read the manuscript in Cambridge and sent back suggestions. The document accused the trustees of doing Gould "a willful and egregious wrong." And pointed out that "the trustees, instead of being actuated by a desire to advance astronomical science, have sought merely the advancement of their own interests and the gratification of personal feeling. They have no idea of an Observatory such as their high-sounding words to us and the public would imply."[120] In this pamphlet of ninety-one pages, the truth often took a back seat to loyalty to Gould. It went through several printings and was widely circulated.[121]

Once again, Bache called Peirce to Albany. He wasn't enthusiastic about going or supporting Gould; only his undeviating loyalty to Bache took him there. He wrote to Le Conte: "As to the Dudley

Observatory it disgusts me. Never did I fight with so little interest, and it has hung like a black cloud over the last two months, driving away all aspects of joy and brightness. But I shall have no more to do with it, if I can possibly help it."[122]

Yet when Peirce arrived in Albany, he, Bache, and Henry met with spectacular, if only temporary, success. Now highly optimistic, the volatile Peirce, wrote:

The Dudley called me to Albany about the closing of my services in college, and all the time of my being there we were obliged to be unintermitting (sic) in our efforts to get our defense of Gould, which you will perceive to be quite long. The generalship of the fight was of course Bache's and I don't think that a contest was ever conducted with more skill and good judgment. Gould has certainly not been over judicious in his administration, and was rather entrapped by Armsby into some mistakes in his intercourse with the trustees, but I cannot ascertain that anything which he did was seriously objectionable nor have they ventured to bring against him the very matters in which he might have been made to appear to any real disadvantage—and the simple reason is that their own conduct in the same affairs was far more contemptible than his own. The real cause of the difficulty is still a mystery which we cannot satisfactorily solve. The most obvious theory is that Armsby, who has proved himself to be what all Albany calls him, a conceited busybody—who is buzzing about in every dirty intrigue with all sorts of stringing gossip—and who imposed upon us so woefully with his assumption of modesty—could not bear to lose the opportunity of power which he was enjoying over the first time in the direction of the observatory, and was so annoyed to find that Gould, when he came, was actually a man who had a will of his own and meant to be the actual as well as nominal director, that he turned against him and that the whole storm was brewed in his little brain. This story is simple enough but the cause does not seem to me to be adequate to the effects. I am rather inclined to the belief that, in one way or another, Olcott has involved the funds of the observatory in a way which will not bear examination—that it was desirable for him, therefore, to reduce the institution to a ten-penny showman's museum of which Peters was to be the Barnum —and which would soon be forgotten by all the respectable portion of society. He expected that we should be turned off without any bother, and is surprised beyond measure to discover that we are not the milk-sops which he took us to be.

During the whole campaign the Chief has been in the most glorious fighting order and is quite delighted at the unanimity and admirable discipline of his forces which consisted of the gigantic Smithsonian as the right wing—The skillful director of the Dudley as the center, and poor old functionary on the left—who carried a key in his hand which frightened much more than it damaged the enemy. The ladies of Albany are wholly devoted to our cause, and it is fortunate for our wives that we are all of us so indifferent to the charms of the sex.[123]

The trustees did not capitulate, and the scientific council's success was not long-lived; in the middle of August, the trustees published the *Statement of the Trustees*.[124] At 173 pages, it was even longer than the *Defence of Gould*, and well done by the capable lawyer and trustee, Judge Ira Harris. Public opinion began to sway against the scientific council.[125]

Bache thought it imperative to write a pamphlet rebutting the *Statement of the Trustees*. Although outraged by the pamphlet, Peirce had doubts as to the wisdom of doing so. As he discussed the matter with friends, he became convinced that it would be a mistake.[126] But Peirce was unwavering in his support of Bache and Gould in the fight. Not so Henry, who was involved in a nasty attack from Samuel Morse over priority in the telegraph, problems at the Smithsonian, and ill health. Beyond being involved with his own problems, Henry would not compromise his principles in support of his friends. He felt that Bache and Peirce were overstating their case.[127] Henry refused to sign a statement that was as strong as Bache desired; Bache refused to sign a statement as weak as Henry insisted upon. And no additional pamphlet was published.

Peirce was unhappy with Henry. He wrote Bache: "As to our trinity of science, I fear that it is in terrible danger of losing its unity. The Great Smithsonian has almost declared himself ready to desert us, and has (behind the scenes) shown the white feather in a way which shows that he is not a thorough bred game."[128]

A month later Peirce expressed stronger disdain for Henry when writing the Le Contes: "Do you suppose that I was jealous about Smithson [Henry]—It was a nice bit of fun. . . . The bitterness of tone which I assumed in regard to him, was upon account of his refusal to sustain the Chief in the continuance of the Dudley controversy, and I was merely following the Chief in referring to him as a moral coward."[129]

Although disagreement over what course to take did not cause an irreparable rift between Henry and either Bache or Peirce, the relations between Bache and Henry never completely healed.[130] The core of the Lazzaroni was shattered.[131] Some members of the Lazzaroni continued to work together to promote American science, but the close camaraderie was gone. Attempts to revive the annual dinners of the Lazzaroni never came off.[132] Although the founding of the National Academy in 1863[133] was the work of the Lazzaroni, Henry was excluded from the endeavor.

The quarrel between the scientific council and the trustees continued. Bache continued to pursue legal channels to obtain control of the Dudley.[134] Gould remained sequestered in the Dudley Observatory writing his own rebuttal to the *Statement of the Trustees*. He wrote to Peirce: "I was indeed glad to . . . learn from you that my friends in Cambridge approve my course. . . . It is a terrible ordeal. And mean time my dear father has been so ill, that it has been only with great difficulty that I have kept to my work here."[135]

Gould was finally evicted on January 3, 1859. Depending upon whose account of the incident you choose to believe, seven or fifteen men accompanied Elisha Mack, an Albany policeman, to oust Gould. In two days, Gould was packed up and out of the observatory.[136]

Gould's *Reply to the "Statement of the Trustees" of the Dudley Observatory* appeared the latter part of February. Longer even than the pamphlets of the scientific council and of the trustees, it was over 350 pages. Mentally and emotionally exhausted, Gould returned to Cambridge. His return to astronomical work was impeded by the death of his father and the necessity of his taking over the family mercantile business. Nevertheless, Gould continued to do astronomical work and work part-time for the Coast Survey. During the Civil War he did statistical work for the Sanitary Commission and was a key player in the founding of the National Academy of Sciences. After the war, Gould worked for the Coast Survey, until he was forced to resign by his former teacher and mentor, Benjamin Peirce.[137]

For some time Gould had been interested in exploring the southern celestial hemisphere, so after being ousted from the Coast Survey, he accepted a position as the first director of the newly founded national observatory in Argentina. There he did his most important work. His achievements were widely recognized in the United States and Europe.[138]

Despite the troubles in Albany, neither the Coast Survey nor the Smithsonian lacked Congressional funding, as Bache had feared they might. Ironically, Congress mistakenly identified Gould with the *Nautical Almanac* and cut its funding.[139]

Mitchel was again asked to be director of the Dudley Observatory; he accepted the position. Before the end of 1859 a thirteen-inch refractor and the meridian circle were in operation in the observatory. The heliometer that Peirce had hoped so much to have was never installed in the Dudley Observatory, although the instrument remained important to astronomical work throughout the nineteenth century.[140]

9

A Stranger in My Own Land

As PEIRCE APPROACHED FORTY-FOUR, HE FEARED HIS BEST YEARS WERE behind him, although he could see some advantages to aging. He wrote to Josephine Le Conte: "My hair is getting as gray as a grisly bear's and old age has stolen a mighty march upon me in the course of a few months. You have seen the best of me, and on the whole I think it better for my reputation in the Athens of Georgia, better be the memory of which I shall leave in your warm heart, that it should be that of what I was, than of what I am. Do not, however, mistake me. I am not afraid of old age with its frosts. I almost welcome it, hoping that it will cool me down to that moderate temperature which is the best adapted to the spiritual ardor of our northern climes."[1]

Three years later Peirce feared that with advancing age his abilities were eroding: "Remember that I am 47 years old and do not like me the less that I have so many cold years piled upon my head. I am not so fit for work as I was when you first saw me, and I cannot help, now and then, feeling a little sad about it, and looking forward with some apprehension to the five volumes of my work which ought to be finished before I put off this terrestrial body and put on nobody knows what kind of body in the other world."[2]

The five volumes that Peirce alluded to were five volumes in mathematics that he hoped to write. Prophetically, Peirce only published one of them, and probably did little work on any of the others. He was currently working on the one volume he would finish, a book on analytical mechanics. Peirce spent his evenings working on this book, so that it would not interfere with his usual work.[3] Although Peirce used it as a text in some of his advanced courses, he did not intend it as a mere textbook, like the volumes he produced in the 1830s.

Despite his apprehension about finishing his planned series, Peirce was quite pleased with the progress of the first volume. He wrote to Bache telling him how his methods compared with those of European mathematicians: "I have just done my chef d'oeuvre. I cannot give you any idea of it in a letter, but do say that it belongs to the analysis of differential equations. Euler had a multiplier. Jacobi found a new multiplier which was his greatest work by far, of his life, and was the fundamental principle of his new principles of mechanics. I have discovered a system of multipliers which includes, at either extremity the Eulerian multiplier on one side, and the Jacobian on the other. I call my system, that of Partial Multipliers. It is in the new work, and will be sent to the Journal."[4]

Since this was an advanced treatise on mathematics, its audience was limited. Peirce had trouble finding a publisher, and the book was printed only after a sufficient number of people had subscribed for it.[5] Peirce's sister, Elizabeth, commented, "We are all very glad that George has undertaken the printings off of his [Uncle Ichabod's] book, which I have no doubt will be very interesting, much more so than Ben's last work, which none of us can read understanding a single page. I believe there are not more than three or four in the country who can."[6]

There may have been more people in the country who could comprehend the book than Elizabeth thought. Peirce was delighted two years later, when stopping at a hotel: "At Springfield, the landlord . . . carried me into the back portion of his sanctum sanctorum and what do you think was the sole volume lying on his table! his especial study! Peirce's Analytic Mechanics. This is no myth—but a positive fact—the landlord of the Ranapoit Hotel at Springfield rejoices in the perusal daily and nightly—of my big quarto. When I offered to pay—Oh! No! it was impossible. He could take nothing from me and my ladies—they have only taken a little tea at his invitation."[7]

Although the date of publication in the book is 1855, it was not actually printed until late 1857 or early 1858.[8] Peirce complained about the delays, but apparently they were not caused by a lack of subscribers. He was happy with the way the subscriptions were going.[9]

The most striking feature of the book is Peirce's masterful discussion of determinants, a recent discovery.[10] Many mathematicians have praised the work.[11] It was well received in Europe: "A year or two later [after the publication of the *Analytical Mechanics*] an American student in Germany asked one of the most eminent professors there, what books he would recommend on analytical mechanics: the answer was instantaneous, 'There is nothing fresher and nothing more valuable than your own Peirce's recent quarto.' "[12] At home Peirce's former student, and soon to be president of Harvard, Thomas Hill, praised the work as "the . . . most valuable original mathematical treatise as yet written in America."[13]

Φ^n

One of Peirce's colleagues at Harvard was the botanist, Asa Gray. Like many naturalists of his day, including Louis Agassiz and John Le Conte, he had attended medical school. Gray practiced medicine for only a short time, before turning to botany. He studied and worked for eight years with John Torrey, the leading botanist in the United States, before joining the Harvard faculty in 1842. He was progressive and dedicated to the progress of American science. Peirce's audacity in questioning Leverrier pleased him, and Gray sent Peirce a short note saying so.[14]

Seemingly, Gray was an ideal colleague for Peirce, but alas they were usually at odds. One reason may have been that Gray and Agassiz seldom agreed upon scientific matters, and Peirce tended to side with his good friend. As a student in Europe, Agassiz had been strongly influenced by the *Naturphilosophie*. The excesses of this philosophy soon discredited it, but it had a strong influence on such men as Agassiz. Proponents of this mode of scientific thought insisted upon putting all of nature into a unified and absolute paradigm. To Agassiz this paradigm, or idealism, came first; empirical facts were manifestations of it, not evidences. Gray, on the other hand, was a thorough empiricist. To him, theory or generality must follow from observation, not precede it.[15]

But beyond supporting his friend, Peirce, too, had a mind that took to ideality.[16] Peirce and Gray were very different types of men. Their sole instance of scientific collaboration illustrates this point. Phyllotaxy, the study of the arrangement of leaves on a stem, fit well into *Naturphilosophie*.[17] Agassiz was interested in this topic and generated interest in it among the Harvard faculty when he first came to the United States. Gray and Peirce worked together on the topic for a short time, but Gray could not see Peirce's point of view; he thought Peirce "ran off into the sky dementedly."[18]

Gray was shy and preferred scholarship to politics. Unlike Peirce, he preferred local scientific societies to national ones, and soon became well ensconced in the leadership of the American Academy of Arts and Sciences, becoming its corresponding secretary. Eventually Gray was elected president. Peirce may well have resented Gray's prominence in that organization; many of his complaints about Gray centered around the American Academy. But whatever the cause, Peirce developed a strong disliking for Gray and often described him as "cloven hoofed," a typical Peirce disparagement.

Φ^{n}

Much of the scientific and intellectual life in Cambridge centered around the American Academy of Arts and Sciences. It was there that Peirce boldly proclaimed that Leverrier's discovery of Neptune was a happy accident, and many years later it was the venue for the debates, between Gray and Agassiz, on Darwin's *Origin of the Species*. The American Academy was founded in 1779, lest the American Philosophical Society of Philadelphia should take undue luster from Massachusetts. Peirce was elected to the American Academy on January 29, 1834, soon after becoming a professor at Harvard.[19]

Although the American Academy supported science, many prominent citizens of Massachusetts who belonged to it were not even amateur scientists, but patrons.[20] Peirce resented their patronage and their meddling in scientific affairs. He wrote: "[The] Am. Ac. is a creation of patronage, and it fauns at the foot of wealth. I despise it with all my heart, and I regard it as our duty to put it aside and fill its place with a society worthy of the country and of science."[21]

When Peirce's friend Bache was elected president of the American Philosophical Society, of which Peirce was also a member, he wrote: "And so you are president of the Philosophical Society. . . . Can you make it what you want? Do you wish us to aid you in giving it life? Do for heaven's sake try to rescue it (if possible) from the imbeciles of patronage—which has as bad as killed the American Academy."[22]

Peirce's long-standing frustration with the American Academy came to a head in the spring of 1858. Peirce was outraged with the proceedings at a meeting and resigned from the academy. When he relented a week later, Agassiz proposed that he be reelected. "Agassiz said . . . he would not refer to past unpleasant circumstances but hoped a full ballot would be taken. A Rev. Scholar present said he should like to know if Prof. P. would visit the Acad. or consider the election favorably & respectfully if elected. Agassiz said such a question was an insult to Peirce and his friends who wished his election & would not reply to it."[23] Upon Agassiz's insistence a vote was taken:

On Monday last the vote was taken on the re-election of the Functionary to the American Academy, to which he was nominated by the names of Josiah Quincy, Louis Agassiz and Joseph Lovering. For the election a three-fourths vote is required. When the vote was declared, it stood eighteen ayes to eleven nays! And so the functionary is rejected. This is the way in which the Lazzaroni are honored at the north. . . .

Bye the bye—I ought to have said that Agassiz is so wrathy he has resigned his place in the academy. He is angry enough to please you—but he is a little too much dejected at the result. As to the Functionary he is rather amazed by it and especially to see how terrified the doers of the deed are at the sound of their own act. But it is quite fair to use it against them and to make the most of it in our favor. The Lazzaroni cannot afford to lose the advantage which this false move gives us. We shall be defeated in the end. Then right never prevails in their world.[24]

Despite the bravado, this was a troubling personal blow for Peirce.[25] He exclaimed, "Ingersoll Bowditch voted against me!!!!! G. P. Bond voted against me."[26] Yet he put the blame at Gray's feet. "My sincere belief is that the end is near at hand to Gray and his friends."[27] Ingersoll Bowditch had been a close friend since grade school. Bond had been Peirce's student and was now assisting his father at the Harvard Observatory, and was the heir apparent director. Peirce was irate with Bond, but as with the whole proceedings, saw his action as an opportunity. Peirce wrote to Bache: "As to G. P. B. he has made the mistake of his life—perhaps? B. A. G. will probably be driven from Albany, with his council in contempt. . . . In the course of time W. C. B will go to the stars. When the vacancy occurs, I shall be relieved of all allegiance to George, who loves me so sincerely. I shall take the field in earnest. I shall call the court jurors and the Smithsonian to my aid. I never was in better opportunity of action upon the corporation—and it will be a hard fight—but I hope to gain it.[28]

Bache was also disgusted with Bond's actions, but he cautioned Peirce, "Pity for Geo. B. that he was one of the 18 nays. . . . Now Φ^n [Peirce], darling Φ^n [Peirce], make sure that Geo. B. did that, for it is the mistake of his life. If he did it, no honest Lazz[aroni] can trust him now or hereafter."[29]

Writing to Bache about the prospects for his eventual reinstatement, Pierce wrote: "The American Academy matter greatly amuses me— . . . Agassiz and Peirce will both be nominated. Peirce will be elected, Agassiz will be rejected because he undertook to assume a dictatorial tone which even Bowditch would not endure and voted against my admission although he exceedingly desired it. Peirce will decline . . . and the discord will not be resolved—Is the world upside down or do I walk on my head? I seem to be a stranger in my own land—and I cannot understand the language of my neighbors nor those who are called my friends."[30]

Both Peirce and Agassiz were reelected.[31] Peirce relates, "The men in the American Academy who voted against me have now renominated me for selection and I find that Boston seems to have found out that it has a most extraordinary regard for my popular assassination! Vox Populi Vox Diaboli."[32] Peirce at least feigned ambivalence as to whether he should accept reelection. He told Josephine that he accepted it reluctantly, and partially due to the Chief's encouragement. Only two people voted against him, and it would have been an insult had *they* voted otherwise.[33]

Φⁿ

Peirce had other problems at this time besides spats with members of the American Academy. Among them was the government's refusal to pay him:

> One of the comptrollers of the Bureau at Washington threatened to decide that I am not entitled to receive any compensation for the work which I have done for the Coast Survey because I am at the same time an employee of the Nautical Almanac. Would it not be the height of injustice for Uncle Sam to benefit by my work and then refuse to pay for it. In case this decision is made I may be called upon to refund what I have received already—and since I have it not—I may possibly pass my vacation in prison, instead of Columbia. Would it not be jolly? Should the pay be stopped and the return of the back pay not demanded which is not likely—I shall not have the wherewithal to get away from home—even to the Baltimore meeting of the Scientific Association. However I cannot think that such a gross injustice will be committed at the caprice of this low-lived block-head whose chief object seems to be to give annoyance to the Chief. But I hear queer stories about his doings in other cases, in which he has disturbed the equanimity of some of the most distinguished members of his own party.[34]

Obtaining his pay was not resolved as quickly as he hoped; he wrote, "All the family are well, but the pecuniary troubles with which we are surrounded are most appalling and have involved us in much affliction."[35]

Φⁿ

Peirce also had troubles with the Bonds that started long before George voted against Peirce's reinstatement into the American Academy. Well before this time, bad feelings had developed between Peirce and both the Bonds, despite Peirce's earlier high regard for the senior Bond. After William Bond's son, George, graduated from Harvard in 1845, he became his father's assistant in the observatory.[36] The great refractor was powerful enough for him to see a previously unnoticed gap in the rings of Saturn. This observation, made on November 11, 1850, led him to hypothesize that the rings of Saturn were not solid, as was then believed, but fluid. He made mathematical calculations to verify this, communicated his findings to the American Academy, and sent papers to the *American Journal of Science* and to Gould's *Astronomical Journal*.[37]

Excited by Bond's discovery, Peirce immediately made his own mathematical analysis of the problem and concurred with George Bond that the rings must be fluid. Peirce concluded, however, that the planet's satellites were necessary to keep even fluid rings stable. He then shared this research at the next meeting of the American Association for the Advancement of Science in Cincinnati.[38]

Peirce's results were enthusiastically received. A Boston newspaper lauded his presentation as "by far the most important communication yet presented, [to the meeting of the Association] and . . . the most important contribution to astronomical science made since the discovery of the planet Neptune."[39] Peirce not only gave Bond credit, but praised him lavishly for his discovery.[40] Nevertheless, both Bonds felt that Peirce had stolen George's thunder. William Bond complained both to scientific colleagues, like Alexander D. Bache, and to the press. This resulted in an unfortunate and bitter priority dispute.[41] Peirce was supported in this dispute by his former student, Benjamin Gould, who gave Peirce the lion's share of the credit for the discovery.[42]

The Bonds' bad feelings widened when Peirce sided with his friend O. M. Mitchel, concerning a dispute between Mitchel and Bond over the priority and the accuracy of Mitchel's new method of determining right declensions. In addition to supporting Mitchel, Peirce suggested that the dispute between Mitchel and the Bonds should go before a committee of the AAAS, a procedure that Peirce requested in several other disputes. He suggested that the matter be brought before the association in order "to discuss this question fully, and adopt a course which shall settle its worth beyond all future controversy. Would it not be a good plan to propose a committee of reference to be selected by the President of the Association after free and open correspondence with all parties interested?"[43] George Bond resigned from the AAAS that year (1851), apparently in protest.[44]

There was a fundamental difference in the scientific orientation of Peirce and the Bonds that was likely at the heart of their conflicts. Although both of the Bonds were excellent observers, they had little understanding of mathematical astronomy. Peirce, of course, did. Both Bonds were eager for scientific recognition and had several squabbles with other scientists over what they felt was their due recognition. No doubt the Bonds, especially William, felt that the observational discovery regarding Saturn's rings should be given the most credit and failed to understand the important contribution that Peirce made and for which he received so much recognition.[45]

Φⁿ

The rift between Peirce and the Bonds was not so wide as to prevent Peirce's giving a eulogy before the American Academy of Arts and Sciences on the occasion of William Bond's death. While Peirce was generally generous in his praise of Bond, he was careful to note that Bond was no mathematician: "In his original investigations he naturally restrained himself to those forms of observation which were fully within the reach of his own resources, He did not, therefore, seek those inquiries, which could only be accomplished by long, intricate, and profound mathematical computation, but preferred those which were merely dependent upon the thorough discipline of the senses. He consequently availed himself less of the remarkable capacity of his instrument for delicate and refined measurements than of its exquisite optical qualities. But when observations were required which must be passed to the computer his skill was not wanting the occasion."[46]

Although Peirce praised William Cranch Bond's contributions to astronomy, he had no intention of supporting his son as William's successor as director of the Harvard Observatory. In a letter to Bache, Peirce wrote, "The time is come. Bond died yesterday morning and I did not hear of his death till this morning. Should I be asked with regard to George—I shall distinctly declare that I regard him as incompetent to the directorship. But the men in power think that he is good enough for their purposes and he will be elected. If it is my duty to speak out without being asked, I am ready to do so."

At first Peirce supported Gould for the position. Although he found Gould "wanting in the qualities of an administrative officer . . . as to his scientific fitness for the office, I shall and must speak with all the power of which I am capable. I regard him as head and shoulders above any other man except our Dear Chief in this America. . . . But as to his administrative capacity I shall not speak—I shall declare my incompetence to judge and I shall strive not to injure his cause by reticence."[47]

Yet before Peirce finished writing the above letter, Gould was out of the running. Agassiz had asked Peirce how Joseph Winlock, another of Peirce's protégées, would do. But the two of them decided on Peirce, himself. "I have just seen Agassiz. I am to be the candidate for the Directorship. Agassiz will see each member of the Corporation and the chance is greatly in our favor."[48]

One of the reasons that the Lazzaroni dropped Gould as a candidate was that his election to the directorship was too uncertain.[49] Peirce assured John Le Conte that he was concerned about Gould's state of mind over the position "so that I would not consent that my name should be used if I did not know that there was not the shadow of the possibility of his election at the present time."[50] Peirce was also sympathetic with Gould's father, who came to him hoping to garner his support for his son. Although Peirce could give him no encouragement, he stressed that if there came a time when he could bring Gould forward he would withdraw in his favor.[51] Peirce assured Le Conte that he did not covet the position. "It will be of no selfish value to me—but on the contrary will add to my labours more or less, and that more to my responsibilities."[52]

Gould, upon hearing of Peirce's interest in the position, after expressing surprise, withdrew himself from the running, not wanting to compete with his former professor. Peirce attempted to negotiate some sort of compromise with Bond and the college, but in the end, Bond was elected to the post.[53] Although Peirce may not have coveted the position in the beginning, losing it to Bond was a bitter blow.[54]

After Bond was appointed director, he wrote Peirce a most conciliatory and generous letter, attempting to heal the rift that had grown between them. After noting that their relationship had deteriorated, he said:

> My object now is to propose and to open the way for a return to a better state of feeling.
>
> No one can appreciate more highly than I do the advantages which would accrue to the observatory from your cooperation with it. Both your position and attainments are such as to enable you to render most valuable aid, and they must always ensure a respectful consideration of your opinion and advice.
>
> Any assistance which I can give in aid of your scientific investigations, by furnishing the results of observations, or in other ways, shall be freely extended, but a cordial and earnest cooperation can be brought about only by a mutual consent to give up past differences. On my part nothing shall be wanting to accomplish that end.[55]

It is unlikely that Peirce ever responded in kind. The following unfinished letter is in the Peirce Papers. Whether a copy of it was sent is not known, but none is in the Bond Papers,[56] where it likely would be if it had been sent: "I agree with you that the time is come when the dead past should be buried in oblivion, when the director of the observatory should take the Perkin's Professor by the hand, and when an effective cooperation should be established between them. Whatever plan you shall propose in this direction I shall rejoice to accede to, and shall desire to confirm by investigations & any system of observations . . . from the observatory which will tend to elucidate to mankind the laws which the Creator has written upon the firmament."[57]

Peirce's feeling towards Bond had not softened some years later when he and his Lazzaroni cronies selected the initial membership for the National Academy of Sciences. Despite Bond's international reputation as an astronomer, he was excluded from the list.[58] Nor had Peirce's estimation of Bond changed when Bond was elected to the Royal Astronomical Society. He wrote to Bache, "I perceive that G. P. B. is just elected an associate of the R. A. S. Better late than never. Better to deserve it than be elected."[59] Later he added, "You are right in all respects[:] the R. A. S. is not a high honor—it was rather the opposite—that G. P. B. was not worthy even of that."[60]

$$\Phi^n$$

Peirce continued to be troubled by old age and especially feared turning fifty. He wrote to John Le Conte: "I intend to let myself quietly subside into old age and obscurity—and the universal obligation of living men and women. My notions of my coming paradise will be as follows. I shall get up at nine o'clock—breakfast—read my newspaper—make believe lecture—dress for dinner—dine—take a nap— a short walk—indulge in half hour reverie . . . —drink a cup of tea— play a hand of whist—go to bed. Is not this a charming prospect for a happy and useful old-age?"[61]

Several weeks later, after having lost his bid to become director of the Harvard observatory, his mood was no longer playful: "I am indeed most heavy at heart and the weight increases as I near my fiftieth birth-day. . . . The two days which I spent with the Chief, were bright and his presence seemed to drive off the storm—but the blackness has returned upon me, and I am too sad to write to the Queen. I am all out of tune with all mankind and womankind too. There is nobody at my side, who can read me, and give me the food which I need to return me to myself. I am not fit to write the queen, and this fiftieth birthday! Look upon it as the end of my life. Henceforth I am a mere . . . machine, and never desire again to be roused to life. Let me for the rest of my material existence, be buried in the darkest abysses of geometry—in the unfathomable cavern beneath the feet of the tortoise which sustains the elephant who holds the world upon his back.[62]

John responded to his letter, saying that both he and Josephine were concerned about his depressed state. Fortunately Peirce's spirits rose; when he wrote again, his tone was upbeat:

In the course of the last month, I have received a succession of blows which almost broke me down, and made me feel as if the world were hollow, and fear that all my friends were just like the world. But the storm is blowing over and I cannot tell why but it seems to me that a clear sky is soon to open upon me, and that the evening of my life will be serene and joyful. Do not suppose that I have felt disturbed at the idea of becoming fifty years of age. Blessed with good health, a loving wife, dear children, a beloved mother and sister and aunt and some of the most glorious friends, that ever surrounded a man with a halo of affection, the Queen and her John, Agassiz, the Chief, Henry Davis, Charles Mills—oh! I have much to live for! And may God forgive me that I have ever permitted the sin of despondency to cloud my atmosphere! But for some reason or other I have always been afraid that at the age of fifty, I was to be visited with some dire calamity. When then I found that within a month of this time I was, as if by some mysterious combination of the powers of evil against me, mortified by the most unprovoked insults from those whom I had served most faithfully—and my means of living all of a sudden were down much below my actual expenses—I felt that all my worst anticipations were about to be realized—and that I was given over by the dispenser of events to be the sport of the devils. Let me enter more into detail as to the college and the Nautical Almanac—so that you may realize that I have not been forgetting my friends in a weak indulgence of a peak of selfishness.[63]

Some years later Peirce, now comfortably past fifty, extolled the virtues of old age: "Youth is <u>not</u> an institution. It is a destitution of age. It is an unripe cherry, an uncooked lamb, a blank sheet of paper, an unperformed promise, the flat surface of a mountain lake. A sour bunch of grapes is better for dessert than insipid youth, and its poverty of thought and experience, and its waste of words. Youth is a disease, which is to the part of the wise philosopher to avoid as a pestilence or to cure for the sake of humanity and out of pure love to one's neighbor. Use is its antidote. Age ripens it, cooks it, writes upon it, performs its promise—and brooding over it, furnishes the sky and the clouds, which become, in its reflections the picture of heaven."[64]

Φ^n

Although Peirce had always worn a full beard, he now grew a moustache. At first it wasn't well received: "This is the whole story. Silently, I stopped shaving—and one day one of the babies exclaimed—Why!

Benjamin Peirce after growing a mustache. By permission of the Houghton Library, Harvard University, MS CSP 1643.

Papa is growing a moustache." The wife attacked me—the friends attacked me—the world looked sharply at me—Even Agassiz said one day—I don't like that black upper lip—it changes you so. My defense was that I was getting old and did not own a barber. Bye and bye the climate began to shift the other way and the old gentleman has even received some complements upon the becoming aspect of the additional whiskers. But now that the thing is did (sic.)—I rather think that it will be submitted to, without much complaint. However the dear wife has the final decision of the question."[65]

Φⁿ

Local scientific meetings, even those that featured Asa Gray, were not always unpleasant in Peirce's eyes. In May 1859, Peirce attended a meeting of the Cambridge Scientific Club, where they had a good dinner and Asa Gray had talked about Mr. Darwin's new theories. Peirce was captivated:

> Dr. Gray gave an account last night of a Mr. Darwin and a Mr. Wallace's speculations as to the possible changes of species under the influences of natural agencies. Those writers profess to approve the Lamarckian system of changes arising from internal desire, but inject upon it, that nature can produce just such kinds of changes as are effected by humans already only to a much greater extent. If, for instance, a class of animals was formed which fed upon the leaves of trees, the taller individuals of the class could be so much favored in the amount and degree of food, that they would become decidedly stronger and more numerous than the others, would be more likely to propagate, and thus lead to an increase in the size of all the living members of the family. With a new generation, the same kind of process would lead to a still further increase, until giraffes might be deduced from horses, etc. With them, the life of the wild animal is a perpetual struggle against the opposing influences, which are constantly preventing the increase of their numbers so that they are in the most favorable state for the exhibition of their localities to change.[66]

Peirce tended to side with his friend, Agassiz, who opposed Darwin's theory, however. He wrote to John Le Conte: "What do you say to Darwin? What does Joe say to him? I cannot think that his observations, however curious and useful they may be in themselves and however they may tend to the elucidation of the laws of change to which species are subject, have anything to do with the larger and more radical transformations which have taken place in the transitions from one geologic age to another. Agassiz insists that the geologic changes are thorough and complete, and that there is no instance of a species common to two successive epochs, whatever may be the appearance to a careless and inaccurate observer."[67]

Agassiz and Gray had several debates before the American Academy of Arts and Sciences. Gray, though lacking Agassiz's charm and charisma, was the winner.[68]

Φ^n

A couple of years later, Peirce himself gave a series of lectures at the Lowell Institute. They were well received. His wife, Sarah, wrote, "The lectures are more than successful enough—I have been to all and am surprised at my Φ^n [Peirce] in his, to me, new character. His lectures are splendid—I was in fits at first because he would not write them but I see that they are far better extemporized. He is much complimented upon them & has a large and choice audience."[69]

10

This Holy Cause

ALTHOUGH ANTISLAVERY SENTIMENT IN MASSACHUSETTS WAS STRONG, it did not generally extend to Harvard College.[1] Neither did it predominate before the Civil War. The abolitionists were a minority, if a growing one. Well before the onset of the war, Peirce and his family strongly opposed them. The abolitionists were radicals, who threatened the pleasant life of privileged Massachusetts. He wrote to his wife in 1844, after attending a political rally at which a popular abolitionist, Cassius Clay, spoke: "I am myself of the opinion that the abolitionists have made great headway in this quarter and will soon govern the state. A man like Huntington is much wanted to throw himself boldly into the breach, and if he would step forward he would do a great deal of good to his country. . . . Now is the time for him to take a stand for which every freeman will thank him after he has run the gauntlet for a time. If I were in Northampton, I would say everything I could think of to urge him to it."[2]

While visiting Washington several years later, Peirce wrote to his wife, "Bye the bye our servant is a slave, tell it not to Cousin Sophia, but the fact is not more horrid than true that our meals are served by a slave."[3]

Benjamin's brother, Charles Henry, shared his feelings toward the abolitionists. He wrote Benjamin of his unhappiness with the integration of the Salem schools: "The school committee voted to abolish the black school and admit the blacks to the white school. . . . As soon as I heard of the course of the school committee, I said that the abolitionists had obtained a victory. I cared not whether the black children amalgamated or not with the whites, but I could not bear the idea of the city of Salem securing to countenance abolitionism."[4]

Φ^n

Thus, in August 1851, when Peirce met the Le Contes in Albany, he already had considerable sympathy for their holding slaves. On the other hand, the Le Contes had many ties to the North, despite their strong Southern loyalties and their championing of slavery. John's father was born in Shrewsbury, New Jersey, and graduated from Columbia College. He did not move to Georgia until he was about twenty-eight years of age.[5] John himself had studied medicine at the College of Physicians and Surgeons of New York, and later taught there.[6] While studying in New York, John met Josephine, who was herself a New Yorker. Le Conte later sought an academic position in the North. Furthermore other members of his family lived in the North. His uncle, nicknamed "the Major," lived in New York, and his cousin, whom he visited often, resided in Philadelphia. And the Le Contes sent their daughter, Lulu, to Miss Easton's School, near Philadelphia.[7]

Neither were Peirce's contacts with the South limited to the Le Contes. Jefferson Davis was an ally of the Lazzaroni and a close friend of Bache. He was one of the key senators who had come to the defense of the Coast Survey when it had been attacked in 1849 and 1850 by Senator Thomas Hart Benton.[8] Just before the Civil War, Peirce was present at the christening of one of Davis's sons, "and the darling Chief blessed the child by standing as his god-father; and after the services we were invited to the house of the young baptized where there was a nice little party, with cake and wine, and President Buchanan toasted the youthful Mississippian and hoped that he would be as great an honor and blessing to his country as his father."[9]

$$\Phi^n$$

In 1858, Peirce visited Columbia, South Carolina, and lectured at the newly founded Columbia Athenaeum, at John Le Conte's invitation.[10] Peirce liked what he saw in the South, and his pro-slavery sentiments were strengthened: "Most beloved of wives—Here I am in the capital of South Carolina surrounded by the horrid institution of slavery and for the life of me I could not as yet suspect that I was served by such a downcast race—for the countenance is cheerful, eager—and having no external aspect of servitude about it. But I suspect that I am stupid and perhaps when the day clears up and the sun breaks out and I am able to wander about I shall then see the evils which are now hidden from me."[11]

The train ride back to Cambridge provided a contrast between North and South:

What a change from the cars of the South in which are gentleman and servants, to those of the North in which are to be found two gentlemen and forty blackguards. From the conductor of the South, who does not disturb your meal by the way . . . to him of the North whose chief glory is to . . . wake you from the first sleep into which you fall in spite of your uneasy position, by a punch in the ribs with a stentorian demand for "Your ticket sir?" If the separation takes place, remember that I am the other side of Mason and Dixon's line. . . . I need the genial clime to warm my cold body. Take me to your hearts, my true Southern friends. My constant text now is I have seen slavery and I believe in it. You may believe that some works are averted—but I am persuaded that the earnestness of my sincerity gain at least many listening ears, and I am confident that my arguments make more or less of an impression. And they are not met unkindly. The time may come however, when I shall become the object of public attack.[12]

Φⁿ

Peirce's strong Southern sympathies did eventually separate him from many of his Northern friends. Peirce severed his association with the prestigious Saturday Club, whose members were among the most distinguished intellectuals in the country. Such men as Emerson, Hawthorne, Longfellow, and Whittier belonged. Peirce declared, "I have left a club in which some of the most accomplished scholars of Boston were accustomed to meet and have a good dinner once [a month]—because as I honestly told them, they had become such desperate abolitionists."[13]

Φⁿ

One of the reasons that Peirce hated the abolitionists was that he feared they would destroy the Union he loved. Peirce became increasingly despondent as the divisions in his country widened. He feared that Buchanan was incapable of maintaining the Union and upon visiting Washington said, "I found things at Washington much gloomier than I expected. Fears for the dissolution of the Union are gaining strength on all sides."[14] In Peirce's mind, slavery was essential to the salvation of the Union: "More than ever do I feel the immense importance of southern slavery to the permanence of the American Republic. You, at the South, have been the first to be tried in the fire of discussion, and to be purified in the trial. I see clearly that our time is coming[,] and as a result of careful and deliberate assurance I state it as my solemn conviction that the salvation of the north is finally to depend upon this very institution which is now the object of much irrational attack, which the Christian influences have been working at the south, now warming you into a hope of perfect organization, and are in which the laboring classes have become more and more sympathetic with the higher classes."[15]

Φn

As the separation of North and South loomed on the horizon, Peirce's Southern sympathies were so strong that he began to wonder where his loyalties were. He wrote to Josephine, "Is there any danger that this spirit of separation will ever gain the heart of the South? On which side of the line will the Queen be? On which side shall I!"[16] Feeling despondent about not being readmitted to the American Academy, Peirce wrote to Josephine, "And so the Functionary is rejected. This is the way in which the Lazzaroni are honored at the North. The Functionary stands ready to accept the first decent offer which comes from south of the Mason and Dixon's line."[17] But it was not a new sentiment; he had expressed the same sentiments only a few weeks before.[18]

The South looked increasingly more attractive to Peirce; he wrote to Josephine:

> You have been the inspiration of the most inspired portion of my life—the greatest blessing of my existence. You have taught me to know the south and for this I thank you. Like yourself, though born at the north, I was made for the south—for the society of gentlemen and if it were possible for me in the necessary arrangements for living, I would remove my dwelling to the land which holds you. . . .
>
> This North—it is infected by a race of money changers—who know nothing but a pecuniary standard—but she is our mother and she must not be deserted by her children without a strength for her rescue. . . . Because the south can take care of herself, shall she therefore desert the north? This is not worthy of South Carolina. . . . No. The South will never separate in such a ungenerous and ungracious form. It will not or can't desert the union for which it shed its most precious blood.[19]

Φn

Peirce did actually consider taking an academic position in the South. He and the Le Contes had for some time dreamed of a great university in the South. When John promised more news of the new university from his wife,[20] Peirce replied, "Your professorial plan is most delightful—the only drawback is that before the time comes—I may have received too many of the wanings of old age to be acceptable as a professor—to a new and thriving institution—where all will have to be young. But if there is anything which can make me young, it will be these long letters from the darling Queen."[21]

A great university in the South remained John Le Conte's dream; no steps were ever taken towards its fulfillment.[22] Sarah nixed Benjamin's further involvement in the scheme. She hadn't been keen about her husband running up to Albany pursuing the phantom of a great university; she certainly wasn't going to have him leave Harvard, to chase a fantasy in the South. She wrote to Josephine:

> The momentous questions contained in your kind letter of the 21st could not of course be hastily answered & even now after three days & nights of anxious deliberations I am still perplexed with doubts & misgivings—Were the Professor alone in the world, I feel no doubt that even at his ripe age he would be ready & willing to make the exchange—but unfortunately for him, he has ties which are not easily loosened or severed & he is not willing rudely to break them. For my own part, if our children were not also [to] be considered, I would willingly, as he knows well, share his fortune in a land of strangers, let that fortune be what it might, but I feel that they are at an age when such a change of their whole future, such an extreme transformation in this course of life, & in all their relations to those about them—Such a disruption of every thing, could not but have a most dreadful, and perhaps dangerous, influence upon them, in many ways. I honestly believe that it is purely on their account, without reference to any selfish regrets, that I hesitate to say that at all risks I am ready to take this step. No, I think for the present, while the affairs of the University remain undeveloped, & the Observatory a thing of the future, the Prof. ought to content himself with saying that an "Extraordinary Professorship[,]" if such are established, which he could hold in conjunction with his place here, lecturing during the months of our summer vacation would be <u>highly acceptable to</u> him—afterwards, as the Institution becomes established & the Observatory in process of time takes its place there & is equipped for action—he could, if his interests & position would be benefited by it, occupy a post there if desired—without pledging himself to it so many years in advance. I hope all this seems reasonable to you—to me it appears the only course to be taken considering the happiness & best interests of all concerned. A place as Director of an Observatory, with at the same time an active Professor's duty would hardly be the "graceful repose upon laurels" you so glowingly paint. I fancy there would be work to be done even there, dear Josie.[23]

Still, even Sarah remained somewhat ambivalent about going to the proposed Southern university. Benjamin wrote:

> You can have no idea, what uncomfortable times these are for those of us who love the South. And now let me tell you that there is one great comfort to it, for Sarah is quite warmly upon our side and fights the battle of

the South against everybody who ventures to insinuate a word against you or rather us; she even gave her Dear Dr. Wyman a heavy propounding last Monday, because he ventured to suggest that the South could not leave the Union if it would. But she mixed up her defense of the South, with indignant attacks upon the College for having done me injustice and said, "We mean to leave and go South." I find that most of my friends approve of my plan of going to the great university of the south—all at least who have any true feeling in regard to our duties to the country. . . .

Sarah is most anxious to hear from you again about the University, and she is every day more and more favorably disposed to leave the north and go to your noble university.[24]

$$\Phi^n$$

Although happy about the prospect of a Southern university, the nation's events disturbed Peirce. He was particularly troubled by John Brown's raid on Harper's Ferry. By 1855, Brown had done little to distinguish himself in life. He had lived in Kansas during the conflict between pro- and antislavery forces during the 1850s. Enraged by violence of proslavery forces, he formed a guerrilla band, of which six of his sons were members, to strike back against those with Southern sympathies. After assessing that proslavery forces had killed at least five Free-Soilers in Kansas, Brown, with four of his own sons and three other men, kidnapped five proslavery settlers from his own neighborhood of Pottawatomie. None of these men had any involvement in the murders Brown was avenging. Near the end of May 1856, Brown and his men split open their prisoners' skulls with broadswords. Despite the savagery of this massacre, it went unpunished.[25]

Several years later, Brown organized another guerrilla band. He hoped to invade the southern Appalachians and attract large numbers of renegade slaves to his ranks. Although initially most abolitionists had adhered to nonviolence, violence as a means to end slavery was becoming increasingly acceptable, and Brown was backed by the "Secret Six," six abolitionists of wealth and rank.[26] Brown planned to attack an arsenal at Harper's Ferry, Virginia, and distribute its arms to slaves that would join his ranks. He met with disappointment in recruiting blacks for his band, and had only five black men and seventeen whites in his group. Despite grim chances for success, on October 16, 1859, Brown led eighteen of his men in an attack on the arsenal. Brown took the arsenal, but a company of United States Marines under the command of Colonel Robert E. Lee quickly quelled the insurrection. Brown and all of his men who survived the counter-attack were taken prisoner.[27]

Although Peirce was incensed by John Brown's raid on Harper's Ferry, he hoped Brown's radical actions would awaken the country to the dangers of the abolitionists. To Peirce, such excesses must surely mark the end of the abolitionist movement:

What say you in South Carolina to the Harpers Ferry insurrection. We think here that it will have a good effect to wake up the good sense of the North, which has been strangely dozing for the last few years, and expect it will be a heavy blow to the republican party. It would be absurd to suppose that any of the leaders of that party were privy to this special design. But it is apparent that the abolitionists, who have always been more or less associated with them and have held a rod of control over them, have for a long time been stimulating just such enterprises, and that Brown had received means for the purpose of such damnable treachery. It is true that the details of the designs were reserved to the abolitionists themselves, and that the republican leaders were aware of the "impossibility of their being consulted in relation to them." But it is the old story of the thieves, one of whom did not do the stealing, and the other did not do the keeping of the stolen goods. They are a pack of knaves all of them, and the air of the North is rampant with their abomination. As to Brown himself he shows a pluck, which his backers have not got—and would that he could hand over a half dozen that I could name as substitutes for him—men of white-livered-courage.[28] The abolitionists have spread their vile principles all about, and the republican party has endorsed them. Both are guilty of the successes of their principles. But how greatly they have misunderstood the Southern slave! They seemed to think that if you could only put arms into the hands of the blacks—the work was done, and the matter would be made at once to disappear from the face of the earth. How little they know the fidelity of these cherished servitudes! to whose very feelings I have observed that the Southern gentlemen were so nobly considerate.[29]

Peirce later wrote: "But how can the abolition's leaders be looked upon here after as anything but murderers? Stimulators of murder, assassination, robbery and treason? That devil of a Wendell Phillips has been permitted to address an audience in Brooklyn since the affair, in which he has had the impudence to boast of it and to call it the 'beginning of the end.' That it will be the end of all such beginnings & fini I believe. I feel, though, that the northern atmosphere has been greatly cleansed by this little thunder storm, and that the effect of it is upon the whole, beneficial with us. At any rate, it has killed Seward and Republicanism."[30]

Peirce soon realized that Brown's raid would not mark an end to the abolitionists. He was outraged; given the events in the North, to Peirce it was once again time to think of moving south:

When Wendell Phillips said the other day in Brooklyn, N. Y. that Brown was about to die for the truth, and that eighteen centuries ago another had died with great glory for the truth, was it not blasphemy? And was not every cheer which was given him an insult to our divine religion? Was it not the same when Mr. R. W. Emerson said at the meeting of a society in Boston, the other day, that if Brown should be hung, it would make a scaffold as glorious as the cross. Oh! That I had been present! I would have stood up and hissed till I had been dragged from the crowd, even if it had been as a corpse. I am ready today or any day to sacrifice my life upon the altar of my country, in this holy cause. Let there be fanatics upon the other side—There shall be a fanatic on this side also. I am grieved most of all to hear that my classmate Oliver Wendell Holmes, the author of the "Autocrat of the Breakfast Table" is uttering in private these same horrible sentiments with Phillips and Emerson. Let us wait in patience, however, for what the day is to bring forth. I am now fully prepared to look disunion in the face as a most probable necessity—but then I <u>must not be left</u> on the Northern side of the line. Given the Harper's Ferry affair, I have talked in confidence with two of our wealthiest citizens, worth not less than half a million of dollars; and both of them assert that if worst turns to the worst, they will go to the South—and one of them distinctly said to me "Come, if you will go with me, I am leaving." Oh! The North does not know what it is doing, and what must be the harvest. I believe that in the case of dissolution the number of valuable citizens and the amount of wealth, which will emigrate south, will astonish every nod; I find that the merchants are well aware that the advantage is with you. But as to myself, I am influenced by no pecuniary motive. My heart yearns for the South.[31]

Φⁿ

As the country came closer to war, the Republicans, who had largely absorbed the old Whig Party, became increasingly vocal against slavery. The Democratic Party, at least in the North, was not so much proslavery, as pro-Union. It played upon the fears of having a free Negro race, which would compete with whites for jobs. The Democrats also played upon the fears many whites had of blacks becoming social equals, and having black men marry white girls. Peirce was decidedly in the Democratic camp. He wrote, "Now when the Republican Party has been showed their <u>black</u> powder of discord and dissension all over their country, so that it strained the powers of the Democratic Party to the utmost to preserve our glorious union from a fatal explosion, . . . so that the union may never again be exposed to this danger from firebrands."[32]

Peirce also became increasingly involved politically. He reported to Bache that he and Joseph Winlock had been to an abolitionist meeting. Benjamin wrote, "We have seen Washington treated with contempt, the American God unauthenticated, and the Supreme Being brought to the bar of judgment and condemned for having destined man for eternal slavery."[33]

But Peirce was not content to just go and listen at political rallies; he began to take an active part in them. Later that year he presided at a Democratic meeting. His political debut was successful; one gentleman told him, "I profess to be a judge of such matters, and I regard Professor Peirce's speech as the best political speech that I ever heard." "However it did no good," Benjamin lamented, "We were beaten and still to be represented by that goose of a Burlingame ."[34]

John Le Conte was quite pleased with his friend's foray into politics. But he warned, "In most cases, it is very hazardous for Scientific men to touch politics. They are almost invariably soiled by the contaminating contact."[35]

Benjamin remained active politically, serving as one of the vice presidents of a Union meeting in Faneuil Hall,[36] where "there was no lack of plain out speaking against the enemies of the country."[37] Yet Benjamin doubted if there was enough patriotic fervor even here to put out the fires of abolition. Nevertheless, he was dedicated to serving his country: "There was a Benjamin Peirce of Salem who was the only man of that time shot at the battle of Lexington. He was my father's uncle, and through my father his name has descended upon me. I feel that I shall be true to it and that I shall be able to sacrifice my life too upon the altar of my country. Oh! My dear Professor, the times are at hand, which will try men's souls—and God grant that I may not be found wanting when the occasion comes."[38]

$$\Phi^n$$

Benjamin's strong political views seeped into his professional life. When asked to lecture at the Smithsonian, Benjamin was reluctant to go at all. Yet the Chief had requested that he deliver the lectures, and his wife assured him that it would do him good. So Peirce decided to give four lectures, "two upon the various powers for mathematics of the different races and nationalities of man, and two upon comets."[39] Warming to the assignment, Peirce wrote, "Perhaps I shall do them good for I shall lecture, if I go, upon the 'Diversities in Mathematical Power of the Various Races and Nationalities of Man.' Who ever heard of a nigger mathematician? He might do for a black board."[40]

Peirce explained to his son Charlie, "I specially spoke of the fact that no African had ever yet become a mathematician. This can not be from want of opportunity for they have been in the most intimate relations with the Egyptians, Carthaginians and Romans in ancient times, with the Arabs in the Middle Ages, and in modern times in our own or other country."[41]

Peirce made it clear, however, that he did not wish to embarrass "our poor timid giant of science, the venerable Smithson[42] in Party strifes."[43] Peirce was to make a careful distinction; he was to discuss the diversity, "not <u>the inferiority</u> of the races."[44] Yet Peirce would point out "that no man of the African Races has ever shown himself capable of any advance in the mathematical <u>sciences.</u>"[45]

While in Washington, Peirce was especially well received by the Southern gentlemen. The lectures drew excellent audiences, especially the ones on comets.[46] Peirce was in the Senate "when Douglas made his last bid for the presidency." Peirce surmised, "I am rather inclined to the opinion that he will be nominated at the Charleston convention, in which case he will surely be elected."[47] Peirce was also introduced upon the floor of the Senate by Senator Pryor of Virginia, a great honor.[48]

$$\Phi^n$$

Sarah became seriously ill in the spring of 1860. Her mother was living with them at the time and she also took sick, so that the whole care of Sarah and the family rested upon Benjamin and his sister Elizabeth. According to Elizabeth, "Ben fortunately excels in nursing almost if not quite as much as in mathematics & always seemed to do just the right thing."[49] But it was a touch-and-go affair: "One day, Sarah gave me in the coolest way, full directions what I should do when she was dead. Fortunately her eyes were so closed with disease, that they could not witness my conditions and I was able to command my voice, when it was necessary to speak."[50]

Fortunately Benjamin was able to find the right antidote. He wrote to Josephine:

What a terrible fortnight it has been! The Φ^{na} [Sarah Peirce] has been at Death's door with encephalitis in the head. It is a frightful disease, with . . . its horrid distortion of the face—But how the Queen is always the blessing of her friends—In the instant of greatest danger when her pulse was hardly sensible, and the doctor was anxious lest she was right in her assertion that she was fast sinking out of the world—it was the Queen's glass of hock which saved her. There was nothing else in the house which she could take and it was long past midnight when she first began to recover glimmers of reason, the Queen's letter came, and <u>we both bless it</u> and Sarah's heart was never so warm toward the Queen whom she calls "the sweetest creature in the world." And wanted to send me at once to town to see a new photograph of [her] which has just been taken and send it to the beloved queen. As soon as I can get out, I shall most religiously perform her wish.[51]

At this time, Benjamin himself had an accident and became seriously ill:

Sarah & Ben are really better at last, but they have been so sick that I had not the heart to write & besides I have been there every moment I could spare from home duties & occupations. The doctor says Ben must go to Europe, therefore he is going. Yesterday the Corporation had a meeting & decided that he should have leave of absence for six months; his foot is still lame, & he has to keep it very quiet—but the doctor says he wants him to go away as soon as his foot is well. When Sarah was very ill, a piece of coal fell on his foot; he did not pay any attention to it at the time; but when Sarah began to get well, his foot began to pain him; the Doctor said that the nail would come off & then the toe would soon be well, which all happened just as the Doctor said. Then Ben walked about too much—he went into Boston last Friday & when he came out I observed that he looked very pale & he complained of pain in his foot; the pain increased till it became so violent that he sent for the Doctor; all that night he had to keep inhaling ether, besides taking a great deal of laudanum; the next day the doctor ordered leaches, as the foot was swollen; to day he is much better. It is now more than thirty years, that Ben has been hard at work—& the Doctor says he absolutely requires a rest.[52]

The doctor prescribed a trip to Europe. Furthermore, Peirce was invited to attend the meeting of the British Association for the Advancement of Science. His son Charles agreed with the doctor in thinking that the trip would do him good. He wrote, "How anyone so fond of traveling should persist in refusing all invitations to Europe I really can't see."[53] Benjamin did decide to go to Europe: "The Corporation of the college have taken the matter in hand and given me (without being asked) a furlough of six months, and have appointed James to be tutor, and perform his father's duties (at his father's expense for the college is too poor to pay). Some of my friends have offered me the aid of five hundred dollars and I am going to Europe! I shall leave my native land sometime in the last week of May, and shall attend the meeting of the British Association in Oxford. I shall then go to Paris and after a few weeks there, expect to find myself reshipped for home in restored health and vigour."[54]

Bache furnished Peirce with letters of introduction to European scientists,[55] and went to New York to see his friend off.[56] At first Benjamin found the voyage on the *Amazon* exhilarating, "Here we are half way across the broad Atlantic running away from home at the rate of nine miles an hour under full sail, rather close to the wind.

It is the most magnificent merchant ship that sails from New York, our Amazon, and we have got greatly attached to the noble animal. . . . The beautiful ship rides the waves like a swan."[57] Peirce's nautical heritage did not serve him well, and despite calm seas, he was seasick for over a week. Nevertheless, after recovering, his spirits were high, "But I now feel so well, that I thank the dear Doctor with all my heart for his sending me upon this sea voyage. I believe that it has saved me from severe disease, and that if I were now at home my Dyspepsia would have been worse than any attack of illness, which ever befell me."[58]

By the time he reached shore, however, he was weary of sea travel, "You cannot think how happy we were to step again on Terra Firma. I dread the thought of again getting on board a ship—even for two hours, to cross the Channel from Dover to Calais—and nothing but home would be inducement enough for me again to undertake the passage of the vile sea, which separates me from my wife and children. . . . I hate the sea; I was never intended for it."[59]

Peirce found the train ride to London delightful; he found all of England "to be a garden from end to end."[60] Peirce was no less pleased with Oxford, "This venerable and aristocratic University, imposes upon you at first sight, but the more you see it, the more you are impressed with its ponderous architecture, and the weight of the centuries, which have piled up their learning in its vast stone houses."[61]

Peirce stayed at Exeter College while he attended the meetings of the British Association for the Advancement of Science. He was excited to see His Royal Highness, Prince Albert, who opened the meeting. He enjoyed the food, found the waiters "more supurb in their bearing than Abbott Lawrence," and was pleased to learn that American scientists were "a great deal better known" there than he had supposed. He renewed his friendship with James J. Sylvester, and met other British scientists who "treated him as a distinguished guest."[62] He was particularly gratified that the British scientists thought well of Bache. He gave several papers at the meeting, which were well attended and well received.[63]

Peirce then went on to London, where he enjoyed the opera and the city's parks, finding them far superior to city parks in the United States.[64] Quite by accident, he ran into Michael Faraday, the English physicist who did pioneering work in electricity and magnetism. Peirce found him "a most pleasant and great man."[65]

Upon crossing the English Channel from Dover to Calais, Peirce was pleased to report that it was "wonderful to tell I was not sick." While staying in Paris, he visited the garden of the Tuileries, Versailles, and the Louvre. He liked France even better than England. "I found Paris grander and more superb than my most magnificent castles in the air,"[66] but he missed Sarah and his family. He wrote, "Oh that Sarah could be here to admire these elegant things with me!" Although Peirce was homesick, he saw many people; he visited his friends the Greenoughs. Lillie Greenough was engaged to marry Charles Moulton, son of a prominent American banker who lived in Paris. Peirce visited the Moulton's estate and was very impressed with it. On one visit to the Greenoughs, Peirce recorded, "In the evening Lillie Greenough sang and she is the wonder of the world. Every body here is infatuated about her music. It is really a fact and all that her mother boasts of her is true. The Moultons seem quite to idolize her. Her voice has acquired every good quality which is most admired in the great singers, and wholly free from defects."[67]

Peirce also attended several meetings of the Academy of Sciences. There he met the mathematician, Joseph Bertrand, and the mathematical astronomer, Charles-Eugene Delaunay. He prepared an abstract of a paper on comets for the French Academy of Sciences and was pleased to hear that it would be published shortly in the *Comptes Rendus*.[68]

From France, Peirce proceeded to Switzerland. Although he enjoyed seeing the sights, he remained homesick. Seeing Europe impressed upon him his love of his own country; he wrote: "Oh America is the land of freedom! It is the only country in the world where the mind of man can hope to expand to its full powers. In less than half a century if civil dissension does not ruin our glorious country, we shall look down with wonder on the ignorance of Europe and the diminutive stature of its intellect."[69]

Unfortunately, Peirce also suffered from seasickness on the trip home. He wrote shortly after returning, "I hate the sea and when I see that it covers two thirds of the surface of the globe—I cannot help doubting whether the earth was actually intended for man. Poor fish! How I pity them! They must bless the fisherman—for, to get upon land—even though it be to die—must be a blessing."[70]

Φ^n

Peirce's trip brought new vigor to his body and boosted his patriotism, but his country's situation was even more perilous when he returned than when he left. John Le Conte wrote frequently, defending the South's point of view:

But all candid minds must admit, that the threats of South have not been arrogant or <u>offensive;</u> she merely threatens to withdraw from a confederation in which her <u>rights of property</u>, secured by the Constitution, are not practically recognized. . . .

God grant that we may be spared the horrors of civil war.[71]

Writing again, he said:

In relation to the prospects of our country remaining united, it seems to me that they are gloomy and forebode nothing but revolution. . . .

The truth is, the people have borne injustice and aggression as long as they could be endured, and the conservative men are now the <u>leaders</u> in the secession movements. . . .

There is not the slightest doubt, . . . abolition emissaries have been clandestinely attempting to alienate our slaves, and peddlers and other Northern men of no character have been furnishing poison to them for the most fiendish purposes.[72]

Peirce replied that although he saw great danger of the country splitting apart, he was not without optimism. He reported that even Republicans in Massachusetts were now declaring, "Within the last week I have heard men of the Republican party speak of the duties of the north, as I never heard them before. They declare that the fugitive slave law should be carried out in good faith, the personal liberty laws repealed, and that there should be no impediment for the south to carry slaves into the territories. I was amazed to hear such doctrines from Republican lips—but the men who spoke were those who have a real influence in the affairs of their party, and are not likely to say what they do not mean!"[73]

11

Every Drop of Blood

JUST BEFORE THE WAR, JOHN LE CONTE WROTE TO PEIRCE, DESCRIBING the beauty of springtime in Columbia, South Carolina, where he lived: "You cannot imagine what a beautiful appearance our little town presents at this time. The thousands of Water oaks, Willow oaks, and Elms, which fringe our streets, are now in the verdure and freshness of their fully developed foliage; while numerous gardens, attached to the private residences, are completely embowered with roses and other flowers."[1] By the next spring the Le Contes were comfortably settled in their new house where they looked forward to entertaining their Northern friends.[2]

Although initially Le Conte had opposed the South's secession, by early 1861 he fully affirmed it.[3] "The Formation of a Southern Confederacy is now a <u>certainty</u>,"[4] he wrote to Peirce. Le Conte was hopeful, even exultant, about the Confederacy: "It is true, Fort Sumter is not yet in our possession:—but we shall lose nothing by forbearance, while it <u>will</u> and <u>can</u> be taken whenever it is deemed necessary. The preparations for a successful assault are so complete that we hope Maj. Anderson will not tarnish his military reputation by ultimately surrendering the fort, and thereby avoiding the effusion of blood. If otherwise, its capture will probably cost the lives of his whole command, as well as those of 200 or 300 of our best citizens:—but it will be taken whatever the cost."

Le Conte saw the formation of the Confederacy as the potential fulfillment of not only his political, but professional dreams. He added: "We hope that the Southern Congress which meets at Montgomery Ala. on tomorrow will organize a provisional Southern Confederacy on the 22nd of February; and that but a short time will elapse before <u>all</u> the slave holding states will be with us. As soon as this is consummated, it will give an immense impulse to the cause of the great University of the South. The Queen says she expects to live to see the "good Chief" as Chancellor of the great University with his true and faithful friends around him!"[5]

Despite the cataclysmic political events, Le Conte foresaw no reason why his personal relationships with Northern scientists should change, "What say our Northern friends in relation to attending the Meeting of the "Scientific Association" at Nashville next April? . . . Science is <u>universal</u>—she is not circumscribed by geographical lines." A few months later Le Conte realized that he had been overly optimistic about scientific meetings proceeding as usual in the divided country. Now anticipating a cancellation of the Nashville meeting, he wrote Peirce, "I sincerely hope, that when matters become amicably adjusted between the two American Confederacies, the more catholic spirit of science will maintain the <u>unity</u> of the American Association for the Advancement of Science."[6]

Le Conte was concerned about how the rupture of the nation might affect national scientific institutions and scientists associated with them, like the Coast Survey and Bache, but he remained overwhelmingly pleased with the course of events: "Northern people have been perversely and persistently <u>blind</u> to the inevitable result of their systematic crusade against our system of slavery. It is extremely doubtful, whether many of them are, even at this time convinced, that our people are <u>seriously in earnest</u> in this matter. . . . I suppose we will have to abandon the idea of seeing you this month. This is a most sad disappointment to all of us. It seems almost an age since we have seen each other."[7]

On April 12, 1861, ten days after Le Conte wrote this letter, the Confederacy fired upon Fort Sumter. It would be many years before John would see his friend again. Le Conte had no way of knowing this, of course, and his excitement was not deterred. His brother Joseph's daughter, Emma, recorded the family's feelings when they heard of the fall of Fort Sumter: "The joy—the excitement—how well I remember it. For weeks we had been in a fever of excitement. On the day the news came of the fall of Sumter we were all sitting in the library of Uncle John's. The bell commenced to ring. At the first tap we knew the joyful tidings had come. Father and Uncle John made a dash for their hats—Jule and Johnny followed. We women ran trembling to the veranda—to the front gate eagerly asking the news of passers-by. The whole town was in a tumult."[8]

Φⁿ

News of the war troubled Peirce. "Our hearts are sad at the fall of Sumter," he wrote Bache.[9] A day later, after receiving a telegram informing him that the Baches would have to cancel their visit to Cambridge, Peirce wrote, "We were depending upon your visit and that of the dear Chiefess to feel that in this reign of hate, there was still some love left in the world. The capacity of this young blood for intense hatred is more wondrous than its power of love; and the whole community has wheeled into line with such suddenness. The God of War!"[10]

Bache was saddened by the war: "What a mess! But for Country—alas—alas—alas! And things are at such a pass that liberty, life, & property are all feared at the hazard of war—Civil war! God help us and save us!" He added that he was irritated by the letters that John Le Conte had been writing him; "He indulges in politics & expresses sentiments with which he should have felt I could have no sympathy."[11]

Although Peirce was unhappy about his country being at war, he immediately shed his Southern sympathies and became fiercely loyal to the Union.[12] When his sister complained of the New England mosquitoes, "Ben proposed that Mother should catch all she can & pack them up to send against the rebels; also flies."[13]

Less than three weeks after the South began firing on Fort Sumter, a somber Peirce wrote Bache: "All Massachusetts is ablaze for the war, and there is no resisting the current which is as deep as it is swift. All men are ready to embark in the service of the government, lives and property and all which they hold most dear. There is a selflessness and generosity of patriotism which is truly heroic, and the individual excitement is less noticeable than the calm and resolute determination to devote everything to the service of the country. No man withholds his hand and his heart from the cause—and the government need not fear to exhaust the heart of the state as long as it has one drop of blood or property which can be shed in the war."

He exclaims his change of heart when he adds:

> God save the country! And we may be thankful that General Scott is in command and not some visionary civilian like the professor of Mathematics in Harvard College.
> I have a sty in eye and not pigs in it.
> Goodbye my beloved Chief—If I can be of any service, I belong wholly to the country; and you may use me as you please.[14]

South Carolina's attack on Fort Sumter greatly affected the attitudes of all the Peirce family, but initially they were confident that the war would be of short duration. Benjamin's sister, Elizabeth wrote, "Ben thinks that Scott is a truly wonderful man & is the greatest general in the world. I consider him a sort of Moses & have a notion that he has been prepared, educated & kept for us just as Moses was for the Israelites. . . . I understand that Scott says that the war will be all over by the first of next May."[15] After the Union loss at the first battle of Bull Run, however, Benjamin, like many Americans, realized the war would be long and bloody. Elizabeth wrote, "Aunt Saunders said that there was something in yesterday's paper about peace & asked if he [Benjamin] thought we could have peace soon. Ben said, not at present—especially now that we have just lost a battle."[16]

Elizabeth was against slavery: "I hope, dear Aunt, that there will come an end to slavery in our country. I consider it a great evil, but I think we must have faith with regard to this as to all other evils—& leave them in the hands of our all-wise Ruler. I must say too that in my opinion the South have behaved in a most mean, wicked & awful manner. Especially would I like to get hold of Floyd Toombs & Co. The Cheats! & Pickpockets."[17]

She, as most Americans, remained a racist, however: "As to the blacks—they <u>can not</u> mingle with the whites—they must divide apart from them—& if the reign of love, of Christianity is only established amongst us & we will have patience, all the evils of slavery & of infidelity will disappear from the North & South. We <u>must</u> love each other, love even our enemies so Christ says. If all your people, dear Aunty & Uncle & Cousins were like you & my darling brother & had their hearts full of <u>love,</u> the questions that now disturb our country would soon be settled. If I were a poet I would write a National Hymn & every other line should end in "Love." This beautiful land—so smooth & convenient for passing to & fro—was meant to be inhabited by <u>one</u> loving nation of Christians & the blacks could live on the islands."[18]

Even Benjamin's attitude toward slavery changed, as his sister observed: "Ben talked about the war & I wish you could have heard him. He is firm for supporting the government—he says he thinks that we should all unite steadily. I asked him if he should vote for Andrews if he were put up again for governor? He said Yes. He seems to be of the opinion that if a slave owner is a rebel his slaves at once must become free & being free once—can <u>never</u> become slaves again. He says, that the more united we are at the North—& the more convinced the South are that we are united—the more quickly all these troubles will be settled. As I said at the beginning of this war—We are fighting for our sovereign The Constitution—And as to Slavery—Cotton, & c. God is all-wise—all-powerful & good—We will therefore trust to Him."[19]

Φⁿ

Less than a month after the fall of Fort Sumter, Peirce's former student, Charles Russell Lowell, approached Peirce, asking him if he would ask Bache to help him acquire a commission, preferably in the artillery. He was becoming impatient with normal channels. Peirce wrote to Bache, "When such men as he are ready to offer themselves to the public service, I must believe that the country is safe, and that treason will fail of its end."[20] That Peirce would respond so favorably to a man who himself, and whose family, had been such strong abolitionists, is additional evidence of his change of heart.[21] For example, before the war, one of Peirce's former students, an abolitionist, had attempted to rescue a fugitive slave. He told Peirce that if he were imprisoned for his actions, which he feared he would be, he would have time to read, Laplace's *Mécanique céleste*. Peirce responded that in that case, he sincerely hoped that he would be.[22]

Lowell hardly required a recommendation from Bache.[23] He was quite capable of recommending himself. He arrogantly wrote to Senator Sumner: "I speak and write English, French, and Italian, and read German and Spanish; knew once enough of mathematics to put me at the head of my class in Harvard, though now I may need a little rubbing up; am a tolerable proficient with the small sword and the single-stick; and can ride a horse as far and bring him in as fresh as any other man. I am twenty-six years of age, and believe that I possess more or less of that moral courage about taking responsibility which seems at present to be found only in Southern officers."[24]

$$\Phi^n$$

None of Peirce's sons served in the war. At its outbreak, James and Charles were both in their twenties; Benjamin Mills was seventeen, certainly old enough to get into the action before the war was over. Only Herbert was too young to serve. Charles, who ignored his father's advice to stick with his chemistry,[25] took a position with the Coast Survey as a computer for his father's Pleiades longitude project.[26] When the draft became a threat in 1862, Charles wrote to Bache:

> Does my appointment in the C. S. service exempt me from draft or not? I have not had any letter of appointment so I am in doubt. . . .

> This town has just raised a full company, tho' considerably above its quota. But I perfectly dread going. I should feel that I was ended & thrown away for nothing.[27]

Bache replied that he was exempted.[28]

Only her age and sex kept Charles's Aunt Elizabeth from serving. "I wish I could go myself & help fight. Well if the women cannot fight they can make songs as Mrs. Sparks has done."[29] Sarah, too, was doing her part for the war effort; her sister-in-law noted, "She is busy knitting for the soldiers & so are most of the Cambridge Ladies."[30]

Peirce strongly supported the government, and served as a delegate to a political convention in Worcester. Although still a Democrat, he now strongly supported Lincoln's Republican government.[31] A year later he wrote:

> Have you seen the action of our Republican convention? And the failure of Dana's attempt to get them to agree to aid the President in his patriotic attempts to save the Union without making conditions. They would not do it. . . .

> I cannot speak for all Massachusetts but I am sure that Dana fairly represents the portion of his party which is in Boston and its vicinity. I am not one of them—but I can bear testimony to their sincere loyalty, and that they (the last requisite of the Republicans here) are willing to make any sacrifice to the union—and of course all conservatives not in the Republican ranks are ready to do the same.[32]

Typically, Peirce was not beyond going to extremes in placing the ultimate blame for the catastrophe of war: "God bless the dear Chief and Chiefess, and carry them in safety through these terrible times of trial and affliction. But after all, dear Chief the country has been so false and rotten that these times are needed to slough off the mortifying and dead parts. The state of all our scientific bodies, academies and societies, observatories, etc., etc., all are due to falsehood which pervades society."[33]

During the course of the war, Peirce regularly discussed matters with his sister when he went there for breakfast, as he often did. His daughter, Helen, said that she saw her "father and Agassiz talking over some bad news from the front during the War of Rebellion with tears running down their cheeks."[34]

Φ[n]

Even before the beginning of hostilities, Bache foresaw that a civil war might "sweep away our organization, or sadly cripple it."[35] In 1861 and early 1862, Congress attempted to suspend the Coast Survey, objecting that it was too expensive a luxury for wartime, and that its cessation would free its employees for military service.[36] Peirce immediately sought political support for the Survey from Agassiz, Gould, Felton, and Ingersoll Bowditch. Bowditch "spoke of the national disgrace it would be to give up the support of a fundamental portion of the government in the face of a rebellion. What greater proof of the power of the rebellion?" Peirce's other friends stressed the military importance of the Survey.[37]

Bache, himself, quickly took steps to make the Coast Survey useful to the military. He increased the production of maps of the Southern coasts from extensive data already collected. With the South's destruction of lighthouses, a naval blockade would have been impossible without these maps.[38]

Bache had already demonstrated the use of the Coast Survey to the war effort, however. As an editorial in the *New York Times* pointed out:

> At every step of our naval operations, we get new demonstrations of the immense utility of that great National undertaking, the United States Coastal Survey. . . .
>
> It is not, we think, too much to say, that without the data which the Coast Survey has supplied, it would have been impossible for the Government to have undertaken the system of coast operations, which have already been so brilliantly inaugurated. . . . And the neatest thing of it all was, that while the whole Southern coast had been surveyed, only a small portion of the results—and they mere primary generalities—had been published up to the time of the breaking out of hostilities. . . .
>
> If the Coastal Survey should render no further service than that which it had already rendered, it would still have repaid tenfold all it has cost the nation.[39]

The bill to suspend the activities of the Coast Survey was defeated by a three-to-one margin.[40] Bache's adroit administrative and political skill saved the Coast Survey from its enemies.

Bache did not confine himself to the Coast Survey, but took on other responsibilities related to the war, including work on the Permanent Commission of the Navy and on the Sanitary Commission. He was also one of a four-man board that planned the coastal blockade. During Lee's campaign in the summer of 1863 Bache worked eighteen hours a day planning the defense of Philadelphia. Peirce warned him that he was overextending himself and that he was worried about his health. Although Bache was aware that his Herculean workload was affecting his health, he found it hard to moderate his work, and his health continued to decline.[41]

$$\Phi^n$$

As terrible as the war was, there were other sources of tragedy that touched Peirce and the Cambridge community. On July 9, 1861,

Henry Wadsworth Longfellow's wife, Fanny Appleton, was sealing up cuttings of her children's hair when her dress caught fire. Flames quickly engulfed her. "She ran from her children that they might not be burned, into the front room where Longfellow was sitting." Hearing her screams, he attempted to extinguish the fire himself with a carpet, but it was "too brittle" and Longfellow failed to save his wife's life. He only succeeded in badly burning himself. At Longfellow's request, Fanny's funeral was held the next Saturday, the anniversary of their wedding day. Longfellow was too badly burned to leave his bed and listened to the funeral service through an open doorway. Peirce wrote: "Oh this dreadful tragedy! How the fire is burning into the heart of our Cambridge! Mrs. Longfellow was so beautiful that you cannot understand how the fire dared to burn her. But it spared her face, and I am told that she is looking this morning most beautiful in death."[42]

This was the second wife that Longfellow had lost, his first wife having died from complications due to a miscarriage. He never fully recovered from this second tragedy.[43]

$$\Phi^n$$

Peirce was involved with the war only in minor ways, such as doing work for the Sanitary Commission and suggesting plans for the defense of Boston harbor.[44] He continued to fight valiantly for the advancement of American science. Several of his battles involved a former student, Charles W. Eliot, who entered Harvard in 1849. Eliot was from a prominent Boston family. His grandfather was perhaps the richest Bostonian of his day; his uncle was Harvard's treasurer; and two of his uncles by marriage, George Ticknor and Andrews Norton, were former Harvard professors and leading American scholars.[45] Although, as a student at Harvard, Eliot's principal academic interest was chemistry, he studied mathematics with Peirce, bravely opting to take the elective courses. During one class, Eliot remarked that what Peirce had said about functions and infinitesimals seemed to him to be "theories or imaginations rather than facts or realities." Peirce looked at him gravely and said gently, "Eliot, your trouble is that your mind has a skeptical turn. Be on your guard against that tendency or it will hurt your career." This surprised Eliot, but on reflection, he agreed that Peirce was correct.[46]

A year after graduating from college, Eliot became a tutor in mathematics at Harvard.[47] In 1858 he was made assistant professor of chemistry, and, when Eben N. Horsford, the Rumford professor, relin-

Charles William Eliot, ca. 1865. By permission of the Harvard University Archives, call # HUP [20a].

quished supervision of the chemistry laboratory in the Lawrence Scientific School, Eliot took it over. But it was with administrative duties, not chemistry, that Eliot truly sparkled. He quickly assumed many such tasks, and he became acting dean in 1862, when Henry L. Eustis, Dean of the Lawrence Scientific School, left to serve as a colonel in the Tenth Massachusetts Volunteers.[48]

Eliot moved at once to bring more structure to the teaching in the Lawrence School and to introduce general studies. Much of the teaching at the Lawrence School, especially Agassiz's, was often informal. Agassiz was convinced that students should learn natural history from experience, and that lectures and close supervision were obsolete. Joseph Le Conte, one of Agassiz's first students, was surprised that Agassiz had no interest in his previous education and that he seemed to have little interest in what he was doing. Agassiz gave Le Conte a collection of a thousand shells and told him to separate them into species. When Le Conte had finished the task, his instructor informed him that he had made a significant improvement over the existing authority on this subject. Agassiz's students had a deep veneration for their teacher, who was giving them "the most valuable education in natural history yet offered in the United States."[49]

Although Agassiz's students revered him for his teaching style, Eliot abhorred it. He was also concerned that it took only one year to complete a degree. A committee was formed to make suggestions for improvements in the Lawrence School. The proposed plan, largely Eliot's, would require written examinations, two years of general studies before embarking on specialized work, and four years to complete a degree. Although Agassiz, himself, was chair of the committee making the report, he quickly opposed the proposal. Peirce joined with him. Both of them feared that such innovations would deter Lawrence's traditional attraction of offering the opportunity to study with a single, outstanding professor. Many of the students came to the Lawrence School with college degrees and extensive knowledge; they were seeking advanced training. To Peirce and Agassiz, Eliot's proposals were aimed at strengthening undergraduate, not graduate, education and as such, were a move away from the university ideal.[50]

Peirce no doubt considered himself to be the nation's leading mathematician. Agassiz certainly considered himself to be the nation's, and perhaps the world's, leading naturalist. Criticism from a young man of twenty-eight, with no scientific distinction, did not sit well. As an alternative to Eliot's innovations, Agassiz suggested the establishment of University Lectures, where Harvard professors and visiting experts would offer series of advanced lectures. Agassiz and Peirce felt that this was a step towards a true university. Although Agassiz's and Eliot's plans were certainly not mutually exclusive, President Hill sided with Agassiz and Peirce, and shelved Eliot's plans.[51]

Φn

Thomas Hill was Peirce's former student and loyal friend. He was inaugurated president of Harvard in April 1862. While a student at Harvard, Hill had distinguished himself in science and mathematics. But after graduating from Harvard, he entered Harvard Divinity School and became a minister. Although a successful minister, he never gave up his scientific interests.

Agassiz and Peirce campaigned for his appointment as president, and his inauguration was for them a day of celebration. Peirce invited all his classmates to his house for a banquet on that day. His sister, Elizabeth, said, "Being a very uncommon occasion, I wanted to contribute my share to the entertainment—& so I told Ben I should make some jelly & some cheese-cakes for him."[52]

Peirce wrote to Bache: "I shall not be surprised to find Hill getting us into some exciting discussions about modes and plans of education, and advocating changes favorable to science at the expense of the ancient languages. Such changes I shall not totally approve, but nevertheless I think that the greatest good will come from their discussion, and the advocacy of so powerful a mind as Hill's must give us much new light upon the subject, and will now lead to serious improvement in the scientific configuration of the college."[53]

Peirce liked Hill's plans for the college: "Our new president told me yesterday, that his most earnest desire was to associate with the college every man in the country of <u>first rate</u> scientific and literary ability, who could be got. He proposed to establish three grades of professorships, regular, honorary, and associate—and will also have lectureships and privat-dozents. We want you here to consult with us in these great matters."[54]

Although Hill proved to be a weak president and was ineffective in realizing the reforms both he and Peirce desired for Harvard,[55] Peirce did not become disappointed with him: "Our new president Hill is more after my own heart than I had known him—and I hope that his administration will last beyond my day. His aspirations towards the scientific reforms of the University are all in the right direction. And mum be the word—but everything looks as if Wolcott would now be secured for us and perhaps Gould and perhaps Chauventet. I fear that G.[eorge] P. B.[ond] is playing opossum about his consumptive attacks—but if ever the devil should take his own from the mighty watches of the stars—I believe that there will be a reform in the observatory. But heaven forbid that I should wish harm even to the devil's own!"[56] Peirce gave Hill the highest praise possible when he wrote, "I have been to talk with Hill this morning about Gibbs. . . . I think Hill worthy to be a Lazzaroni. He is high enough and it is not easy to get to the top of him—he is Alps on Alps. But the most delightful of his qualities is his incomparable sincerity."[57]

Φⁿ

Early in 1863, Peirce wrote to Bache, "<u>Mum</u> must be the word until he accepts. But the word is, Horsford has resigned and Wolcott is offered the Rumford Profship. Hurrah! Hurrah!"[58] The resignation of Horsford, the Rumford professor, gave Peirce and Agassiz an opportunity to move Harvard towards the university they envisioned it to be. Again the principal stumbling block in their efforts was Charles Eliot. Eliot's expansion of the chemical laboratory, his leadership in the Lawrence School, and his family connections, made him, in many people's minds, his own included, the natural successor of Horsford. In previous years he would have been a shoe-in.[59] Peirce feared, "Public opinion says that Eliot is to have the Rumford Professorship."[60] But times were changing, and the leading chemist in the nation was fellow Lazzaroni member Wolcott Gibbs. Gibbs had been trained in Europe and had an established research record. The Lazzaroni had failed to secure a professorship for him at Columbia a decade earlier; they had no intention of failing again. Eliot, although acknowledging Gibbs's preeminence as a chemist, still sought the Rumford professorship. He also pointed out that he had expanded the chemical laboratory sufficiently that another professor was needed to oversee it.[61] Rather than hire Gibbs, however, Eliot hoped the university would hire his friend, Francis Humphrey Storer.[62]

To Peirce, Gibbs's appointment to the Rumford Chair was of extreme importance, much more so than Eliot's proposed changes in the Lawrence School. Peirce lamented, "Oh! If we fail of Gibbs! The prospects of science in this vicinity is very dark in the next generation, and poor Gibbs, if he comes, will have to fight a giant's fight. Above all, there is such a total indifference to honor and honesty."[63] Becoming despondent about the Rumford professorship, Peirce wrote to Bache, "I am anxious about the Rumford Professorship. I fear that things are not going right. Suppose that the election is given to Eliot; what would you advise me to do? Is it worthwhile to be laboring to elevate the University standard, when we see that as soon as Agassiz and myself are gone, the whole thing must fall into the hands of imbeciles?"[64] Peirce's gloom was unusual; although he wrote to Bache repeatedly about the Rumford professorship, he usually was hopeful, but cautious; as in this more typical note: "And the Lazzaroni have carried the Corporation[,] and Gibbs has the election to the Rumford Professorship. His confirmation is all but certain, and nothing but the importance of it makes me feel any anxiety as to the result."[65] Peirce's anxiety was needless; Gibbs was appointed.

Hill was aware of Eliot's abilities and was anxious to make an accommodation for him. He first offered him the professorship connected with the chemical lab, at a reduced salary, but with the possibility of the difference being made up by fees from students using the laboratory. Eliot quickly rejected this second-class professorship. Hill then attempted to raise money for the professorship from wealthy members of Eliot's family, but Eliot did not want a professorship his family had bought for him. Wartime economic conditions prohibited Hill from raising endowment for a new chemistry professorship, and Eliot resigned.[66]

Peirce absolved himself of all blame in the matter. He wrote to Bache:

There is at present much feeling in Cambridge on account of what is thought to be the injustice done to Eliot. But I find that those who complain are ready to listen to a reasonable discussion of the subject and when they hear the whole case admit that it is not so bad as they thought—and they even think that some of Eliot's plans were inadmissible to the University. . . . I freely admit Eliot's administrative abilities and regret that he should leave the college, but maintain that it is his own act that removes him, and that at any rate Agassiz and myself are not accountable for it, for we were very careful not to exert any influence with the Corporation against Eliot, and that our advocacy of Gibbs was undertaken when we had no idea that Eliot was in the field. That after we knew of Eliot's earnest desire for the place we could not withdraw what we had said of Gibbs anymore than Eliot could take back the exalted admiration which he expressed for Gibbs when he was advocating Lovering as President in opposition to Hill, and Gibbs as successor to Lovering. But we said not another word to the President or any member of the Corporation about the appointment and that it was Eliot's own weak exertions, which had lost him the place.[67]

Although Eliot's father had earlier met with a business reversal and lost his wealth, a fortunate investment allowed Eliot to travel extensively in Europe. After returning, he wrote several influential articles on education for the *Atlantic Monthly* and took a professorship at the newly founded Massachusetts Institute of Technology.[68]

Φⁿ

In 1863, the Lazzaroni was engaged in another endeavor, which was on a national scale. As early as 1851, Bache had called for a scientific body patterned after the French Academy. This organization bestowed public recognition upon scientists, advised the government on scientific matters, provided governmental support for research, and issued publications. Bache believed that the time had now come to found a national academy that would serve the same purposes.[69]

The American Association for the Advancement of Science was open to anyone with scientific interests, without regard to their credentials. Peirce recognized that the all-inclusive AAAS was limited in establishing professional standards, and had long acknowledged the value of the kind of organization Bache proposed. "In general as to the Association [AAAS], it does much good. The positive is remembered and its negative is forgotten; but then we need the other society for science, and this for sociability in fair weather. But when the storms arise, we must have a frigate at hand in which the wise men may rest in safety, while the spumes go to the bottom."[70]

Bache never lost sight of this dream, which his friend Agassiz shared with a passion equal to his own. In the summer of 1858, Agassiz set down a specific plan for the organization and membership of such an organization.[71] By January 1863, Agassiz had found a crucial political ally in Senator Henry Wilson of Massachusetts.

One key member of the Lazzaroni opposed the measure. In early 1863, Bache asked Henry for his opinion of such an organization. Henry replied that he did not think such a proposal could be "passed with free discussion in the House." Even if it were passed, it would be impossible to "obtain appropriations to defray the necessary expenses of meeting and of the publication of the transactions." He also feared that "it would be a source of continued jealousy and bad feeling . . . on the part of those who were left out." And, it would "be perverted to the advancement of personal interests or to the support of partisan politics."[72]

Bache appeared to be convinced, and as an alternative to a national academy, Henry, Bache, and Davis then proposed the formation of a board to advise the government on the scientific questions that had been arising with respect to the war. As a result the Permanent Commission in the Navy Department was formed on February 11, 1863. This commission functioned until the end of the war. Peirce occasionally served as a consultant on this board.[73]

Bache and the other members of the Lazzaroni had no intention of heeding Henry's council, and excluded Henry from further discussions about the academy. Senator Wilson met with Agassiz, Bache, Davis, and Peirce at Bache's house in Washington, on February 19 to draft a bill. Both Davis and Bache had written preliminary drafts. Together they hammered out a single bill that evening.[74] The bill proposed establishing a group of scientists who could be called upon by the government, at no expense to the government, to give scientific advice. It also included a list of the fifty members who would comprise the academy.[75]

In the frenzy of the closing session on March 3, Wilson rose and proposed "to take up a bill, which, I think, will consume no time, and to which I hope there will be no opposition. It is a bill to incorporate the National Academy of Sciences. It will take but a moment, I think, and I should like to have it passed."[76] As there was no objection, Wilson explained that the first section of the bill was merely a list of the names of the incorporators, names of prominent scientists of the country. He then read the rest of the bill, which gave the incorporators the power to organize the academy and stated that their sole obligation was to give free scientific advice to the government.[77]

The bill passed by voice vote. Several hours later the House passed the bill, and President Lincoln signed it into law before midnight.[78]

Agassiz was exultant. He wrote to Bache, "Yes there is a National Academy of Sciences, and we may well rejoice. It inspires me to see how young you feel about it."[79] Agassiz saw the National Academy as a means of validating scientific credentials. He wrote to Bache a few months later: "It [the Academy] has already accomplished one great thing. We have a standard for scientific excellence, whatever our shortcomings may be. Hereafter a man will not pass for a Mathematician or a Geologist, etc. because an incompetent Board of Trustees or Corporation has given an appointment. He must be acknowledged as such by his peers, or aim at such an acknowledgement by his efforts and this aim must be the first aim of his prospects."[80]

Although many scientists were not happy with the clandestine way the Academy had been formed or with the list of incorporators, the initial meeting went well. Henry, now hoping to "give the association a proper direction and to remedy as far as possible the evils which may have been done," raised no opposition.[81] Bache was elected president; Dana, Agassiz, Wolcott Gibbs, and Fairman Rogers[82]—all sympathetic to the Lazzaroni—were the other officers. Peirce was chosen to be the chairman of one of the Academy's two classes, the class of mathematics and physics.[83] Later when Bache became ill in the spring of 1864, Peirce "was unanimously elected to preside for the meeting."[84]

The only opposition at the first meeting came from William Barton Rogers, and that did not amount to much. Peirce reported, "All things have gone to our most complete satisfaction in the National Academy, which is completely organized and adjourned till next January. . . . W. B. Rogers could not get up the least opposition and lost all his influence. All the indications are that the Academy will be a thorough success."[85]

Rogers was unhappy, nevertheless. The meeting had been held on the campus of New York University, where John William Draper was a professor. Rogers felt awkward in running into Draper, who had not been included among the fifty original incorporators, on the campus. Draper was a chemist with an international reputation that far exceeded that of all but a few of those who had been selected for the Academy. Also notably absent from the list were George P. Bond, director of the Harvard Observatory, and Spencer F. Baird, the naturalist, who was Henry's assistant at the Smithsonian.[86] It is not clear why Draper was excluded; perhaps the Lazzaroni feared his anticlerical writing would reflect poorly on the Academy.[87] Baird, though highly regarded by the country's naturalists, dealt primarily with the classification and identification of species. Agassiz in particular felt he represented the old and amateurish school of American natural science and was vehemently opposed to including him. Bache, Gould, and Peirce were enemies of Bond. Bond had failed to support Gould and the Lazzaroni during the Dudley controversy. Peirce had lost the position of director of the Harvard Observatory to Bond, and Bond failed to support him in his bid to be reinstated in the American Academy of Arts and Sciences. Peirce thought little of him as a scientist and disliked him bitterly, even lacking sympathy for his terminal illness. Early in 1863, he wrote Bache, "The . . . scheme is quite separate from the observatory matter, which is dependent upon George's game of consumption—which some people play for many years without being consumed."[88]

Resentment about the way the National Academy was founded and the way the names were selected was not limited to the few worthy men who had been overlooked for membership. Newspapers raised objections to the founding of the Academy, and even fellow Lazzaroni member, James Dwight Dana, criticized the selection of its members in the *American Journal of Science and Arts.*[89]

Peirce sensed resentment to the founding of the National Academy within the American Academy; he wrote to Bache: "Yesterday, met the American Academy, and to show their hatred of the National Academy, all its opponents combining to elect Gray as President and William B. Rogers as Corr. Sec. of the Amer. Acad. and refusing to elect Lovering to either office because he has not been thought fit for election to the National Academy."[90]

When Peirce's temper cooled down, he saw less of an insult to the National Academy in the American Academy's action: "The election of the officers of the American Academy is less unpleasant to me the more I think of it. Agassiz and myself had positively refused to be candidates, so that it left Gray as the first . . . choice."[91]

Unlike Agassiz and Bache, Peirce was ambivalent about the National Academy, and expressed his doubts about the course they had taken: "Recent occurrences lead me to fear that we have made a mistake in founding the Academy—and that it may not be so great a misfortune if its enemies were to succeed in overturning it. Nevertheless I shall do my utmost to sustain it. Perhaps the naval commission was the best working plan, and might quietly have grown up into a more substantive organization."[92]

Serious dissension within the Academy first manifested itself with the election of Spencer Baird. In the minds of those who saw injustices in the selection of the incorporators, the obvious solution was to elect those omitted when vacancies occurred. Peirce was adamant in the necessity of rejecting Baird. He grabbed the moral high ground in a letter to Bache:

> As to Baird—I wish to be able to begin debate upon the election of new members, with a statement of principles. I wish to say that the Academy is not merely an academy of sciences—it has been entrusted with a charter by the government for the performance of certain duties—involving trust and responsibility. The opinions to government must be above the shadow of suspicion. This is its only hope of stability and permanent influence. We must refuse to receive into its body any member, however gifted, who is not sound to the core. The bad leaven must be rigidly excluded. When the said leaven is not only bad, but weak, its objection cannot be a matter of an instant's hesitation. I have no serious fears, however that the Agassiz's will ask us to take Baird when they recall his history. It is too repugnant to their natures. Perhaps they may be impelled by generosity. But that will not do. You have no right to be generous with what belongs to another. The Academy is not ours, but America's; and it is our duty which we have sworn to perform, to be faithful to it in the election of members, and in all other respects.[93]

In this case, Henry broke ranks with his friends, Agassiz, Bache, and Peirce. He joined with Gray and Dana, and they elected Baird.[94] Henry wrote to Agassiz, admitting that he voted for Baird, and explaining that by overplaying their hand, the Lazzaroni had once again aroused the animosity of outsiders:

The feeling also exists, that the few who organized the Academy intend to govern it; and I think this was the animus which excited the determination to elect Professor Baird. He was the choice of a larger majority of the cultivators of natural history; and although your opposition was honest in intention, and your position correct in general principle yet I think that had you prevailed in your opposition, a majority of all the naturalists would have resigned, and a condition of affairs would have been produced deeply to be deplored. I fully agree with you in opinion, and I presume the philosophical world also concurs with you, that as a class of investigations those which relate to Physiology and the mode of production and existence of organic forms are of higher order than those which belong to descriptive natural history. The good however which two persons may have done to science in these classes will depend on the relative amount, as well as, on the character of their labours.[95]

Since membership in the National Academy was limited to fifty, no additional members could be elected until the astronomer, James M. Gilliss, died in 1865. The Swiss geographer, Arnold Guyot, hoped the vacancy would be filled by Bond. He wrote to Henry, "Our Cambridge friends had their own way in the last meeting. Let justice have its own way in the next."[96] Unfortunately Bond died of consumption before he could be elected.

The government did call upon the Academy for studies of various things, but it received no government funding, and remained purely an honorary organization until World War I.[97] Although its instigators had hoped for much more, they were pleased that a national scientific honorary organization now existed.

The Academy did hold sessions of scientific papers, in which Peirce participated, but, as Henry anticipated, they lacked funds for publication, and, in fact, were unable to publish Peirce's most important mathematical work.[98] With no federal money supporting the National Academy, except for reimbursement for expenses, and lukewarm support from the scientific community, the National Academy might well have drifted into oblivion, except for two things. First, Bache left his estate to the Academy for the "prosecution of researches in physical and natural science by assisting experimentalists and observers."[99] Endowment from other sources followed. Another boost to the Academy came when Henry reluctantly accepted the presidency, in deference to his deceased friend. Henry made original research the principal consideration for membership, shifting the purpose of the Academy from practical service to the government to recognition of scientific accomplishment. He also increased the number of members beyond fifty.[100]

Φ^n

Unlike Peirce, his friends, the Le Contes, were deeply involved in the war itself. Their son Julian fought for the Confederacy, being away from home for periods of more than a year at a time, his safety always in doubt.[101] South Carolina College closed in June 1862, because nearly all of its students had joined the Confederate army. John Le Conte spent the rest of the war working as superintendent of the extensive nitre works in Columbia. He was given the rank and pay of major, but never wore a uniform.[102] His brother Joseph first worked for a company manufacturing medicines for the army, and then joined his brother in the Nitre and Mining Corps.[103]

Life was hard in South Carolina during the war; John's seventeen-year-old niece, Emma complained, "How dreadfully sick I am of this war. . . . No pleasure, no enjoyment—nothing but rigid economy and hard work—nothing but the stern realities of life. These which should have come later are made familiar to us at an age when only gladness should surround us."[104]

The last months of the war were particularly devastating. Hearing of Sherman's march through Georgia, Joseph Le Conte traveled to the family plantation in Liberty County, Georgia, to rescue family members there. He hoped to reach the plantation before Sherman, but did not succeed. Fearing capture, he had a harrowing experience reaching his family while evading Sherman's troops.[105]

After Sherman completed his march to the sea, he cut a narrower, but even more devastating path through South Carolina. It was a campaign with a vengeance. South Carolina had started the war and Sherman's army was determined that South Carolina should be punished for it. Columbia was in Sherman's path.[106] Anticipating the invasion wore heavily on the Le Contes. Emma wrote, "They are preparing to hurl destruction upon the State they hate most of all, and Sherman the brute avows his intentions of converting South Carolina into a wilderness. Not one house, he says will be left standing, and his licentious troops—whites and Negroes—shall be turned loose to ravage and violate." A few days later she wrote, "The uncertainty is very horrible. But how accustomed we have grown to what is horrible!"[107]

John attempted to save his scientific papers, including the manuscript of a book on general physics. He left them with a Catholic priest for safekeeping, but despite his precautions, all his papers were destroyed, as were his brother Joseph's.[108] Attempting to save personal property, as well as supplies and equipment at the nitre works from the invading Northern forces, John and Joseph packed it in wagons and left Columbia. Union soldiers intercepted them, and John and his son were taken prisoner, although they were soon paroled.[109]

Sherman claimed not to have given orders to burn Columbia; but it was largely destroyed by fire. Emma Le Conte described the spectacle: "Imagine night turned into noonday, only with a blazing, scorching glare that was horrible—a copper colored sky across which swept columns of black, rolling smoke glittering with sparks and flying embers, while all around us were falling thickly showers of burning flakes." Union soldiers "would enter houses and in the presence of helpless women and children, pour turpentine on the beds and set them on fire. . . . The wretched people rushing from their burning homes were not allowed to keep even the few necessaries they gathered up in their flight—even blankets and food were taken from them."[110] Over 360 acres and almost fourteen hundred homes were burned.[111] The Le Contes' homes fared better than most, probably because they were close to the college, which was being used as a hospital. John and Josephine's house caught fire numerous times, but the flames were always extinguished. Joseph and Caroline's home came through the night unscathed.[112]

After Sherman's invasion, the Le Contes lived by the charity of their former slaves, who supplied them with food, the source of which they never knew.[113] John and Joseph found a flatbed boat that had been used for the nitre works. The Union garrison commander, Col. Haughton, generously agreed not to confiscate the boat, and it was given to the Le Conte brothers in lieu of unpaid salary. They used the boat to transport food into the city, by which they were able to make a living.[114]

Despite the weariness of the war, its end brought no joy, only sorrow and occupation. Emma said, "Why does not the President call out the women if there are [not] enough men? We would go and fight too—we would better all die together. Let us suffer still more, give up yet more—anything that will help the cause, anything that will give us freedom and not force us to live with such people—to be ruled by such horrible and contemptible creatures—to submit to them when we hate them so bitterly."[115]

Josephine wrote to Benjamin, complaining that earlier letters to both him and Bache had received no reply. "I wrote with the same feeling you have always inspired in me—perfect confidence—Perhaps too much confidence for the times and without taking into consideration the possibility of change. . . . I am equally certain that if you received ours [a letter] no matter how we might differ upon the events of the day, you would certainly have sent some reply by the Dr. He reached here on Friday and brought nothing for us."[116] Peirce did send a message to the Le Contes through Col. Haughton, the federal commandant at Columbia after Sherman's occupation. He offered the Le Contes an interest-free loan with indefinite time for repayment. Emma was indignant: "Of course his offer was declined. . . . Father would not borrow from anybody, but to be under obligations to a *Yankee!*" Apparently John and Josephine felt the same way; they too declined the offer.[117]

The Le Contes were saddened by the devastation of the war and the domination of carpetbaggers. In 1866, the college at Columbia was reopened as the University of South Carolina, but its status and support for its faculty remained uncertain. The board of trustees, now consisting of freed slaves, was taking steps to "convert the University into a school for illiterate negroes."[118] The poor conditions at the University of South Carolina and the larger problems with reconstruction in the South made the Le Contes look elsewhere for a place to live. The old South was gone and they did not wish to live in the new one.

Yet finding employment outside of the South proved difficult. After having worked through the war supplying the Confederate army with nitrate for gunpowder, John and Joseph found it impossible to obtain faculty positions in the North.[119] Another blow to John and Josephine at this time was the death of their daughter, Lulu. She died of tuberculosis in March 1868. To the end, John refused to acknowledge the presence of the disease.[120]

Finally Agassiz wrote the Le Contes, suggesting that the newly founded University of California might consider a Southerner on its faculty.[121] Josephine wrote Benjamin complaining of the conditions in the South and telling him they would probably move to California, but expressing her great joy that Jefferson Davis had been freed of his cruel imprisonment. Further, "The second source of individual happiness, was the receiving of your letter to the Prof. which next to your presence was always such a blessed tonic in those days. . . . I need not add how it would delight us to see you here, before there is a general [emigration] of our people made possibly necessary by a revolutionary spirit among our slaves. If left to themselves, they are a very affectionate and docile race, but we cannot yet tell how this leaven of unrighteousness may liven the whole mass and deluge our sorrowing land . . . in blood. Tell Mrs. Peirce I do think of her constantly and will write soon. Say to her that as soon as our captive was released I set to work and made a beautiful dress of the Balmoral she sent some time ago."[122]

With the support of their Lazzaroni friends, including Peirce, both John and Joseph were offered positions at the University of California. John received an offer in November 1868.[123] The Le Contes passed through New York on their way to California, but were unable to see the Peirces, and other Northern friends. The Peirces continued to correspond with the Le Contes, however, until Benjamin's death.[124]

Upon arrival in Berkley, John became the initial acting president of the university, and later served as president from 1876 to 1881. He worked at the University of California as an administrator and faculty member to make educational reforms, much in the spirit of the Lazzaroni. More than any other man, he was the father of the University of California.[125]

John died of severe bronchitis in 1891 at the age of seventy-three. Josephine never recovered fully from his death. Three years later, while dozing by the fireplace, a newspaper fell from her lap and caught fire, which ignited her dress. She subsequently died of burns from the fire. A California newspaper described her as a seventy-year-old lady who was "one of the most magnificent appearing women in the state."[126]

$$\Phi^n$$

In contrast, the war did not greatly interfere with Peirce's scientific or personal life. Harvard was little affected by the war; it remained open and did not seriously lack for students.[127] "Public opinion in the North did not require students to take up arms, . . . draftees who hired a substitute were not despised. President Lincoln himself kept his son at Harvard until he graduated in 1864."[128] Peirce's sons did not fight for the Union, and Massachusetts was not invaded by a hostile army.

In 1859, Peirce's wealthy Uncle Saunders gave money to Harvard to be used to build an alumni center after he died. When he died in 1864, the university incorporated a Civil War memorial into the building.[129] It was built within eyesight of Peirce's home. One hundred and thirty-eight Harvard men gave their lives for the Union, and another sixty-four for the Confederacy. One of those memorialized in the Saunders building is Charles Russell Lowell, the young man who sought Peirce's aid in securing a commission.

In October 1863 Lowell married Josephine Shaw. Just a few months before their wedding, Josephine's brother, Colonel Robert Shaw, died leading the first African-American regiment in an attack on Fort Wagner, near Charleston, South Carolina.[130] The words that her husband wrote to Senator Sumner were not empty bravado. Lowell was "the picture of a soldier," exquisitely mounted, erect, confident, and defiant.[131] He was conspicuous for his bravery and lack of regard for his own safety, wearing his crimson officer's sash boldly to inspire his men, oblivious to its making him a prime target in battle. In recognition of his bravery at Antietam, he was selected to carry captured flags to Washington. While engaged in the raids in the Shenandoah Valley in 1864 he had thirteen horses shot out from under him in as many weeks. He died at Middleton, Virginia, on October 20, 1864, of wounds received at Cedar Creek the previous day.[132]

Lowell's widow, Josephine, gave birth to a daughter six weeks after her husband's death. When the war ended, Josephine devoted herself to public service. She was the first woman appointed to the New York State Board of Charities. For many years she led the New York Charities Society, an organization founded through her efforts. She also founded the Consumers League of New York. A friend said, "One could not be with her . . . without seeing the halo upon her brow."[133]

12

We Shall Always Be Boys Together

BY THE SPRING OF 1864, BOTH PEIRCE AND BACHE WERE SERIOUSLY ILL. Peirce wrote, "Oh! my Dearest Chief! We have both of us been deep in disease and death. Our souls and health have been together although our bodies have been separated."[1]

Peirce had started complaining of bad health as early as the summer of 1863. He begged off going to Bache's summer Coast Survey camp, saying "I have had another of my old attacks. . . . I hardly now undertake to move from the house."[2] The "old attacks" were some form of kidney disease, probably kidney stones. Peirce wrote to Bache: "God bless you. I am stupider than ever—for I am just waking up from a blow which I have been having for the last three days—a most dissipated blow in which I have been in a regular state of intoxication from opium and ether for the whole time-and I am still so permeated with the direful forms of the intoxicating phials that I nearly hate myself. It was to save me from suffering during one of my ancient attacks from one of the seven devils who has taken up his abode in my kidneys. Poor fellow! He is more to be pitied than I am."[3]

In Peirce's day, the only way to relieve the intense pain from passing a kidney stone was with ether or laudanum (tincture of opium). Sometimes his attacks were less severe and he did not need to resort to using these drugs. He wrote Bache "last night [I] came as close as possible to one of my attacks. But I escaped with a few hours of suffering & without being compelled to resort to opium or ether."[4]

The attacks continued. Several months later, Peirce, again writing to Bache, feared that he was "just upon the point of breaking. My attacks are more continuous and, every night, I feel quite discouraged." Perhaps being reminded of his own mortality, not only by his own bad health, but by the death of several friends, including his uncle Charles Saunders, he told Bache, "I am trying to arrange all my affairs in the best way to be suddenly left, and if I should be taken, dearest Chief, exert all your influence to save me from eulogistic biographers. Before the Academy, let my account be merely a list of the works which I have written and the most important memoirs, without criticism or comment. Let posterity decide for itself in what estimation it will hold me."[5] But Peirce's fear of death was premature; by May his health was largely restored.[6]

Some months before Peirce recovered, Bache suffered from a series of strokes; contemporary accounts describe his condition as a softening of the brain.[7] As Peirce's health improved, he became increasingly concerned that his friend's was not. He wrote to Bache, "As to myself I am daily getting in better and finer health and hope that the beloved president of the National Academy is assuming his duty to the country and science in this most important particular."[8] A few months later, Peirce reiterated his concerns: "God bless you dear old Chief. In our hearts the love of you is ever young. We shall always be boys together. How I long again for one wise look, from that trusting eye of yours. Ah! Dear Chief! I miss you. There is no one to take your place with me. All our science misses you. You are the chain of love, which has held us together in such fine harmony. You must hold a little longer yet. The cement of mutual faith and trust is not yet strong enough."[9]

Bache's illness was a terrible ordeal for Peirce; he wrote to his wife, "It almost made me faint when I heard of Bache's illness and I can hardly sit still now when I think of it."[10] When Anna Cabot Lodge[11] wrote him, sympathizing about his prolonged illness, he responded: "Your allusion to the physical suffering which I have borne is kind and sympathetic. But I can say from experience that the severest pain is a little compared with the sympathy of such a friend. I have not been much tried in this world, perhaps not enough for my good—but my hardest trial has been the failing mental power of my dear friend Bache. I have been much with him in his illness this summer and he is coming today to make us a second visit."[12]

Peirce visited Bache often during his illness. Henry wrote him, "I learn that like a good Samaritan you visit the Professor almost every fortnight."[13]

Φⁿ

In addition to diminishing Bache's memory and intellectual ability, the strokes incapacitated his right hand. When Bache could no longer write, his wife, Nancy Fowler Bache, or Ency, began writing Peirce. She wrote long letters every few days describing her husband's condition. At first they were full of hope:

2 July 1864

I was to have written you this morn in behalf of the Prof. to beg you to come up and & see him, but the sight of your hand-writing delighted him so much that he could not resist the desire to visit you himself. . . .

You will find Mr. Bache quite feeble, tho much better than when he came up here.[14]

Mt. Wooster, 4 July 1864

Mr. Bache is a great deal better; gets stronger every day, but wants to have his friends around him & and he has been talking about your coming, constantly for the last five days.[15]

Mt. Wooster, 8 July 1864

He was not quite as well as usual yesterday.[16]

Danbury, 2 August 1864

You have been so kind to me since Mr. Bache has been sick, as well as to him, that I am getting to feel as if I had a right to call upon you whenever I need help.[17]

Danbury, 8 August 1864

Mr. Bache begs me to write & tell you how much he wants to see you. He is so ungallant that he pines for you even while he has two charming ladies with him, a dear little girl & any quantity of boys. But all together they cannot fill your place. . . .

The other page was written for the Prof. to read—but I must tell you I have been very much worried. His mind has wandered so much today.[18]

Danbury, 9 August 1864

I feel so much worried about Mr. Bache, that I cannot help turning to you for advice. His mind is getting very much weaker & wanders more & more. I feel that I ought not to sit still & see this without letting a Dr. know—could you see Dr. Séquard & tell him about it? If there is any thing to be done for him now is the time & every day is of importance—if not— it is better I should know it, don't you think so?[19]

New York, 29 September 1864

I have the most wonderful news to give you—there has been a most wonderful change in our invalid for the better as regards his mind, for the last week he has been brightening up so much that I have felt uneasy, as his bodily health did not keep pace with it. . . .

It seems as if a veil was being lifted from his mind.[20]

The Baches went to Europe, hoping that the trip would restore Bache's health, but it offered no therapy beyond that of a distraction. Bache dictated the following words to his wife:

> Naples, 27 December 1864
> I do not think the oysters are particularly nice, certainly not worthy of the Lazzaroni. . . . I wish you could be here enjoying this with me but I know you are more useful where you are.[21]

Nancy wrote:

> Naples, 3 January 1865
> This is the 3rd day he has been so much like himself than he has been since the summer. I feel the more encouraged that he has never gone backwards, tho he seems at times to stand still-it is true that the weakness affects him a great deal. . . .
>
> I have been very much worried about our money matters. . . . The difference between traveling for health & for pleasure. I find it much more expensive than I had expected.[22]

Now back in the United States, Nancy Bache wrote:

> West Hill, 11 August 1865
> Mr. Bache continues to improve & is gaining strength daily, walks more without any assistance for a little while at a time, but is still too frail & uncertain on his feet.[23]

> New York, 2 March 1866
> Our invalid has had a nice quiet night & I cannot but think he will be more comfortable today than usual. Yesterday he was very restless, very much as he has been for some days past, but with only one distressed turn.[24]

> New York, 2 April 1866
> Mr. Bache has had three comfortable days in succession! A rare thing, but only one good night during the time.[25]

> Newport, R. I., 14 September 1866
> A great change has taken place in Mr. Bache within a few days, he has been in bed for three days from weakness. I have seen for some time that his bad nervous times were returning, slowly, but steadily on the increase. On Sunday afternoon he had a fearful one, like those he had in New York.[26]

> Newport, R. I., 14 October 1866
> For the last week Mr. Bache has not been as well as usual, indeed I think he gets more & more feeble every day, he is now so weak that I dare not

take him from one room to another without help & his mind has gone even much more since you saw him. He notices less & talks less—is more restless & has the alarming turn more frequently.[27]

> Newport, R. I., 4 November [1866]
> Mr. Bache is getting so much more feeble & so often has very bad days, that I sometimes feel as if we should not be able to move him. He had not been out of his room for more than two weeks before the day we brought him here—he does not sit up very long at a time, gets up late in the morn & by six in the eve—is so tired that we give him his tea & put him to bed. I think he must suffer a good [deal], tho' it is impossible to find out in what way. There are times when he is sweet & gentle & like himself. At others he moans constantly. Obviously he does not feel comfortable, but it is when the nervous times come on that he must suffer the most. . . .
>
> Mr. Bache does not talk or take much notice, yet he always seems pleased when his friends come.[28]

> Tuesday Night, [late 1866, or early 1867]
> Mr. Bache has had another very bad day. . . . He would be quiet all day from the effects of the Anodyne of last night but between 1 & 2 the nervousness came on again & we have been giving him Hoffman's Anodyne at intervals all day—this eve he had the worst attack he had had all day & it was some time before we could quiet him—it is wonderful how he retains so much strength! You can form no idea of his sufferings unless you have seen the dearest—It is heartrending to witness them.[29]

Bache died in Newport, Rhode Island, on February 17, 1867. A few days later, Peirce attended his funeral in Washington.[30]

$$\Phi^n$$

When it became apparent that Bache would not recover, Henry was concerned not only for his friend, but for the future of the Coast Survey. Under Bache it had grown into an important, perhaps the most important, institution for American science. With little graduate work available in the United States, it served as an essential training ground as well as source of employment for American scientists. Henry feared that all of this might be lost to patronage unless a distinguished scientist and dedicated administrator replaced Bache. As early as 1865 "many applicants were in the field for the office."[31]

Henry felt Peirce was the only scientist in the country capable of maintaining Bache's scientific standards for the Survey.[32] He wrote Peirce, reporting that he had told the secretary of the treasury "that the condition of the professor, was such that he might be taken away at any moment, that a number of persons are preparing to make an effort to obtain the office, as soon as the death of Mr. Bache should occur, and, hence the necessity of determining in advance, on the proper candidate, in order to insure the efficient and faithful prosecution of the work. I repeated the statement, which I made last summer, that you were the man, of all others, for the position—that no one else could compete with you in regard to the necessary mathematical attainments, and that your appointment would command the approval of the scientific world."[33]

At this point Henry was under the impression that Peirce was willing to take the position. He wrote to Peirce: "In this I think you have done wisely, because you have done that which is right, without being influenced by personal considerations, and I doubt not that you will find your reward, not only by an approving consciousness, but also in the successful completion of the great work in which our dear professor has been so deeply interested, and in the increased prosperity of yourself and family. Without being subjected to the continued interruptions incidental to my office or the study of the minute details necessary in the primary organization of the Survey, you will be in a position of great power and influence, in which you can do much good in the way of advancing science, adding to the prosperity and reputation of your country and of ameliorating the condition of humanity by facilitating the intercourse of nations."[34]

Either Henry was misinformed or Peirce quickly changed his mind; Peirce firmly refused to accept the position. Shortly before Bache's death, Sarah wrote, "The people of science in Washington wished your father to be at the head of the Coast Survey—but he says he shall not take it."[35]

Peirce went to Washington to attend Bache's funeral, with no intention of replacing his friend as head of the Coast Survey.[36] Joseph Henry did not relent in his efforts to put Peirce in the position. The day after Bache's funeral Henry wrote:

> I have just returned from the Treasury Department, and have the pleasure to inform you that the Secretary will nominate you without asking as to your acceptance. He is convinced that your appointment alone can prevent a struggle which will be damaging to all engaged in it, as well as to the work. I beg, therefore, that you will not decline, even though you do not intend to retain the office.
>
> A very important principle is involved in your acceptance, namely that of retaining the Coast Survey and the Smithsonian Institution in the charge of science, and of not suffering them to be engulfed in the vortex of partisan politics. Let it for once be conceded that the head of the Coast Survey need not be a man of Science, and the character of the work will

be changed. The Patent Office was at first under the charge of a man of
Science, Dr. Jones, the editor of the Franklin Journal. General Jackson
thought that the office merely required administrative ability, and since
then not only the head of the establishment, but the examiners under him
are appointed on political grounds.

. . . I doubt not that on proper representation the authorities of Harvard will allow you to retain your professorship, and that such an arrangement can be made as to render your duties in the survey agreeable.[37]

Despite Peirce's protestations, he accepted the position before he
left Washington.[38] Upon receiving a letter of congratulations from
John L. Le Conte,[39] an embarrassed Peirce replied:

I have suffered an almost ignominious defeat. I had said when I left
Boston, that nothing should induce me to accept the Coast Survey, and
the first thing my friends read in the newspapers that I am sworn into
office. The multitude hurried me on against my will. I was powerless. . . .

But ever so—many thanks for such kind congratulations. . . . All the
assistants of the survey have received me with open arms and with a cor-
diality which is quite affecting—but I am not deceived. They delude themselves with the idea that I am dear old chief revived again. How little they
know my inability to fill his place. He was a perfectly rounded sphere. I
am a pyramid of the fewest number of sides and they will find it out, too
soon, alas! But I shall try to do my duty to carry out his plans in his spirit—
and if I fail, it will not be from lack of heart but from lack of brains.[40]

Peirce's wife was no less surprised than his friends when Peirce
agreed to head the Coast Survey. Sarah confided to her son "after all
your Father's protestations that he could not and would not—he has
been & gone & taken the Coast Survey! I for my part was completely
astonished & hardly reconciled. But he is to keep this place here. . . .
He is in grand spirits."[41]

$$\Phi^n$$

Other members of the family, as well as his wife, were concerned
about Peirce's being head of the Survey. His sister, Elizabeth, feared
that he would no longer be able to come over to her house for breakfast, and Peirce's Aunt Saunders hoped that he would still be able to
keep track of her dividends. But Peirce took hold of the new position,
and things went on much as they had before. Soon his sister was
happy to report that Benjamin "comes as usual to breakfast."[42]

His wife observed: "I feel more and more as if it [Peirce's becoming head of the Coast Survey] was a fortunate change for him to make

although at first I shuddered at the load of responsibility & work he seemed to be assuming—but he seems in so good spirits about it & to have taken hold with such energy & zeal & life that I begin to think that he needed 'fresh fields & pastures new' after browsing over those columns of figures for a quarter of a century. We expect to go to Washington about the 1st of April. . . . He has 70 men under him & 20 ships—which gives you some idea of what he has to do. . . . We shall only be in Washington about 3 weeks & your father thinks he will never need to be there long at a time."[43]

Peirce had insisted that he be able to remain in Cambridge and not sever his ties with Harvard. During his entire tenure as the Coast Survey superintendent, Peirce retained his professorship at Harvard and lived in Cambridge most of the time, traveling to Washington as necessary. One of the reasons he decided to remain in Cambridge may have been that Washington was a wretched place to live. Six years earlier a congressman described the nation's capital: "It was then as unattractive, straggling, sodden a town, a wandering up and down the left bank of the yellow Potomac, as the fancy can sketch. Pennsylvania Avenue, twelve rods wide, stretched drearily over the mile between the unfinished Capitol and the unfinished Treasury building on Fifteenth Street, West, where it turned north for a square, and took its melancholy way to Georgetown, across the really once very beautiful Rock Creek. Ill paved with cobblestones, it was the only paved street of the town. The other streets, which were long stretches of mud or deserts of dust and sand, with here and there clumps of poorly built residences. . . . Not a sewer blessed the town, nor off of Pennsylvania Avenue was there a paved gutter."[44]

A few years later, Peirce's daughter-in-law, Zina, also found Washington to be a deplorable place. She wrote, "if ever magnificent possibilities were cruelly marred, mutilated, and mangled, those of the National Capital have been, and every day are being, so treated. . . . [Washington is] the most disappointing, disheartening conglomerate that ever shocked the pride or patriotism of order-loving, beauty-worshipping woman."[45]

Despite Washington's shortcomings, Sarah found the city pleasant to visit as the wife of the superintendent of the Coast Survey:

Darling Children—We arrived here last night at about six o'clock—found Hoover at the station with a carriage & splendid fires all over the house to welcome us when we arrived at the house. The gas also was brilliant & showed to advantage the delightful arrangement made here for your father's comfort. The furniture of the two chambers is very nice & pretty—with a Brussels Carpet & chintz curtains in our room—& Kiddiminster carpet in the others. The wash stands are as large as the one in our room & very elegant—also the bathing rooms—which are exactly alike—In fact it is quite luxurious here—This morning we went over to the senate restaurant for our breakfast & I went all over the capitol which is much improved since I saw it last when much of the finish of the interior was quite incomplete—The Marble work of the outside too is finished & the blocks & stones cleared away which have lain so many years in confusion about the yard. I am going over to hear the speaking by & by, & expect to spend nearly the whole morning there.[46]

On a later trip to Washington, Sarah had an equally grand time visiting her sister: "I have as you know been to Washington with your father where I had a charming time. It was nice to breathe for a while a milder air & to walk forth reasonably shod upon dry pavements under a southern sun. I saw the Davis's (sic) a good deal, & Charley & Annie Huntington who are luxuriously established at Welker's where they fare sumptuously every day, as they like so well to do—we were invited to dine with them one day with all the Davis's (sic). Dinner at six in their private parlour, beautifully served—courses—12— wines varying with every course—& after coffee we were taken to the theatre to see Jefferson in Rip Van Winkle."[47]

Since Peirce was usually in Cambridge, he depended upon Julius Hilgard, the assistant in charge, to oversee the Coast Survey affairs in Washington. Although Peirce publicly praised Hilgard, he was not entirely pleased with him.[48] When Peirce took over as superintendent, he wrote that the Survey "had fallen into a slovenly state in many matters and even in the office, where Mr. Hilgard might and should have prevented it."[49] Nevertheless, Hilgard continued to manage the day-to-day affairs of the Coast Survey at the Washington office, while Peirce was in Cambridge.

Hugh McCulloch, the secretary of the treasury, nagged Peirce to make Washington his permanent abode, saying that he didn't want Peirce "to be the merely <u>ornamental</u> head of the Survey."[50] Peirce took exception to such a title: "If the Secretary thinks that I have hitherto been the mere <u>ornamental</u> head of the Survey—where does this absurd idea come? My whole time and thought have been absorbed in the Survey ever since I took the office. So much so that I have not even undertaken to give the little attention required for printing the memoir presented by me to the National Academy at the last meeting. And it is the opinion of the Chief clerk . . . that my predecessor had not for very many years devoted as much time to the Survey as I have done."[51]

To Peirce, being away from Washington was not merely a matter of personal preference or convenience. If he was going to be the scientific, as well as the administrative, head of the Survey, he needed time away from the distractions and interruptions of Washington. This was not a position Peirce coveted, and he wrote Henry, "if the Secretary disapproves this plan, he has only to say so, and I will resign the office with many thanks to him for the kindness which he has hitherto manifested to me, and shall avoid all necessity for an unpleasant rupture."[52]

It was not merely the problems of squaring himself with the secretary that annoyed Peirce, but the nature of the work itself. He complained to his friend Lewis Rutherford:

> As to myself—I feel somewhat as Mrs. Rutherford said she should feel about me if I accepted the Survey—almost ashamed of myself. I am no longer the mathematician, with my heart and head full of science, and beauty and God's works, not excluding fair Eve and her fair daughters, But I am a chief of topographic and hydrographic parties, with a soul ground down to estimates, and writer's work, and this man here and that man there, and men with jealousies—and members of Congress to be properly worshipped. There—get thee behind me, Satan. Every now and then the tempter rises and puts before my eyes a glass in which I see myself covered with chains and lying in a dungeon. How many does he betray to their destruction by this false reflection! We are, after all, every man of us in a world of duty—and if we do our duty the chains fall off. The dungeon becomes a palace, and we are kings instead of slaves.[53]

With Peirce's new job came an unexpected reevaluation of the capabilities of African Americans. Peirce's sister, Elizabeth, described an assistant assigned to Peirce: "It is only when I see his black servant 'John' or his secretary that I realize his changed position as in other respects all seems the same. John takes care of his clothes, study, papers, pamphlets, books & c. & goes his errands. He is very intelligent, writes a handsome hand & seems to have a clear head for accounts. He arranged all your father's pamphlets according to the subjects & labeled them. . . . When your father and mother went to Washington John packed your father's trunks, & even repaired the covering of one of them. He is I should judge from what I see & hear a general favorite in the family."[54]

Sarah was equally impressed with John, noting that he was "very handy about the house & willing to do anything."[55]

$$\Phi^n$$

One of the first things Peirce did when he took over the Coast Survey was to force Benjamin Gould to resign. Gould had been in charge of the telegraphic longitude department of the Coast Survey since

Benjamin Apthorp Gould. Courtesy of the NOAA Photo Library.

1852. This work was of great importance to him, as he explained to Peirce: "To this work I gave the almost exclusive labor of the best years of my life, working unremittingly, & trusting to the future for whatsoever recompense might be rewarded me, when the results should be published."[56] Currently Gould was anticipating the completion of the Atlantic cable. Using telegraphic transmissions with this cable, astronomers could compare American longitudinal coordinates with those of European observatories. This was important and highly visible scientific work that would bring prestige to the man who accomplished it.[57]

Although Peirce had long been frustrated with Gould, he was particularly unhappy with Gould's actions in 1865 regarding the replacement of George Bond as director of the Harvard Observatory.[58] Upon Bond's death, Peirce's former student, Truman Stafford, was made acting director; many felt he should be given the post.[59] But Peirce supported Joseph Winlock. Peirce had met Winlock at the 1851 AAAS meeting in Albany, and hired him to work on the *Nautical Almanac*. Gould ostensibly joined Peirce in supporting Winlock, but in fact was campaigning for the position himself. His good friend Wolcott Gibbs had been organizing support on his behalf, and it is extremely unlikely that Gould was not aware of his friend's actions.

Back in 1856, when Gould agreed to be the acting director of the Dudley Observatory, he angered the trustees by seeking a position at Columbia University shortly after agreeing to be the director of the Dudley Observatory. Gould protested that he had no knowledge that others, again most notably Wolcott Gibbs, were working on his behalf for the position. The trustees did not believe Gould's protestations. This time Peirce didn't believe them either. Peirce thought that Gould was deliberately deceiving Winlock.[60]

So, despite past friendship, and the fact that the transatlantic telegraphic connection was nearly completed, Pierce removed Gould from all connection with the telegraphic work and demoted him to a reducer of numerical observations. Gould felt that his only course consistent with self-respect was to resign from the Survey.[61]

He wrote to Peirce complaining of his treatment: "In September last the completion of the Atlantic cable seemed to render it my duty, from loyalty to the Survey & its Superintendent, then seriously ill, to take personal charge of the transatlantic operations; & without a moment's hesitation I left everything at home in order to attend to this matter, at very great personal inconvenience, & a pecuniary loss of more than my whole year's salary."[62]

In a second letter written the same day Gould protested that:

> I cannot send you my official resignation without a couple of private words, for the bitterest part of all this is that it should have come through you,—whom I have always loved so much.
>
> The intimations which escaped you yesterday were however the deepest stabs of all.[63]

Gould protested his innocence of allowing his friends to oppose Winlock's confirmation, while he was professing to be his friend and supporter. He then closed: "And if ever you should find reason to change your present estimate of my character I shall be glad to know it,—as enabling me to resume the affectionate relations in which I have stood toward you, & which I have supposed mutual."[64]

Not content to let matters rest, Gould wrote to one of the Harvard trustees, the Reverend Artemus B. Muzzey, requesting him to write Peirce, confirming that he knew nothing of Gould's designs on the directorship of the observatory. Mezzey complied, stating that indeed he was unaware that Gould had opposed Winlock's nomination, and added that there was never any possibility that Gould would have been considered for the position.[65] Gould forwarded the letters to Peirce:

> I enclose copies of two letters, which for the sake of our long friendship I beg you to read.
>
> Should they convince you that you have been mistaken in believing me capable of the action attributed to me the other day, this will be a source of more satisfaction to me than I can express.[66]

Two days later, Gould again wrote to Peirce, lamenting the demise of their friendship:

> One more note I must send you, & that is to ask that so far as may be, our personal relations may rest unchanged now that the official ones are no longer subjects of discussion.
>
> You tell me you are my friend, & I have never doubted your desire to be so; you cannot doubt my affection for you, I am sure; whatsoever soreness I felt at your supposed doubt of my integrity ought to be removed by your note; & the severance of my connection with the Survey was due merely to what I regarded as a change in my position there, though in the absence of documenting evidence to the contrary you did so regard it. . . .
>
> If you will permit me to feel that we can meet & care for each other as before, it will be a source of great comfort to me.[67]

Peirce was not moved by Gould's painful remonstrations. His terse reply did not offer any hope of reconsideration, and did little to heal the breach between them: "I respond to your note with all my heart & hope that your separation from the Survey will not interfere with your early occupation in some position more purely astronomical, & you will always find me ready to cooperate as one of your devoted friends. Give my love to your generous wife & a kiss to dear little Lusie & sweet Julie."[68]

Two years later Henry was able to convince Peirce to smooth things over between himself and Gould. Peirce responded to his promptings: "After I left you, I reflected long and deeply upon all which you said. Your purity, truth and loveliness entered into my heart and sanctified it. I perceived that you deserved to have your desires for the preservation of the harmony of science gratified. I resolved to write a new letter to Gould, such as your spirit would have dictated—and such as I would have written, had there never been any unkindness between me and him. The result was that which you would have anticipated. He wrote a grateful and sincere letter in reply; and I am sure that when you have seen the whole correspondence, you will feel that your earnest advice was not thrown away. May I always have so good a friend, and may God open my heart to follow his advice."[69]

Henry made the following gracious reply: "Your characteristic letter has just been received and I hasten to say in reply that your words have excited the warmest throbs of my heart, and the sympathetic moisture of my eye. It was a very easy matter on my part to give the advice I did, but a difficult one under the circumstances on yours to accept; I rejoice at the result."[70]

$$\Phi^n$$

In September 1868, Thomas Hill resigned as president of Harvard. Andrew Peabody, the highly regarded professor of Christian morals and preacher to the university, was appointed acting president. When the university gave additional grants to all the professors, except for Peirce and one other, the volatile Peirce sent Peabody his resignation. Peabody wrote Peirce, urging him to withdraw his resignation. He explained that while the college had been in financial straits for some time, he felt that the faculty members were in dire need of income beyond their salaries. He then pointed out: "The reason for such a provision did not, however, apply to two of the professors, yourself & Mr. Lowell, both of you having other & lucrative employments, & having, in order to pursue those employments, virtually resigned the duties appertaining to the headship of your respective departments into the hands of your assistant professors."[71] Peirce relented and asked that his resignation be withdrawn.[72]

A more serious problem facing the university was the selection of its new president; this responsibility fell to the six remaining members of the corporation. After the corporation nominated Hill's successor, his appointment had to be ratified by the overseers. Finding a new president proved to be difficult. Since there was no consensus as to the direction in which the university should go, many favored

Peabody's becoming the permanent president, especially those with traditional views of college education. But, hoping to fill the position with a man of national stature, the corporation offered the position to Charles Francis Adams, who had just returned from his ambassadorship to Great Britain. Adams immediately turned it down. Then, quite surprisingly, the position was offered to Charles Eliot,[73] the man who had lost the Rumford Chair due to Peirce's and Agassiz's lobbying.

Eliot had retained close ties with Harvard and become a member of the university's board of overseers in 1868. But, happy with his professorship at MIT, Eliot was not anxious to assume the duties of the president of his alma mater. Yet after consulting with friends and associates, he told the corporation he would accept the position. Early in 1869, Eliot had published two articles in the *Atlantic Monthly* on higher education. In the articles he had championed practical education and severely criticized Harvard's Lawrence Scientific School.[74] His ideas offended many people, including Peirce and Agassiz. Despite Peirce's earlier work in developing engineering at Harvard, science meant much more to men like Peirce and Agassiz than technology. They had also resented Eliot's interference in the Lawrence School when he was acting dean. Agassiz, in particular, lobbied hard to prevent his appointment. But unlike the controversy over the Rumford professorship, this time Peirce and Agassiz had no candidate of their own to put forward in place of Eliot. Opposition to Eliot was sufficiently strong that the overseers returned his name to the corporation, an act tantamount to rejection. But nevertheless Eliot was appointed.[75] As the psychologist and philosopher William James put it, "C. W. Eliot was confirmed President yesterday. His great personal defects, tactlessness, meddlesomeness, and disposition to cherish petty grudges seem pretty universally acknowledged; but his ideas seem good and his economic powers first-rate,—so in the absence of any other possible candidate, he went in. It seems queer that such a place should go begging for candidates."[76]

Eliot's appointment brought the end of the Lazzaroni's influence at Harvard. Peirce and Agassiz had passed from academic liberals to conservatives, having been replaced by younger men.[77] Peirce did not realize that Eliot was leading Harvard toward the institution he had always wanted it to be. Eliot, on the other hand, bemoaned the fact that the current educational system in the United States was failing to produce replacements for men like Peirce.[78] In actuality, Peirce's son Charles not only equaled his father's intellectual abilities, but even exceeded them. After Eliot became president of Harvard, he and Charles Peirce had an unfortunate squabble over support for the work Charles had been doing for the Harvard Observatory. As a result of this misunderstanding, Eliot was vehemently opposed to Charles working at Harvard, either for the observatory or as a professor.[79] Although there is great irony, even tragedy, here, it does not tell the whole story. Under Eliot's leadership, Harvard did become a great university. Peirce, too, was far from an aging, bitter man whose time had passed him by. At this same time he was in the process of significantly extending the scope of the Coast Survey, and he was also producing not only the best mathematics of his lifetime, but by far the best mathematics produced by an American to date.

$$\Phi^n$$

One of Peirce's new friends was the astronomer Lewis Rutherford. Rutherford married Margaret Stuyvesant Chandler, who, like Rutherford himself, came from a prominent and wealthy family. Although Rutherford was trained as a lawyer and practiced for twelve years in New York City, he eventually devoted all his time to his scientific interests. Rutherford built an observatory in the garden of his New York City home and equipped it with an $11\frac{1}{4}$-inch equatorial telescope, a large telescope for the time. He eventually replaced it with a larger telescope. He did pioneering work in astronomical photography and spectroscopy, which won him an international scientific reputation. He was a founding member of the National Academy of Sciences, a foreign associate member of the Royal Astronomical Society, and was awarded the Rumford Medal by the American Academy of Arts and Sciences.[80]

Peirce and Lewis were not only scientific associates, but personal friends. The Peirces visited the Rutherfords in New York, and the Rutherfords visited the Peirces in Cambridge. Just before a visit from them, Sarah wrote, "They are very nice people—cultivated, scientific —rich, aristocratic—everything a lady & gentleman should be but somehow not quite congenial. They are too rich and too fashionable."[81]

They were not the Le Contes; they were not the Baches.

13

A Public Functionary

ALTHOUGH CHARLES ELIOT WAS KNOWN TO HARBOR A GRUDGE, HIS DISagreements with Peirce did not prevent him from speaking generously about Peirce's accomplishments as head of the Coast Survey:

> When Professor Bache retired from the superintendency of the U. S. Coast Survey, he procured the appointment of his intimate friend Benjamin Peirce as his successor in the superintendency. Those of us who had long known Professor Peirce heard of this action with amazement. We had never supposed that he had any business faculty whatever, or any liking for administrative work. A very important part of the Superintendent's function was to procure from Committees of Congress appropriations adequate to support the varied activities of the Survey on sea and land. Within a few months, it appeared that Benjamin Peirce persuaded Congressmen and Congressional Committees to vote much more money to the Coast Survey than they had ever voted before. This was a legitimate effect of Benjamin Peirce's personality, of his aspect, his speech, his obvious disinterestedness, and his conviction that the true greatness of nations grew out of their fostering of education, science, and art.[1]

Peirce was remarkably successful as superintendent. He was able to garner Congressional support for many of his Coast Survey projects, exceeding even Bache's success in having the United States government support basic scientific research.[2] Peirce, himself, credited much of his success to his naïve forthrightness. He wrote to Hugh McCulloch, the secretary of the treasury:

> I sometimes think that I have not been quite deficient in practical ability, but again every body that I meet seems so much shrewder than I am, that I feel as if any success of mine was a favor granted to my weakness. Almost every body has such a wonderful gift for diplomacy and such an artistic skill in managing man, that my excessive simplicity, which has no power of penetrating below the outer surface of events and persons, feels to me to be a weakness. I have got into a habit, of which I cannot break, merely of speaking the real facts when I speak at all. This does not seem to be the

custom of the world, or at least of those who govern the world—except in
a few cases—such as that of my secretary. Almost all the others set about
in a way which would overset me—and, to tell you the honest truth, I
would rather be direct than follow their customs. I know that my day of
oversetting will come, and when it comes, I shall drown without a mur-
mur provided I retain the approbation of yourself and Agassiz and a few
other men and women whom I could name.[3]

Although lacking administrative experience, Peirce was neither a
stranger to Washington nor to lobbying for the Coast Survey. He had
often done the latter to support his friend Bache and also visited the
capital on Coast Survey business during Bache's administration.
Peirce had been well received in Washington; twice he had been introduced on the floor of the Senate—a great honor. He was also well
acquainted with many of the leading politicians from his home state,
including Senators Charles Sumner and Henry Wilson. Peirce's friend
Ebenezer Hoar was appointed attorney general shortly after Peirce
began serving as superintendent of the Coast Survey. And while on the
Scientific Council of the Dudley Observatory, Peirce had stayed at the
home of, and become a friend of, J.V.L. Pruyn, who returned to the
House of Representatives the year after Peirce's appointment.

There is no evidence whether or not any of these contacts aided
Peirce in running the Coast Survey. But there is no doubt of the
importance of Joseph Henry's friendship. Henry had urged his
friend to accept the position, and he did not abandon him when he
took it. After Peirce had been the superintendent for almost a year,
he was becoming increasingly frustrated with the problems of the
Survey. He wrote Henry a long letter detailing his difficulties. In particular, the secretary of the treasury wanted him to be in Washington
more, preferably to reside there. But Peirce protested, "By the course
I have marked out I am actually in Washington as much as my predecessor was, and therefore as much as the administrative interests of
the Survey require."[4]

Henry postponed his plans to travel to Princeton, so that he could
talk with the secretary of the treasury. After the visit, Henry wrote
Peirce assuring him that he was convinced that the secretary had both
the Coast Survey's and Peirce's best interests at heart. Henry explained,
"He was anxious that you should become a little better acquainted with
the members of Congress."[5]

Henry was correct in assessing that Peirce had exaggerated his
troubles with McCulloch. The secretary was pleased with the new
superintendent of the Coast Survey: "I deserve no thanks from Prof.
Agassiz for anything that I have done for his friend Prof. Peirce, as in

Joseph Henry. Courtesy of the NOAA Photo Library.

calling the services of this gentleman into requisition, I have only done what everybody who is acquainted with his high character and splendid attainments has approved. In making Prof. Peirce Superintendent of the Coast Survey, I have served the country more than I have served him."[6]

Henry suggested to Peirce: "If you could remain in Washington say two weeks immediately after the holidays and do as Prof. Bache and myself did, viz. make a regular business of calling upon Senators and Members, I think it would be well. We went together and devoted four evenings a week until we had visited all of the most prominent members of both Houses."[7]

When Peirce arrived in Washington in January, he wrote Henry, "I am here and ready to do whatever you advise in regard to making calls."[8]

Peirce also maintained good relations with many scientists outside of Washington. Fortunately, he was not only cursed by past altercations, but blessed by past favors. In particular, Peirce had supported the candidacy of Joseph Winlock for the directorship of the Harvard Observatory. Winlock, Peirce's protégé, had become director in 1866, just before Peirce became head of the Coast Survey. Peirce's son Charles began making observations at the Observatory in 1867 and was appointed Winlock's assistant in 1869. Working relations between the Coast Survey and the observatory had never been better.[9]

$$\Phi^n$$

When Peirce became head of the Survey in February 1867, the State Department, led by William Seward, was negotiating with the minister of Russia to purchase Alaska. The treaty ceding Alaska to the United States was signed on March 30 and ratified by the Senate on April 9. Before the terms of the treaty could be fulfilled, however, the United States had to pay Russia $7,200,000. This required a Congressional appropriation that had to originate in the House of Representatives. Although the first stages of the purchase had gone smoothly, the appropriation from Congress was in doubt. In addition to the ardent anti-expansionists in Congress, President Johnson's political enemies saw the purchase as a means of embarrassing the administration. Opponents to the purchase called Alaska a "worthless iceberg" and "Seward's folly." A small but vocal segment of the press launched a vigorous campaign denouncing the purchase. The administration responded with a publicity campaign of its own.[10]

Despite a previous scientific expedition to Alaska, little was known of the territory.[11] Merchants and sailors who worked in the area generally regarded their knowledge of the territory as proprietary and were unwilling to share it.[12] Well before strong opposition to the purchase had been apparent, Senator Sumner had called for a scientific expedition to the territory in his speech in favor of the purchase. The New York *Herald*, which supported the purchase, saw an exploring expedition to Alaska as the means to settle "vexing questions of climate and the general worth of Russian America." The *Herald* recommended that: "Our government should without delay dispatch a steam revenue cutter from San Francisco for special service this summer in exploring that portion of the coast which half a century of experience has taught the Russians is available for general commercial purposes. . . . with an active and wide awake little corps of the proper men, untrammeled by red tape restrictions, the expedition would bring back more live information on all important points than could be obtained by years of fumbling among the accounts of ancient navigators and musty books of travel."[13]

Seward quickly organized an exploration party, asking Peirce, through the secretary of the treasury, to find someone to lead the scientific corps for the expedition. Peirce was excited about his part in acquiring the new territory. His wife wrote: "There is already issued a map of the new Russian Territory which looks quite tempting & we are thinking of taking a party out to survey it. It is a new world to conquer. One needs to be in our house but a day by the way to feel what a thing it is to be a 'public functionary'. Red tape abounds on the study table—& the room is thronged with men who want some great work assigned them & rolls of paper & voluminous letters are coming all the time—We are all well thank God. Your Father never better."[14]

Captain W. A. Howard was chosen to lead the expedition on the steam cutter *Lincoln*. Peirce requested George Davidson to lead the scientific corps of the expedition, comprised of men from both the Coast Survey and the Smithsonian. Davidson had been in charge of the Survey work on the West Coast before the Civil War, although he was currently in Pennsylvania, recuperating from a fever he had acquired while in Panama. Davidson had taken a leave of absence from the Coast Survey to be the chief engineer for an ill-fated surveying expedition making explorations for a canal there.[15] Citing his poor health, Davidson declined, but Peirce urged him to go.[16] He wrote: "I could not avoid great disappointment at your declining the reconnaissance of the New Coast of the Pacific. It is, you know, to be sent out under the special patronage of the State Department, and the Commander will have diplomatic office and functions. It will be his duty to collect all materials important to influence the House of Representatives in reference to appropriations for the purchase, among which it will be his special duty to develop the facts as to the maritime and military importance of the islands which are just upon the great circle track from California to China and to show whence the coal can come from for the purposes of steam navigation. Such information can only be collected by a man of comprehensive mind and it will require the authority of an influential name to give it weight in Congress."[17]

While sympathetic to Davidson's poor health, Peirce pointed out that robust and vigorous health was a necessity for the principal assistant to the Coast Survey on the West Coast, and if Davidson's health prevented him from performing his duties, he would have to be reassigned.[18] Peirce succeeding in persuading Davidson to take the job, and sent him further instructions:

> It will be your duty to prepare a descriptive memoir of the coast, which you will compile from the best authorities, and to combine therewith all the results of your own reconnaissance.
>
> It is highly important that you should examine the chain of islands, as far at least as Unalaska, in order to find a suitable place for a coaling station for vessels in the Japan and China trade. Hence, good harbors, clear entrances, absence of fog, and if possible, presence of coal on the islands, must be amongst the objects of your special investigation.
>
> It is stated by Mr. Bryant[19] that fishing banks exist, running parallel with the islands, in latitude about 50° N. The determination of that question is of especial interest.
>
> Local surveys of places which appear favorable for maritime purposes are quite desirable.
>
> It is understood that the subjects of Timber, agricultural capacity of land, the existence of Coal, of Gold, &c, and report on the general geological features will be in the province of the Smithsonian observers. You will thus be left to study the essential features which come within the usual range of operations in the Coast Survey.[20]

Peirce was now optimistic about the venture; he wrote Davidson, "I am very much pleased at the very promising aspect of your expedition to Alaska, and especially that you have undertaken it with such hearty good will."[21]

Davidson's journey to Alaska was indeed arduous. He took a steamer from New York to Panama, where he crossed the isthmus. From there he sailed to San Francisco, where the treasury department made the *Lincoln* available for the Alaskan expedition. The scientific party left San Francisco, under Davidson's direction, on July 21, 1867.

The party's assignment, to explore such a vast territory in such a short time, was even more daunting than the journey itself. Its mission was complicated by bad weather, especially persistent fog. Davidson collected existing documents and relied heavily on them in his report. After a tour along the Alaskan coast, the *Lincoln* returned to San Francisco at the end of November.[22]

George Davidson. Courtesy of the NOAA Photo Library.

Peirce considered the expedition to be a success; he wrote to Davidson: "Welcome back to the Atlantic. You come laden with treasures of information and I am thankful that your health has not suffered. When your report is ready, it is well that we should peruse it together, and that then you should be permitted to lay it before the Secretary. . . . I am also desirous that you should come before the National Academy, which will [be] on the 22nd of January, and give a full account of your observations and results of all kinds—and we will have a nice discussion of them."[23]

After thanking Davidson for his exemplary service in Alaska, Peirce made it clear that he had not asked him to make this sacrifice on his own account. Davidson had been Bache's student and had remained devoted to Bache and his wife.[24] Peirce reminded him of their common loyalty to Bache's memory: "I like to have it understood that I am striving to accomplish no personal fame in the administration of the survey. I regard it as the creation of my dearest and best friend and all that I wish is to have it develop itself as it would have grown under his own care[;] that it will be all that he would have made it, I cannot hope—but whatever it is which is praiseworthy, is his and wherever it fails—I am made to acknowledge the failure as my own. In such a game, I have everything to lose and nothing to gain—and nothing but my love of the blessed Chief would have induced me to risk such an undertaking."[25]

Later, Peirce reported to Davidson that, "Your report is in the hands of the President for transmission to Congress and he no doubt will be presented at an early day."[26] Seward hoped that the report would convince the House to pass the appropriation for the purchase of Alaska. In this sense it might be regarded as a failure, for the report was sufficiently objective that, during the House's debate on the Alaskan appropriation, Congressmen on both sides of the question frequently cited the report.[27]

There were many considerations that were important in the purchase of Alaska: its rich fishing beds, its coal and other geological resources, as well as potential agricultural value. But perhaps the most significant factor was that Alaska dominated the northern Pacific and its trade routes. Peirce was aware of this well before the debate in the House began. He wrote to Davidson, "I am much pleased with the graph evinced in your letter, of the important question in regard to the sailing lines between San Francisco and Japan."[28] On the House floor, several congressmen expressed this view. Among them was Representative Orth of Indiana who said : "With the possession of Alaska and her hundred islands we can command and control the commerce of that ocean [the Pacific]. With Alaska in the possession of a commercial rival such as Great Britain our commerce will be crippled if not destroyed."[29]

Φn

Secretary of State Seward hoped to acquire other territory for the United States as well as Alaska. In 1867, Seward had just acquired the islands of St. Thomas and St. John from Denmark and was exploring the possibility of acquiring Iceland and Greenland from the same country. One of his expansionist supporters, Robert J. Walker,[30] requested more information about the countries from Superintendent Peirce. Peirce had his son, Benjamin Mills Peirce, who had just returned from Paris where he had been studying at the École Polytechnique, compile *A Report on the Resources of Iceland and Greenland*.[31] The elder Peirce submitted it to Seward in December 1867. The report was published by the State Department, but there was little Congressional interest, and no action was taken.[32] Indeed during the debate over the Alaskan appropriation in the House, one congressman mocked the proposal by asking if the country was going to annex Iceland and Greenland next. His question was greeted with laughter.[33]

The next year, Peirce had Davidson review M. Hellert's work on the exploration of Panama, looking to build a canal there. Davidson gave suggestions for conducting a survey there in the future. He concluded, "I am convinced that there is enough shown by this exploration to warrant the use of it as a basis for an exhaustive exploration of this region; especially of the heights lying between the head-waters of the Puero, the Paya, and the Tapanaca Rivers, and the Alrato or its tributaries."[34]

Another study was done six years later by the Coast Survey concluding that it would be feasible to reconstruct the harbor in the neighborhood of Greytown, Nicaragua.[35] This would serve as one of the terminal points of a canal through Nicaragua and the isthmus of Darien.[36]

Φn

Peirce did not attempt to disguise the scientific activities sponsored by the Coast Survey that were beyond the strictly practical goals of the agency. Rather, he was pleased to point them out:

It is a distinguishing feature of the service under my charge, that while it has a specific and direct object in its bearing on the interests of commerce and navigation, the performance involves operations and investigations which almost rival in value the primary functions of the survey. The methods and procedures used have been at all times the best afforded by science and art, and the form of publication has gained in accuracy and beauty, so that our charts from the first have been unsurpassed by any which have been elsewhere produced. The methods of astronomical observations employed in the survey are now universally adopted, and have greatly increased the precision with which elements of the relative position of places upon the earth are determined. Our deep-sea explorations have incidentally opened new worlds of discovery to the naturalist and the physicist. The laws of the tides and the distribution of magnetism have been traced with increased distinction, and have been made more intelligible by the observations made in the progress of the survey.[37]

Peirce was quick to point out the nation's obligation to support pure science. For example, referring to the government's supporting parties to view an eclipse of the sun, he stated: "Certain astronomical phenomena of rare occurrence and high importance for the advancement of human knowledge have, in all civilized countries, since modern science has been cultivated, been deemed matters of national importance. Among these are total eclipses of the sun; and for many years it has been customary for the great nations to organize expeditions for the observation of them."[38]

The Survey supported several expeditions that made fundamental contributions to astronomy while Peirce was superintendent of the Survey. Anticipating excellent conditions for observing a total eclipse of the sun on August 7, 1869, Peirce stressed that "it is nothing less than a duty owing to civilization that everything in our power and within our means should be done to make the observations as complete as possible."[39] The Survey sent parties to five different locations to observe the eclipse, along its line of totality. They included: Sitka, Alaska; Des Moines, Iowa; Springfield, Illinois; Shelbyville, Kentucky; and Bristol, Tennessee. Peirce personally supervised the observations in Illinois, and Sarah accompanied him.[40]

After observing the eclipse, Peirce telegraphed home "it was a glorious success."[41] Peirce had much to be happy about. The observations of the sun's corona and perturbances suggested new theories about the composition of the sun. But European astronomers questioned the American theories. The earliest opportunity to verify the American's hypotheses was a second eclipse of the sun, occurring on December 22, 1870.[42]

This time the total eclipse would be visible in southern Europe. Peirce wrote to John Armor Bingham, the representative from Ohio, on March 8, 1870, requesting funds for the expedition:

It is of the greatest scientific importance that competent observers, who may have had the opportunity of observing any total eclipse, should, as soon as possible, upon the recurrence of that phenomenon, be placed where they can again observe it. The phenomena as they appear to the actual observer are so unexpectedly different from any previous conceptions which he may have formed; that which he expected to find clear and distinct is obscure, while the reverse also occurs; so that in the exceedingly short period of duration of totality he has not time for that careful consideration and organization of thought which shall produce the best observations. But after an experience he is quite another man[;] and notwithstanding the extraordinary success of American observers in the observations of the last total eclipse in this country, if they could be sent to the next opportunity for such observations, I am sure that the results will be found worthy of the country and the expedition. Parties will no doubt be sent, as they hitherto have been from England, and France, and I hope that your powerful influence will prevail in securing the presence of American parties also. The next total eclipse of the sun . . . occurs on the 22nd of December 1870. . . . it appears to be the only good opportunity for observation which will occur for a very long time. It would be, therefore, particularly unfortunate if it were permitted to pass unobserved by the presence of American Astronomers.[43]

Bingham responded by submitting a bill to the House to provide an appropriation.[44] Peirce was surprised when he learned that this bill would put him in charge of the expedition: "Your exceeding kindness and most liberal spirit, in undertaking to carry through Congress the bill for the observation of the eclipse of next December, demands the utmost frankness upon my part. When I wrote my letter concerning it, I had no suspicion that it was intended that the expedition should be entrusted to me. Should Congress consent to this mark of confidence in me, I shall feel it my duty to direct the operations in such a way, as will be creditable to the country and worthy of its science. All that the astronomers of the expedition will hope or expect is the payment of their transportation and subsistence, which must I think be estimated for each person at about one thousand dollars."[45]

Peirce then discussed the cost of the expedition, estimating that twenty-eight thousand dollars would be sufficient. Bingham had proposed only twenty, and Peirce assured him that with slightly reduced personnel, that sum would be adequate, "and I shall accept this as a very munificent allowance." Peirce then stressed the importance of the United States taking part in this venture, so as to be a full partner in the world's scientific community: "England is already preparing for the observation of the eclipse on the most magnificent scale, and there is no doubt that the best astronomers of France will be sent to

Algeria with the best equipment which the science and art of Europe can accomplish. But the observers of America are not unwilling to compete with them on their own ground, and I hope that the hearts of Congress will beat like your own in ardent sympathy with this generous rivalry to advance knowledge."[46]

Bingham, either finding or anticipating problems with the passage of the bill, wrote to Peirce seeking a better understanding of why the expedition was of scientific importance.[47] Peirce requested Hilgard to respond immediately. Hilgard stressed the scientific importance of the observations: "The advantages that are expected to accrue to science from the contemplated observations of the Solar Eclipse in December next are additions to our knowledge of the physical constitution of the sun and of the corona surrounding it. There is a class of facts bearing strongly on the nature of the matter of which our planetary system is composed and on the manner of its formation, which can be observed only on occasions of the total eclipse of the sun."[48]

A few days later Peirce responded, again stressing how important the event was to Europeans: "The great importance, which attaches to the observation of a total eclipse of the sun, is evident from the fact of the great preparations being made in Europe for the observation of the eclipse next December. England alone proposes to send sixty observers to the line of the central path under the direction of S. B. Airy Esq., who is the acknowledged head of English science. Why Europe sent no parties to this country for the admirable eclipse of last August is not easy to be understood."[49]

Peirce then pointed out that American science should be able to compete admirably with that of Europe. "The new modes of observation, which American genius has suggested, will probably lead to very valuable results and will throw new light upon the sun, and illuminate the darkness of its mysterious construction."[50]

He then took this opportunity to stress that such expeditions should be administered by civilian scientists and not the military: "Turning to some questions which may have perhaps arisen as to the proper authority to which the conduct of these observations should be entrusted, I will similarly observe that scientific matters will always be best administered in the purely scientific hands. This is the universal experience of Europe—and it is surely not the genius of our institutions and of our people—to place civil service under military control. The principles of civil and military service are essentially different. In the latter, subordination must be rightly enforced, and the inferior must be sacrificed to the commander. But the success of civil and scientific service depends upon the earnest enthusiasm of all the assistants, and this enthusiasm can only be secured where every man engaged in the service receives his due compensation of credit for the work which he performs."[51]

Although it was imperative to Peirce that scientists direct scientific work, he had no objections to having a congressman accompany the eclipse expedition. He invited John Bingham to do so, but the congressman declined.[52]

Since experience in observing an eclipse was vital, Peirce hoped to have as many men who had observed the 1869 eclipse also view this one.[53] He also hoped to take full advantage of the new techniques in astronomical photography being developed by his friend Lewis Rutherford. He constantly sought his advice and told him, "I regard myself as the mere steward of science in this expenditure, and it is preeminently my duty and your right to have that fame which your great experience shall dictate as likely to promote the success of the expedition."[54]

Not waiting for Congress to pass the appropriation for the expedition, Peirce sent his sons Charles and James to Europe to make preparations for the observations.[55] They combined the business of the Coast Survey with a leisurely tour of Europe.

Peirce's sons suggested two locations for viewing the eclipse, one in Sicily and one in southern Spain. Benjamin Peirce followed, taking Charles's wife, Zina, with him as his personal assistant. Peirce, in charge over all, personally conducted the expedition to Sicily, and requested that his friend, Joseph Winlock, now director of the Harvard College Observatory, conduct an expedition to Spain.[56]

When Peirce went to London to make preparations to view the eclipse, he learned that the British government had not provided support for the British parties that had hoped to observe the eclipse in Sicily. Peirce, wishing to aid Britain's scientists, and perhaps embarrassed after stressing to Congressman Bingham how much the English were doing, wrote the British astrophysicist, J. Norman Lockyer: "I have been directed by the Government of the United States to have the best possible observations made of the total eclipse of the next December. If I could aid the cause of astronomy by assisting the observers of England in their investigations of the phenomenon, I should be greatly pleased. I take the liberty therefore to invite your attendance, and also that of other eminent physicists of England, with either of the parties of my expedition one of which will go to Spain and the other to Sicily."[57]

Lockyer explained Peirce's involvement in the British expedition: "The Government had been approached for two men-of-war and for some money by the Royal Society, while the American astronomers had a large subsidy. The American astronomer, Professor Peirce, came over to see me and wanted to know what we were doing. When he heard how we had been treated he asked me if I would go under the 'Stars and Stripes.' I said I would if the Government would allow me, and went to the American Ambassador and suggested that the two of them should see the prime minister, Gladstone."[58]

Peirce, unbeknown to Lockyer, visited Gladstone, expressing his regret that Her Majesty's government had not made any provision for British astronomers to view the upcoming eclipse of the sun. Gladstone responded that he was not aware of the situation, and further, assured Peirce that any request for funding from British astronomers would receive his serious consideration.

The chancellor of the exchequer sent for Lockyer and asked what all the trouble was about. The end result of the matter was that the English expedition went out as originally planned.[59] A gratified Peirce wrote, "Is it not strange that I am exerting an important influence upon English science. It almost seems to me as if they depend upon me here more than at home."[60]

But Peirce's cooperation had helped forge a link between American and British science and scientists. To both Lockyer and Peirce, this was one of the most important aspects of the expedition. Peirce wrote, "The American and English parties were in co-operation, and afforded each other mutual aid. It is hoped that the good feeling thus engendered was not without influence beyond the circle of science."[61]

Unfortunately, bad weather interfered with the observations of both parties. In some cases the clouds parted just in time. Joseph Winlock wrote, "a few minutes before totality, a rift of blue appeared in the west, and we hailed it as affording a gleam of hope. Slowly it drifted along, directing its course straight toward the sun now reduced to a narrow crescent. At last it reached it; in a moment more the moon had completed its event, and there, in clear air, between the dense masses of heavy clouds, hung the beautiful spectacle."[62]

Parties at some locations did better than others, but all of them were able to get at least partial results through breaks in the clouds. Zina Peirce, though not a scientist, was able to record an unknown phenomenon. "She was successful in drawing the corona, and distinctly recognized the dark rifts which have become a subject of discussion, and which were photographed by Mr. Brothers, of the British party at another station."[63]

Despite the bad weather, the expedition was a success, and the Americans' theories of the previous year proved to be essentially valid.[64] Peirce was elated: "The eclipse expedition has been a marvelous success. We have solved problems and brought to light new questions. We have cut off heads of the hydra and new heads have sprung up in their place. The mysteries of science are as eternal as the soul, as infinite as divinity. There is no longer a doubt that there is external to the ordinarily visible sun, an atmosphere at least fifty thousand miles high and perhaps a hundred and fifty."[65]

Φn

Peirce may have been happy with the expedition, but his Aunt Saunders felt he'd been away too long, and told everyone that she would not consent to Benjamin's "going to see any more eclipses."[66]

Φn

Peirce's most significant accomplishment while he was superintendent of the Coast Survey was expanding the scope of the Survey by making a geodetic link between the surveys that had been made on the East and West Coasts. By obtaining an appropriation of fifteen thousand dollars to begin this work, Peirce, in fact, transformed the Coast Survey into a coast and geodetic survey.[67] In recognition of this change, the name of the agency was changed to the Coast and Geodetic Survey in 1878.[68]

The political accomplishment of greatly extending the scope of the Survey pales in light of the scientific accomplishment of actually doing the survey. This triangulation along the thirty-ninth parallel was the longest arc of a parallel ever surveyed by any one country. In making this survey, fundamental progress was made in many areas of geodesy, including "reconnaissance, signal building, triangulation, and methods of computing." This arc also made a more fundamental contribution to science by adding knowledge of the earth's shape and size: "The great triangulation system along the thirty-ninth parallel is probably the greatest single contribution to the world's geodesy that has been made by any one country. It marks an epoch in the scientific history of the United States and in that of the world."[69]

Φn

Under Bache, Peirce, and Peirce's successor, Patterson, the Coast Survey was the most important federal agency for American science not only because it supported extensive work of its own, but because it financed scientific work of others outside the Coast Survey. For example, Peirce aided the geologist, James Hall, in preparing a geological map of the United States.[70] But the best-known example of this support was an expedition the Survey sponsored for Peirce's friend Agassiz that began late in 1871 and continued until the fall of the next year. Peirce received two new ships for the Survey, which he named the *Bache* and the *Hassler*. Although built on the East Coast, the *Hassler* was intended for service in the Pacific. Peirce saw the transporting of the *Hassler* as an opportunity for American science. He wrote Agassiz: "But now, my dear friend, I have a very serious proposition for you. I am going to send a new iron surveying steamer round to California in the course of the summer. She will probably start at the end of June. Would you and Lizzie go in her, and do deep-sea dredging all the way round? If so, what companions will you take? If not who shall go?"[71]

Agassiz was just recovering from a cerebral hemorrhage, which had forced him to be bedridden for many months. Nevertheless he was excited about his friend's proposal, "Your proposition leaves me no rest. . . . I do not think that any thing more likely to have a lasting influence upon the progress of science ever was devised."[72] Two days later he wrote again, telling Peirce that only poor health would keep him from taking such a trip: "But even then I would like to have a hand in arranging the party, as I feel there never was and is not likely soon again to be such an opportunity for promoting the cause of science generally & that of Natural History in particular. I would like Pourtalès and Alex [Agassiz] to be of the party & both would delight to join if they possibly can. . . . I have no doubt between us we may organize a working team strong enough to do something creditable."[73]

Delays in the *Hassler's* construction postponed her departure from Boston until December 4, 1871. Agassiz was able to make the voyage, and assembled an outstanding scientific team to accompany him. One key member was the naturalist Louis de François Pourtalès, an old friend of Agassiz and an employee of the Coast Survey. Pourtalès had had extensive experience dredging to collect samples from the ocean floor. Another important member was Dr. Franz Steindachner, a Viennese naturalist, who had extensive experience in collecting specimens along shorelines. Agassiz's wife, Elizabeth, and Thomas Hill, the retired Harvard president, also accompanied the party.

Agassiz had three goals in making this voyage. One was to collect specimens for his museum. The second was to look for evidence of glacial action in the Southern Hemisphere. Previously Agassiz had discovered somewhat flimsy evidence of glacial action in Brazil; nevertheless he was convinced that much of the South American continent had once been covered by a glacier, just as the North American continent had. Agassiz's third reason for taking the voyage was to reconsider Darwin's theory of evolution. The *Hassler's* voyage would replicate much of Darwin's explorations of South America on the voyage of the *Beagle,* including those of the Galapagos Islands.[74] Agassiz had mellowed greatly with respect to Darwinism, and no longer dismissed it using the metaphysical arguments that had been his mainstay in the past. Now his criticisms were valid scientifically and gave important insights into the problem of evolution, helping followers of Darwin to improve their theories.[75] Agassiz wrote to a colleague: "I had a special purpose in this journey. I wanted to study the whole Darwinian theory free from all external influences and former prejudices. It was on a similar voyage that Darwin himself came to formulate his theories!"[76]

Darwin knew of the voyage and was enthusiastic about it. He wrote to Agassiz's son, Alexander, "Pray give my most sincere respects to your father. What a wonderful man he is to think of going around Cape Horn; if he does go, I wish he could go through the Strait of Magellan."[77] Just before departing, Agassiz wrote Peirce discussing the questions that he hoped the voyage would answer, including Peirce's own theory of continental drift.[78] The dredging was important because Agassiz believed that the ocean's deepest waters were the habitat that would most closely represent the environment on earth millions of years ago.[79] Unfortunately the dredging equipment did not work as well as Peirce had hoped it would. Sometimes it did not work at all, and it was never able to reach the predicted depths.[80]

Although Agassiz gathered a great deal of data pertinent to the question of evolution, and visited the Galapagos Islands, just as Darwin had, he did not become a believer. He wrote to Peirce:

> Our visit to the Galapagos has been full of geological and zoölogical interest. It is most impressive to see an extensive archipelago, of *most recent origin,* inhabited by creatures so different from any known in other parts of the world. Here we have a positive limit to the length of time that may have been granted for the transformation of these animals, if indeed they are in any way derived from others dwelling in different parts of the world. . . . Whence, then, do their inhabitants (animal as well as plants) come? If descended from some other type, belonging to any neighboring land, then it does not require such unspeakably long periods for the transformation of species as the modern advocates of transmutation claim; and the mystery of change, with such marked and characteristic differences between existing species, is only increased, and brought to a level of that of creation.[81]

Agassiz did find conclusive and abundant evidence of Pleistocene glaciation.[82] And he was able to collect a prodigious number of specimens for his museum.[83] He reported to Peirce: "With the results of this last expedition the Museum is put upon a level with the best. I am sure the materials we have on hand are better adapted to advance science in the direction in which progress is most needed than the collections of any other similar institution, and I would not except the British Museum or the Jardin des Plantes from this statement." Agassiz was so encouraged that he now foresaw the dominance of American science: "Our next step must be to prepare such a course of instruction, that European students, capable of appreciating the difference, may prefer to come to us to finish their scientific education, than to remain at home."[84]

Φⁿ

Before the Civil War, the navy made hydrographic surveys, as well as the Coast Survey. The Coast Survey was responsible for surveys along American coasts, and the Hydrographic Department of the navy made surveys along foreign coasts. In addition, the navy had provided ships and men to the Coast Survey. After the war the navy was anxious to resume using its ships and men in the Coast Survey.[85]

From the point of view of the military, there were many advantages to having the Coast Survey under its jurisdiction. It was a natural use of military personnel during times of peace, when there was little else for them to do. In addition, the Coast Survey served as an effective training ground for naval personnel.[86] During the early years of the Coast Survey, Congress had taken the Survey away from Ferdinand Rudolph Hassler, as an obvious cost-cutting measure, and put it in the hands of the navy. Unfortunately, the naval officers lacked geodesic expertise, and the Survey had made essentially no progress after fourteen years of naval supervision, when it was finally turned back over to Hassler in 1832.

Despite the navy's failure, attempts to put the Coast Survey under naval control persisted. The Survey was once more put under their control from 1834 to 1836. Again, little progress was made in surveying the coasts.[87] Some members of Congress attempted to put the Survey under dominion of the navy while Bache was superintendent. But the politically wily Bache was able to thwart off the attacks.[88]

Aware of Hassler's and Bache's problems with the military, Peirce was reluctant to use the navy's resources. To Peirce the only advantage of using vessels for the Coast Survey that were under the control of the navy was that they did not come out of his budget.[89] He wrote, "What shall we say with regard to the navy. Will it cost less to use it or not to use it. Is that the question?"[90] Hilgard reported to Peirce on this matter: "I have carefully looked over the laws and regulations

with reference to the Hydrography, and like yourself I cannot perceive any escape from having Naval officers as chiefs of hydrograph-
ical parties at <u>sea.</u> . . . I presume we can keep the surveys of the smaller
bays & lagoons in the hands of our civil assistants, by not calling the
parties <u>hydrographical,</u> but something else, <u>surveying,</u> if you like. . . .
But for the work at sea we must have Naval officers."[91]

Peirce wanted to have as little to do with the military as possible.
Senator Edmonds of Vermont, unaware that inland lakes were under
the jurisdiction of the Army Corps of Engineers, passed legislation
requiring the Coast Survey to survey Lake Champaign.[92] The head
of the corps, General Andrew A. Humphreys, was fiercely competi-
tive and tolerated no rivals.[93] When he complained to the Coast Survey, Peirce was not intimidated:

> I cannot for the life of me tell what Humphreys has to do with me and
> my work and my appropriations more than I with him. When did he ever
> set me the example of explaining to me any of his peculiar items of appro-
> priation which are such evident encroachments upon the Coast Survey. I
> am surprised that you thought it of any importance to . . . discuss the mat-
> ter with him at all. As to his having any participation which is properly
> applicable to Lake Champlain it is all nonsense. But if he thinks so, and
> undertakes to come upon <u>my ground,</u> he may find that <u>he</u> has made a <u>mis-
> take</u> and not the Coast Survey. It may be that Congress desires to place the
> whole country under a weak military control—but it is yet to be demon-
> strated that the country will acquiesce in such an arrangement or that
> Congress intends it.
>
> However that may be, I believe that we know our duty and will perform
> it without the advice of Army or Navy. We, i. e. the Coast Survey Service,
> is . . . "Small but plucky." At any rate we will hold our heads up as long as
> we have heads on our shoulders. And no kneeling to General <u>this</u> and
> Admiral <u>that.</u>[94]

The economies of having the military doing surveying in times of
peace was tempting to many legislators, and the resources the Survey
expended on pure science was thought to be wasteful by many. The
first indication of an assault upon the Coast Survey appeared in two
newspaper articles. The *Boston Evening Journal* reported that: "A
Department of Science is the last proposed addition to the Executive
Branch of the Government. It is to be composed of the Storm Signal
Corps of the army, the Lighthouse Board and Coast Survey of the
Treasury, and the Hydrographic Bureau of the Navy. With this propo-
sition for consolidation comes complaint that the Coast Survey
Bureau, in becoming an adjunct of Harvard, and the scientific ambi-

tions of Cambridge Professors, is to some extent neglecting the practical side of its work. Great delay in the preparation and publication of charts is asserted, and it is affirmed that years instead of months elapse before changes in important lighthouses and buoys are indicated on the charts furnished to navigators."[95]

The *National Republican* issued the same message, giving as an additional rationale for consolidation that "the Scientific undertakings of the Government have grown wonderfully since the war."[96]

A few months later an additional warning came from Peirce's former student, the astronomer, Simon Newcomb. Charles Saunders Peirce wrote his father: "Newcomb told me very privately that Harkness was seriously moving to get up an attack on the Survey in the winter & have it transferred to the Hydrographic office. He added that every professor at the observatory hoped he would go on & get squelched. . . . If he does get up an attack on the C S would it not be well to court an inquiry & have done with it & squelch Harkness incidentally."[97]

The attempted attacks of the navy's hydrographic office were apparently easily warded off, but there was a serious attempt by Humphreys and the Army Corps of Engineers to take over the geodesic work that Peirce had just begun. When this attempt was made, a number of scientists came to Peirce's aid, including Winlock and Henry, but none were more ardent in their support than Louis Agassiz. Agassiz rushed off a stream of letters asking for support for Peirce and the Coast Survey, making a different appeal in each of them. To Henry Wilson, one of the senators from Massachusetts, he wrote, "You have always been a good friend to me for more than 25 years & I now make an earnest appeal to you to save science from one of the severest blows it could receive. . . . I need hardly add that the C. Survey has been in the hands of civilians since the beginning of its organization for the obvious reason that neither the officers of the military, nor those of the navy possess that comprehensive knowledge & true scientific spirit which is essential to the proper management of the various interests of the C. S."[98]

To Massachusetts's other senator, Charles Sumner, Agassiz wrote, "It is attempted in Washington to give science the severest blow it could receive and to deprive it of the little independence it has acquired in the U. S. by placing it under military rule. It is the vilest plot I have yet heard of; and any thinking man is bound to revile & denounce it."[99] To the German-born senator, Carl Schurz, he complained, "I appeal to you as a friend to come to the rescue of science. A vile attempt is made to dwarf the Coast Survey & put its most important operations under the control of the Army. This is virtually putting an extinguisher upon science in the U. S. The C. Survey is the only national institution which has been conducted upon scientific principles, it is the only scientific school we have, the only resort of scientific men; for Annapolis & West Point are only practical schools for specific subjects. There is not <u>one</u> officer in the Army who has the comprehension of scientific culture necessary for the management of the Coast Survey; from the beginning it has been under the Superintendence of civilians, Hassler, Bache, Peirce."[100] In a letter to a Dr. Heck, he simply stated, "Peirce is in trouble . . ." and then told him, "Now my dear Sir, begin writing, writing, writing, until you have sent a good article to each of the newspapers to which you have access. . . . Stimulate all your friends to write also."[101]

Agassiz reported back to Peirce that the geologist, Josiah Whitney, had prepared a petition, and that both Agassiz and Whitney had solicited signatures for it. "Should you want further assistance Whitney is ready to appear before the Committee with voluminous evidence. . . . Can I be of any use command me also. The evidence of Whitney would be crushing."[102]

Peirce was able to defeat the army in the House committee. He reported to Henry: "The chairman has congratulated us on our success so that we seem to be as safe for the present as it is healthy to be. As chronic invalids never die—so there seems to be salvation in the weakness which never dares to fall asleep with both eyes at once. If I had needed it, your strength would have been given in all its power to save the Survey of Dallas Bache—as it is, you will rejoice greatly in our victory."[103]

Peirce gave a more complete report to Josiah Whitney:

> We have had the fight in the committee—it was forced upon us—it was short but intense—and the victory is ours. In congratulating me upon it the clerk said he had not for a long while witnessed so sharp a fight in the committee—and at one time he trembled for the result. But the principle is now fully settled in our favor in committee and it will not be raised in the House. To bring it up in the House, would be regarded as a breach of faith. . . . It has been a mighty good thing for us that we were thus driven to the wall—and that our adversaries experienced such a injurious and decided repulse. You may be sure that we did not spare words and threats. "The country should be aroused" and all that sort of thing—and the ignorance of the army of geodesy was very distinctly asserted. The bill will be reported tomorrow, and as soon as the House acts, I shall go to the Senate committee—but there is great reason to believe that the Senate committee is strong on our side. . . . We may then consider this campaign as closed—and look out for the future.[104]

Although Peirce was successful in fighting off the army, the appropriation for the Survey that year was disappointing. An angry Peirce
wrote, "But what can we expect from the gang of pickpockets which
now govern Congress. Their first object is to put their hands into
Uncle Sam's pocket and I rate them quite below the pickpockets
who robbed me in New York. . . . I fear that even the immaculate
Sumner and Wilson are not to be excepted from this disgraceful
transaction."[105]

$$\Phi^n$$

Peirce's success in thwarting the army's attempt to encroach upon
the Coast Survey didn't put an end to attempts to combine the Coast
Survey with other agencies. In June 1873, Peirce's son Charles reported
another attempt to combine several federal scientific bureaus, much
like the one the year before.[106] Peirce was ever vigilant to avoid en-
croachments by other branches of the government. Even before Congress had passed the appropriation for the 1870 eclipse expedition,
Peirce was making plans to send out American astronomers to view
the transit of Venus, which could be viewed from the Orient and
would occur in 1874. He was also being careful not to be politically
outmaneuvered by the Naval Observatory, which was under Benjamin
F. Sands. Sands was not a scientist, and thus particularly objectionable
to Peirce as an administrator of scientific work.[107]
He wrote to Davidson:

A committee of the National Academy of Sciences has been appointed
to decide the proper form of a memorial to Congress upon this matter
[the observation of the transit of Venus]. It consists of Admiral Davis (as
chairman) [,] Sands as head of the observatory and myself. . . . The original motion for this committee was made at the Acad. Meeting by one of
the Professors of the observatory who is considerable of a busy body, and
the plan evidently was to have the control of the whole matter given to the
navy, and have the committee under the direction of Sands. But I let
Henry know that such an arrangement was not satisfactory to the C. S. and
might easily be defeated, and insisted upon being upon the committee
myself—and not having any subordinate upon it. When this committee
meets, I may, if it be necessary, throw myself upon my reserved right and
exclaim that I cannot be expected to make arrangements with the Navy
men unless they have the full authority of their department to enter into
an agreement with me—and that they cannot expect one who is himself
a principal to deal with those who are not so. But I hope that I shall only
be obliged to show my teeth and not to use them. But I have already come
to a complete decision in my own mind, as to what I should specially claim

as the part of the Survey and what I should justly regard as belonging to the Navy (as such), and to the observatory as the representative of astronomy. I am the more embarrassed because the observatory is not under purely scientific control and I am quite indisposed to let civil service be ruled by military domination.[108]

Despite his early enthusiasm for viewing the transit of Venus, Peirce became ambivalent about the enterprise. But he told his former student, Simon Newcomb, that he was glad Newcomb had persuasive arguments to justify an expedition, and conceded that the plans for it should go on.[109] Peirce increasingly questioned the value of the undertaking. In 1873, he wrote to Henry: "I shall startle you by the announcement that, in my opinion we should go no farther in the Venus matter. I have come to the conclusion that this mode of determining the solar parallax is unequal to the present demands of astronomy and does not deserve the great cost which it entails. I deem it better to stop where we are—than plunge any deeper with a useless expenditure."[110]

Despite Peirce's misgivings, he did serve as a member of the Transit of Venus Commission, and his active support was crucial to the sending of American parties to view the phenomenon. Eight American parties observed the transit. Two of them were sent from the Coast Survey, one to Nagasaki under George Davidson and the other to Chatham Island in the South Pacific under Edwin Smith.[111]

$$\Phi^n$$

At sixty-five years of age and after seven years of being superintendent of the Coast Survey, Peirce had had enough. He resigned without qualms: "I do not feel myself diminished one tittle by stepping off the chief's platform. I have stepped off—not because things are worrying me or going wrong in any form or shape—but just the reverse. Everything connected with the Survey is in a most flourishing and thrifty state—so that it no longer needs me—and the secretary is such a personal friend that he has been ready to do precisely that which I desired—and so it is that Patterson is delighted. And I am satisfied. . . . I do not think that my crown, if I am crowned, is of such material that it can be removed by any administration's power. If it is worth having—it is of the immortal, which Apollo grows—and which man can not tear from my brow."[112]

With Peirce talking about retiring as head of the Coast Survey, Carlile Patterson was a natural, but reluctant, choice to succeed Peirce. He wrote: "If you do so, I shall probably be appointed, and then they

Carlisle P. Patterson. Courtesy of the NOAA Photo Library.

will have only the Supt. of the Coast Survey and <u>not</u> <u>Benjamin Peirce</u>—
and therein 'lies the difference.' I beg you not to resign and shall
hold your letter to the President. Until I hear from you—better still
<u>see</u> you. Come on here—stay with us.[113] Patterson's diffidence did
not prevent his becoming the new head of the Coast Survey. There
were other likely candidates as well as Patterson. Julius Hilgard, who
had been in charge of the Washington office of the Coast Survey all
the time Peirce was superintendent, certainly expected to succeed
him. But Peirce had never been impressed with Hilgard's abilities.
Peirce's son Charles agreed with him; he wrote his father, "As for the
scientific position of the Survey I think it would vanish forever with
Hilgard for Superintendent, & that henceforth the Survey would be
managed on political principles."[114] Needless to say, not everyone
was happy with Patterson becoming the new superintendent. Sarah
wrote: "The officers are all lamenting the change, excepting Captain
Patterson who is jubilant & assumes the chair with great confidence,
altho' all the men almost without exception dislike him. As for poor
Hilgard he is bitterly disappointed as he has always looked forward
to being Superintendent when Ben retired."[115]

Peirce continued to work for the Survey, retaining the title of "consulting astronomer." By the time Peirce retired, he and Patterson
were close friends, and Peirce remained an influential voice in the
running of the agency. Peirce and Patterson exchanged warm letters,
each addressing the other as "my darling chief."

14

My Most Precious Pearls

IN JANUARY 1857, BENJAMIN PEIRCE WAS PREPARING TO GO TO WASH-
ington to give a series of lectures at the Smithsonian. Three years earlier, when Peirce first contemplated the possibility of delivering these
lectures, his primary motivation was to invent an excuse to go south
to see his friends, the Le Contes. Never regarding himself as a good
public speaker,[1] he was not optimistic about his success in such a venture: "How absurd this would be? Nor who would wish to attend a
mathematician's lectures, even if he could deliver them? What could
I talk about? The mystic power of number is very well in its place with
the dear past or the unborn future—but to the living present it is a
kind of ghost at midday. People would not even be afraid of it."[2]

Despite his expressed diffidence, Peirce had in fact already talked
to Joseph Henry about a proposed topic. He wrote to Alexander Dallas Bache: "He [Henry] seemed to take quite an interest. Perhaps I
will present it to the next meeting of the Asses[3] if you approve. I call
it Potential Science. By this I mean the science which considers the
embodying of power; and which analyzes the acts of creation. This is
a bold subject and a dangerous one, if not discussed with humility
and in the true spirit of Godliness."[4] Peirce gave an outline of his proposed lectures in a letter to Josephine Le Conte: "I sent my subjects
to the majestic Smithsonian.[5] They were, in general, upon Potential
Physics—and the special subjects of the individual lectures were: 1st
the elements of Potential Physics. The material universe regarded as
a machine, as a work of art, or as the manifest work of God. 2nd
Potential Arithmetic or the Abacus of Creation. 3rd Potential Algebra or the Logic of Creation. 4th Potential Geometry or the music of
the spheres. 5th Analytic Morphology or the world's architecture. 6th
the realization of the imagination or the mysticism of matter. 7th The
powers of Justice and Love or the scientific necessity of Sinai and
Christ."[6]

At this point both Bache and Henry had misgivings about Peirce's
proposed lectures and warned him that his topic may not be suitable

for his audience.[7] An indignant Peirce replied to Bache: "If the subjects are out of sight, the lectures will be still more so, and I have no desire to deliver them to an audience which cannot comprehend them. Will you therefore ask the Smithsonian to let me off? If he does not feel that I am conferring a favor by letting him have them. They are the work of my life, and entirely the highest results of all my mathematical investigations. I would rather throw away all else which I have done than these lectures. They are my delight, I say, my most precious pearls, and it would be a great grief to give them where they were not valued. I know that I have a choice circle of friends whom I would collect about me and who would rejoice to hear me. But I almost feel that my best course is to write them out as I can, and leave them for publication by my children."[8]

Peirce confided to Josephine Le Conte:

> These subjects evidently frightened the immortal [Henry] and made him tremble on his throne. He hurried to our dear chief—who wrote to me at once that they were a peg too high—and quite beyond the vision of a Washington audience. To which I replied . . . that they were not the plays of Lancy—but the results of all my life of investigation. . . . Two replies came, one from the Chief, calling me haughty—and saying that it would not be wise to give the marrow of my discourse in such harsh form. I admitted the haughtiness of tone, but not in my heart towards the darling chief. . . . The great and noble Smithsonian—patted me gently on the back—said that my subjects needed pruning, and was quite sure that when he heard my philosophy, he would approve it. How did I answer him? Was <u>Your</u> friend "<u>umble?</u>" I said that my subjects were not so hideous as my lectures were wild and sterile, that I knew nothing about philosophy. I am a geometer and a geometer is incapable of ascending to esthetics and metaphysics. He had placed my lectures upon his program; I begged him to withdraw them and said that I would regard them as withdrawn if I did not hear further from Washington. What will be the result I care very little—less than I ought to care.[9]

Despite the wrangling, Peirce did go to Washington and deliver his lectures.[10] And in spite of Henry's and Bache's misgivings, Peirce's lectures were a huge success. After the first one, Bache wrote to Peirce's wife, praising Benjamin's presentation.[11] Peirce was gratified by the press coverage of the lectures, and was especially pleased with Henry's praise. He wrote his wife, describing a much different Joseph Henry, "Dearest Sarah, when a man—a real man—a great man—and a profound, earnest, sincere philosopher like Henry praises one in this way he must be less of a mortal than I am not to be elated by it."[12]

At the end of the series of lectures, he wrote: "My Darling wife, my lecture last evening, which was my closing one, seemed to be preeminently successful. It was listened to with the most earnest and rapt attention, and was constantly accompanied by a series of successive outward exhibitions of pleasure and satisfaction. The applause was always in the right place, and it was evident that I had the full sympathies of an audience capable of appreciation, even if they did not fully understand, my conclusions."[13]

Φ^n

Peirce usually infused his popular lectures with ample amounts of natural theology. That is, with the evidence of God's existence and character that can be derived from the study of nature, and is not based on scripture or revelation. At that time natural theology was regarded as a respectable academic field. Many eminent scholars wrote on it, and in most colleges a course on natural theology was given to all seniors, typically by the president of the college himself. Neither was it unusual for scientists of this period to talk and write about the theological implications of science. For example, Peirce's associate, Agassiz, saw strong ties between science and religion and often talked about them in front of general audiences.[14] Peirce's former student, Thomas Hill, seized the founding of the American Association for the Advancement of Science as an opportunity to discuss the value of science to religion.[15] One prominent example of a written work that combined science and religion was Edward Hitchcock's *The Religion of Geology and Its Connected Sciences*. This book served the dual purpose of teaching geology and defending the Christian faith.[16]

Peirce found deep religious meaning in science; he believed that "There is one God and Science is the knowledge of Him."[17] Peirce was particularly concerned that society not view science as being opposed to religion. Doing so was not only incorrect; it put science in a bad light in the public mind. Speaking before the American Association for the Advancement of Science as its president, Peirce said: "There are men, and pious men too, who seem honestly to think that science and religion are naturally opposed to each other; than which I cannot conceive a more monstrous absurdity. How can there be a more faithless species of infidelity, than to believe that the Deity has written his word upon the material universe and a contradiction of it in the Gospel?"[18]

Thus any apparent inconsistency between science and revealed truth was illusory: "We may rest assured that Nature is harmonious with herself, and that the inconsistencies by which science may be embarrassed are temporary clouds in the intellectual atmosphere. They will be swept away by the besom of the true wizard."[19]

Peirce was not content to simply extol the harmony of science and religion. To him, as to many of his contemporaries, science not only posed no threat to religion, but was a powerful positive evidence of religious faith. Again, speaking before the American Association for the Advancement of Science, Peirce said: "Is religion so false to God as to avert its face from science? Is the Church willing to declare a divorce of this holy marriage tie? Can she afford to renounce the external proofs of a God, having sympathy with man? Dare she excommunicate science, and answer at the judgment for the souls which are thus reluctantly compelled to infidelity? We reject the authority of the blind Scribes and Pharisees, who have hidden themselves from the light of heaven under such a darkness of bigotry. We claim our just rights and our share in the Church."[20]

Peirce discussed the relationship between science and religion in many different venues. He often interjected religious comments into his classes at Harvard. On one occasion he was in the middle of a lecture on celestial mechanics when he turned to his class and said, "Gentlemen, as we study the universe we see everywhere the most tremendous manifestations of force. In our own experience we know of but one source of force, namely will. How then can we help regarding the forces we see in nature as due to the will of some omnipresent, omnipotent being? Gentlemen, there must be a GOD."[21]

As a young man Peirce gave arguments from design in his review of Bowditch's *Mécanique céleste*.[22] Later he talked about the relationship of science and religion when speaking before the American Association for the Advancement of Education.[23] Even his outstanding book on analytical mechanics is peppered with natural theology.[24]

As early as 1835, Peirce cited spiritual benefits as the most important reason to study mathematics.[25] He was still doing this at the end of his life, when he said, "But the computation of the geometer, however tedious it may be, has a loftier aspiration. It provides spiritual nourishment: hence it is life itself, and is the worthy occupation of the immortal soul."[26]

Peirce gave many lectures like the ones in Washington during his life. Before be gave the series in Washington, he tried them out with the seniors at Harvard, and was happy to report to Bache that they were well received. He continued to give these lectures to the Harvard seniors for several more years.[27] Later he gave lectures before general audiences at Harvard "On the Manifestations of Intellect in the Construction & Development of the Material Universe."[28] He also gave several similar lectures to the Radical Club; one of them was on ancient, especially Mosaic, cosmology.[29] Near the end of his life, he gave a series of lectures dealing with similar topics to the Lowell Institute in Boston. He repeated them in Baltimore the next year. Peirce's ideas changed somewhat over his lifetime, but his general theme of an ordered universe under the direction of a rational deity remained constant.[30]

There was continued interest in publishing Peirce's popular lectures. When he lectured for the Smithsonian in 1857, Peirce anticipated that they would be published before he left Washington. But the cost of publication proved prohibitive.[31] A few years later his seniors were so pleased with his lectures that they asked him to publish them. Since Peirce delivered the lectures extemporaneously, the class provided a reporter to take stenographic notes and then write the lectures out.[32] But once again the lectures failed to be published. A publisher also expressed interest in printing the Lowell lectures Peirce gave in 1879. Nevertheless, Peirce's flippant words to Bache about having his children publish them posthumously proved to be true. The Lowell lectures were published by his family a year after his death in 1881.

$$\Phi^n$$

Peirce's activities in Washington were not limited to giving a series of lectures. He attended the Lazzaroni's annual dinner and met with Washington's elite as well. "I went to the Senate today, was spied out in the gallery by Senator Forester of Connecticut, who introduced me upon the floor." He was also introduced to "Senators Seward, Files and Norman; and Senators Bell and Pearce were very kind in their attention."[33] Later during his stay, Peirce had dinner with the justices of the Supreme Court. He wrote home to his wife, Sarah, "Our dinner with the Judges of the Supreme Court last evening was exceedingly spectacular, and we were very graciously (the Chief and I) received by these high functionaries—the highest in the courts of Washington echelon—after the president and vice president."[34]

Peirce met President Franklin Pierce as well. He sympathized with the president, who had been unable to solve the country's slavery controversy, and had lost the support of his party for renomination. He wrote, "After my lecture, I went to the President's levee, and the aspect of the man was rather a sad one to me, as I hear that his wife is quite sorry that they are going out of office."[35] After a later meeting he wrote, "I went . . . to the President['']s reception. . . . The President was very gracious and he seems to be quite a favorite in society although his political situation is evidently not very elevated. The attendance at the reception was small. Partly perhaps on account of the weather, but still more I fear from the fact that he is so near the end of his administration. Buchanan has just arrived in Washington and I presume that he is now the great center of attraction. How sad this difference between its beginning and the end of an administration."[36]

Peirce was not over-awed by the luminaries of Washington. He felt his own mission there was of the utmost importance; he exclaimed, "O! my dear wife! I feel that I have been permitted to be the medium of conveying important truths to my fellow men and that I ought to be worthy of this high trust!"[37]

$$\Phi^n$$

Peirce's contemporaries knew him primarily as an astronomer. It was in mathematical astronomy that he had first achieved recognition for his work on Neptune and his bold assertion that finding Neptune had been a happy accident. Three-fourths of his published papers dealt with applied mathematics, primarily astronomy, geodesy, and mechanics. His early work in pure mathematics, such as his paper on perfect numbers, had largely been overlooked by his contemporaries.[38]

One of Peirce's best known contributions to applied mathematics is a condition for rejecting unlikely values in a given set of data. Values which are sufficiently different than other data gathered in the same way are likely to be erroneous, due, for example, to the experimental apparatus being out of adjustment or to an error in recording the data. The astronomer Benjamin Gould was "especially anxious to have a numerical standard by which he might be guided in the rejection of doubtful observations,"[39] and asked Peirce about devising one. Peirce quickly formulated what became known as Peirce's Criterion.

Published in 1852, the criterion was the first test for the rejection of outliers. It sparked an international debate on the wisdom of excluding measurements that didn't seem to fit the overall pattern of the data, a debate which has never been satisfactorily resolved. Peirce's Criterion was, however, consistently used in the Coast Survey from 1852 until Peirce retired as superintendent in 1874. Although Peirce's Criterion was discredited by the early twentieth century, for some time it was regarded as one of his most important contributions to science.[40]

Peirce's involvement as a witness in the Howland will case, served as the impetus for one of his most novel applications of mathematics. The plaintiff in the trial was Hetty Robinson, whose family had made a fortune in whaling and foreign trade. Although her parents were living, she had been raised by a wealthy aunt, Sylvia Ann Howland. Shortly before her aunt's death, she and Hetty had had a falling out, because the increasingly niggardly Hetty complained that some of her aunt's expenditures would reduce her inheritance. When her aunt died, she had an estate of two million dollars, half of which she left to charity and half of which was left in a trust, from which Hetty would receive the income during her lifetime. The miserly Hetty was enraged, having expected to be her aunt's sole heir. Just a month earlier, Hetty's father had died leaving her a million dollars outright and the income from an additional five million dollars kept in a trust. So at the time of her aunt's death, Hetty was already a wealthy woman.

Nevertheless, Hetty contested her aunt's will, producing an earlier will in which Hetty and her aunt irrevocably agreed to be mutual heirs. The will was written by Hetty and signed by both her and her aunt. The principal question was whether her aunt's signature was genuine or had been traced by Hetty. The executor of the will did not accept the signature, and Hetty sued; the resulting legal battle dragged on for five years and some of the country's most eminent scientists served as witnesses. Two of Peirce's friends, Louis Agassiz and Oliver Wendell Holmes, examined the signature under a microscope and testified on Hetty's behalf. They pointed out that there were no pencil marks that would likely be present had the signature been traced.

Peirce and his son Charles, however, testified for the executor, showing mathematically that there were too many similarities between the signature on the will and a second signature for them to have both been written out independently of one another. Benjamin was a confident witness dealing with esoteric knowledge that intimidated Hetty's lawyers.[41] In the end the court found for the executor, but Hetty was awarded the amount of money that had accrued from her trust fund during the trial, some $660,000.

Although Peirce's methods would be criticized by modern mathematicians, they were an early and ingenious use of statistical methods applied to a practical problem. Peirce's testimony may well be the earliest instance of probabilistic and statistical evidence in American law.

In 1867, just before the court ruled against her, Hetty married Edward Henry Green, himself a millionaire. Hetty, however, made a prenuptial agreement that kept their money separate. After they married, the couple lived in London for seven years, where they had two children. During this time Hetty had invested her money shrewdly. After returning to the United States, she became increasingly miserly, purposely wearing old clothing to appear poor. She also feared assassination and lived in a flat in Hoboken and other modest dwellings from which she commuted to Wall Street each day. Due to her eccentricities and hard-headed business dealings she became known as the Witch of Wall Street. She used her slovenly garb to avoid making nominal payments. In one instance, her son had seriously injured his leg. Hetty and her son dressed poorly and went to a free clinic for treatment. When the doctors discovered Hetty's true identity, they demanded payment. Hetty refused, and consequently her son's leg had to be amputated.

Hetty continued to make excellent investments in bonds, railroads, and real estate. At the time of her death her estate was valued near a hundred million dollars, making her the richest woman in the country.[42]

$$\Phi^n$$

Despite Peirce's work in astronomy and geodesy, Peirce was a mathematician at heart. Early in his career he wrote to President Quincy, outlining his duties as Perkins professor and explaining how arduous they were. One of the sacrifices he feared he must make to perform these duties was giving up his favorite academic interest: "I have not yet alluded to one source of occupation which is now rapidly increasing & will erelong throw upon my shoulders a heavy weight of labor & responsibility. The Observatory, which has already consumed three weeks of this vacation in incessant labor from early in the morning till late in the evening, must soon engage much of my time by day & night. I shall then be compelled to relinquish entirely the study of the pure mathematics which are my peculiar delight, & my humble attainments in which, constitute my only claim to the title of a man of science."[43]

Over thirty years later, Peirce reiterated his partiality for pure mathematics, when reminiscing about his past work. "I am having my mind diverted, at the present time to . . . questions in pure mathematics, which I have more at heart than any other work."[44]

Although Peirce's favorite study was pure mathematics, he had broad interests and viewed astronomy as the science "best fitted for the application and advancement" of mathematics.[45] He was willing to sacrifice his personal preferences for what he perceived to be the needs of Harvard College and American science. Peirce saw the importance of a good observatory at Harvard, and worked to have one built. Later, doing longitudinal work for the Coast Survey allowed him to escape from the drudgery of teaching freshman and sophomores and aided his friend Bache. Peirce returned to his first love near the end of his life, however, with his last major mathematical work, his *Linear Associative Algebra*. This was not only his best work, but the first important mathematical research done by an American. Peirce also took great joy in doing it: "This work has been the pleasantest mathematical effort of my life. In no other have I seemed to myself to have received so full a reward for my mental labor in the novelty and breadth of the results. I presume that to the uninitiated the formulae will appear cold and cheerless. But let it be remembered that, like other mathematical formulae, they find their origin in the divine source of all geometry. Whether I shall have the satisfaction of taking part in their exposition, or whether that will remain for some more profound expositor, will be seen in the future."[46]

Peirce realized that this was not only his most pleasant scientific work, but his most important, concluding quite correctly that "Humble though it be, it is upon which my future reputation must chiefly rest."[47]

$$\Phi^n$$

It is somewhat surprising that Peirce did his best mathematical work both late in his life and at a time when he was burdened with the administrative duties of the Coast Survey. Peirce explained this paradox to Hugh McCulloch, the secretary of the treasury: "But whatever may be my practical efficiency, you will be perhaps not quite prepared for my statement, that I am sure I have done more science of the purest scientific, ultra mathematical hypothetically transcendental nature than I should have done, if I had not yielded to the irresistible secretary—but had remained entirely devoted to my mathematics. The new labor has relieved my head of the surplus blood, when mathematical analysis was making it boil. It has settled me and quieted my nerves. But when the strength has returned, the problems would be solved, and every now and then, I cover a sheet of paper with diagrams, or formulae, or figures, and I am happy to say that this work again relieves me from the petty annoyances which are sometimes caused by my receiving friends who do but upset."[48]

$$\Phi^n$$

A nonmathematician may have trouble understanding the significance of Peirce's accomplishment. Modern, or abstract, algebra is today one of the principal branches of mathematics, but it holds little resemblance to the algebra widely studied in high school. In mod-

**Benjamin Peirce, ca. 1870. By permission of the Harvard University Archives, call #
HUP [24a].**

ern algebra, mathematicians study the properties of highly abstract mathematical systems. One of the primary impetuses for studying such systems came from nineteenth-century Britain.

Until the second quarter of the nineteenth century mathematics was universally defined as the science of quantity or magnitude. While negative and imaginary numbers had proved to be extremely useful to mathematicians, if mathematics was the science of quantity, neither type of number represented quantities and thus posed a philosophical dilemma.[49]

By the middle of the eighteenth century, some British mathematicians, most notably Robert Simpson, Francis Maseres, and William Frend, questioned the validity of negative numbers, as well as imaginary ones. They pointed out that they were not well defined and that they should not be considered as mathematical entities.[50] Although negative numbers had long been justified by an analogy to debt, these critics observed that if a definition needed an analogy to explain it, it was a deficient definition. Imaginary numbers had no apparent relationship to quantity at all.[51]

These men were in a minority even in Britain; most British mathematicians saw the advantages of using negative as well as imaginary numbers, which were an integral part of the theory of equations. Yet they also saw the need to deal with the problem of negative and imaginary numbers in a precise way. Consequently, several nineteenth-century British mathematicians responded to this attack on negative numbers. Their investigations led to the question of what algebra itself was. The current definition of mathematics as the science of quantity and space and algebra as a universal or generalized arithmetic now appeared inadequate.[52]

One of these mathematicians, George Peacock, attacked this problem by distinguishing between two types of algebra: arithmetical algebra and symbolical algebra. Arithmetical algebra dealt with the basic arithmetical principles of nonnegative real numbers. For example, in arithmetical algebra, one could write $a - b$, but only if $a > b$, so that the subtraction could be performed. In symbolical algebra, Peacock extended the properties of arithmetical algebra, like the one above to any real numbers whatsoever. Peacock did this by defining a negative number to be the symbol $-a$, that could be manipulated according to a list of rules, or axioms, that were analogous to the usual properties of arithmetic, but without the restrictions needed when dealing with only positive numbers. For example, since $(a - b)(c - d) = ac - ad - bc + bd$ in arithmetic, provided that $a > b$ and $c > d$; the same rule applies in symbolical algebra without the restrictions. Other properties of real numbers could be derived from such relationships. For example, if a = c = 0 in the above equation, then the rule for the product of two negative numbers, (–b)(–d) = bd, follows. Thus Peacock took an axiomatic approach to negative numbers by incorporating them in a system of numbers that followed a given set of rules, patterned after the usual ways we combine real numbers.[53]

In his *Treatise on Algebra,* published in 1830, Peacock defined symbolical algebra as "the science which treats of the combinations of arbitrary signs and symbols by means of defined though arbitrary laws."[54] Peacock realized that axioms could vary from those of ordinary arithmetic, thus giving great freedom to mathematics. Yet Peacock was uncomfortable with this new-found freedom; he experimented with no nontraditional axioms and in 1845 he wrote "I believe that no views of the nature of Symbolical Algebra can be correct or philosophical which made the selection of its rules of combination arbitrary and independent of arithmetic."[55] Despite the fact that Peacock failed to take advantage of the new freedom inherent in his ideas, Peacock's approach marked the beginning of an entirely new approach to algebra, an approach that would be followed by others, including Benjamin Peirce.[56]

Augustus De Morgan was influenced by Peacock's work, but recognized more fully than Peacock that algebraic systems need not be tied to the laws of arithmetic. He believed that an algebraic system could be created by devising a set of axioms that the symbols must follow. He also realized that these symbols could represent things other than magnitudes and quantities. Nevertheless, De Morgan denied the value of such systems and failed to create a new system that obeyed laws that were different from those of ordinary arithmetic.[57] Peirce was interested in De Morgan's work as early as 1856, and cited De Morgan's "Triple Algebra" in his *Linear Associative Algebra.*[58]

The first man to devise a system whose properties differed from those of the real numbers was the Irish mathematician William Rowan Hamilton.[59] He attempted to put negative and imaginary numbers on a firm mathematical footing. The impetus for his new system came from his work that gave meaning to imaginary numbers, which unlike negative numbers have no obvious analogy to any concrete concept. Numbers that can be reduced to the form bi, where b is real and $i^2 = -1$, are called imaginary numbers. For example, $\sqrt{-4}$ can be written as 2i. More general numbers of the form a + bi, where a and b are real, are called complex numbers. Such numbers arise immediately when one is attempting to solve equations. For example, there are no real numbers that are the roots of the equation $x^2 + 1 = 0$. The most obvious way to deal with this problem is to simply say that such equations have no roots, and mathematicians were quite content to do this until the sixteenth century when they solved the cubic, or third degree, equation. The general formula for a solution to such equations often gave real roots in terms of $\sqrt{-1}$. The formula that gives the solution to the cubic is called Cardano's formula, after Jerome Cardano, the sixteenth-century Italian mathematician, although it was the work of another sixteenth-century Italian, Scipione Del Ferro.

For example, using Cardano's formula to solve $x^3 - 15x - 4 = 0$ one gets the solution $x = \sqrt[3]{2 + \sqrt{-121}} + \sqrt[3]{2 - \sqrt{-121}}$, although it is easy to see by inspection that 4 is in fact a solution. A straightforward computation gives $-2 + \sqrt{3}$, and $-2 - \sqrt{3}$, as the other two solutions, all of them real. Rafael Bombelli, another sixteenth century Italian mathematician, was able to show that indeed, $4 = \sqrt[3]{2 + \sqrt{-121}} + \sqrt[3]{2 - \sqrt{-121}}$.[60] Numbers that were expressed in terms of the square roots of negative numbers were called imaginary, because they seemed to have no real meaning. Yet, since Bombelli's discovery, using Cardano's formula, expressed real numbers in terms of them, mathematicians could no longer ignore them.

Hamilton devised an ingenious and influential interpretation of complex numbers. To Hamilton, expressing complex numbers as the sum of a real and an imaginary number, that is as a + bi, made no sense. Doing so was like combining apples and oranges. In 1835, he solved this problem by representing complex numbers as ordered or coordinate pairs, (a,b), that describe two-dimensional space. Observing that $(a + bi)(c + di) = ac - bd + (ad + bc)i$, and that $(a + bi) + (c + di) = (a+c) + (b+d)i$, Hamilton made analogous definitions for multiplication and addition of the coordinate pairs that describe two-dimensional space, (a,b). That is, Hamilton defined $(a,b)(c,d) = (ac - bd, ad + bc)$ and $(a,b) + (c,d) = (a + c, b + d)$.[61]

By making this connection between coordinate two-dimensional pairs and complex numbers, and defining meaningful multiplication and addition for them, Hamilton showed that complex and imaginary numbers were nothing of the sort. They were simply the numbers that describe two-dimensional space.[62]

After discovering that complex numbers described two-dimensional space, Hamilton quite naturally attempted to find a system of numbers that would describe three-dimensional space. Hamilton easily found a reasonable definition for the addition of ordered triples, (x,y,z), but failed to find a suitable definition for multiplication.[63]

Hamilton was more successful in defining multiplication for four-dimensional vectors, (a,b,c,d). Even here he was unable to retain the commutative property, that is, a times b was not necessarily equal to b times a, yet Hamilton did devise a useful and meaningful algebraic system. The mathematical system, which Hamilton discovered in

1843, is called the quaternions. Where a typical complex number may be expressed as a + bi, a typical quaternion is of the form a +bi + cj + dk. Much like the complex numbers, $i^2 = j^2 = k^2 = -1$, but i, j, and k, were not simply the square root of -1. Combining i, j, and k, gave the system its novel noncommutative property; for example ij = k, but ji = $-k$. Recognizing that this was a bizarre property, Hamilton was careful to justify his system of quaternions with applications to concrete problems.[64]

Hamilton and others, most notably the Scottish physicists, Peter Guthrie Tait and James Clerk Maxwell, applied quaternions to a wide range of physical problems. The American chemist, Josiah Willard Gibbs, and the British physicist, Oliver Heaviside, independently discovered a less complicated system that was satisfactory for physical applications. This simplification, vector analysis, quickly replaced quaternions for use in applications, so quaternions are seldom studied and little known today.[65]

The significance of the quaternions went well beyond their applications to physical problems. Hamilton had discovered an algebra that had an obvious physical interpretation and applications, and yet had properties that differed markedly from the familiar axioms of ordinary arithmetic.[66] At the time of their discovery many people were fascinated by quaternions, including Benjamin Peirce. They were, in fact, his favorite subject.[67] Peirce incorporated them into his lectures at the Lawrence Scientific School in 1848,[68] the same year that Hamilton himself gave his first lectures on quaternions in Europe.[69] Peirce's position, as the leading mathematician in the country, influenced others to study quaternions and include them in their curriculum. No fewer than twelve colleges in the United States were teaching quaternions during the latter half of the nineteenth century. This is especially surprising in view of the fact that mathematical instruction at American colleges was still on a low level at this time. At some of these colleges, the course on quaternions was by far the most advanced one taught.[70]

Although Peirce was only thirty-four when Hamilton discovered the quaternions, he lamented the fact that he was not younger and could throw himself into the study of quaternions as only a young person could.[71] Peirce's students did a remarkable amount of work on quaternions, considering the small amount of mathematical research being done in the country at that time. Among his students who did work in quaternions were Thomas Hill and A. Lawrence Lowell (who both became presidents of Harvard), Arnold B. Chase (who became chancellor of Brown), and Charles Saunders Peirce and James Mills

Peirce (both Peirce's sons).[72] In addition, Thomas Hill wrote an enthusiastic review of Hamilton's work on quaternions.[73]

$$\Phi^n$$

Other algebraists had been somewhat timid about using arbitrary axiomatic systems. In Peacock's case, for example, he had not veered from the axioms of ordinary arithmetic, and in Hamilton's case he had been careful to justify his system with applications. Peirce was not so squeamish about generating new algebras from sets of axioms, even if they had weird properties that differed markedly from those of ordinary arithmetic or if they had no apparent applications. Peirce's algebras, like the quaternions, have noncommutative elements, that is, elements a and b such that $ab \neq ba$. In addition, Peirce's algebras have divisors of zero. If a and b are positive integers, and $ab = c$, then a and b are both divisors of c. For example $2 \times 3 = 6$, and 2 and 3 both divide 6 evenly. a and b are divisors of zero if both a and b are nonzero and $ab = 0$.

Peirce's boldness sprung from his deeply held religious convictions and from historical precedent. To Peirce any well thought out axiomatic system must come from God, but Peirce believed that with God thoughts are reality, and since God is not capricious or wasteful, God's thoughts will be manifested somewhere in the universe. Thus no matter how bizarre a mathematical system man might create, there must be applications found for it somewhere in the universe.

Peirce wrote: "Wild as are the flights of unchained fancy, extravagant and even monstrous as are the conceptions of unbridled imagination, we have reason to believe that there is no human thought, capable of physical manifestation and consistent with the stability of the material world, which cannot be found incarnated in Nature."[74] Or, as his sister Elizabeth, related it: "Ben says that if a person discovers that a thing <u>can</u> be—it is certain that such a thing really exists somewhere. For instance, Ben can prove that there can actually be such or such a motion—& though <u>no</u> such motion has ever been <u>seen</u> or <u>heard</u> of—still there <u>is</u> such a motion somewhere in the universe."[75]

Peirce found a further justification for his new mathematical systems from history. He pointed out that the conic sections, which for over a thousand years were just geometric abstractions, were found to exist in nature as the paths of planets, comets, and projectiles. He also observed that even though Kepler failed to find that the regular polyhedrons regulate the planets, these same geometrical shapes do manifest themselves in nature in the "modern theory of crystallization." More generally: "The highest researches undertaken by the mathematicians of each successive age have been especially transcendental, in that they have passed the actual bounds of contemporaneous physical inquiry. But the time has ever arrived, sooner or later, when the progress of observation has justified the prophetic inspiration of the geometers, and identified their curious speculations with the actual workings of Nature."[76]

$$\Phi^n$$

Peirce began his *Linear Associative Algebra* by explaining terms and giving definitions. He recognized the importance of two types of elements to the structure of the algebras; both of them have unusual properties. When one of these elements, called a nilpotent element, is raised to a high enough power it is equal to zero. When the other element, called an idempotent element, is raised to a sufficiently high power it is equal to itself. In symbols, an element, a, is nilpotent if there exists a positive integer n such that $a^n = 0$; and an element, a, is idempotent, if there exists a positive integer n such that $a^n = a$.

After discussing definitions, Peirce proved theorems that developed the algebraic structure necessary to generate all linear algebras of a given dimension. In the latter part of, and the bulk of, his treatise, Peirce systematically generated all linear associative algebras of dimension six or less.[77] Peirce gave a detailed description of over 160 algebras.[78]

$$\Phi^n$$

Peirce realized that his *Linear Associative Algebra* was mathematics produced in the new spirit of Hamilton, De Morgan, and the other British algebraists, and that the concept of what mathematics was had changed. Consequently a new definition of mathematics was in order that would broaden the scope of mathematics beyond the purely quantitative. He took pride in pointing this out to his nonmathematician friend George Bancroft, when he presented him with a copy of his *Algebra:* "My definition of mathematics upon the 1st page may not be unworthy of your consideration."[79] Peirce's definition is now well known:

> Mathematics is the science that draws necessary conclusions.
> This definition of mathematics is wider than that which is ordinarily given, and by which its range is limited to quantitative research.[80]

$$\Phi^n$$

Peirce's interest in algebra followed naturally from his fascination with quaternions, and in many ways his work was an extension or generalization of the quaternions, but his immediate impetus for doing the work came from his son Charles. Charles was interested in such mathematical systems in connection with his work on logic, but lacked the mathematical skill to do the work. He said, "I therefore set to work talking incessantly to my father (who was greatly interested in quaternions) to try to stimulate him to the investigation of all systems of algebra which instead of the multiplication table of quaternions . . . had some other more or less similar multiplication table. I had hard work at first. It evidently bored him. But I hammered away, and suddenly he became interested and soon worked out his great book on linear associative algebra."[81]

Peirce's son was disappointed with one aspect of his father's work. He was displeased with his father's using complex numbers as coefficients or scalars. He complained that "he [Benjamin Peirce] was a creature of feeling and had a superstitious reverence for the square root of minus one."[82] One of Peirce's students, W. E. Byerly, confirmed Peirce's fascination with i. In one of his lectures, Peirce had derived the admittedly enigmatic formula, $e^{\pi/2} = {}^i\sqrt{i}$, "which evidently had a strong hold on his imagination. He dropped his chalk and rubber, put his hands in his pockets, and after contemplating the formula a few minutes turned to his class and said very slowly, 'Gentlemen, that is surely true, it is absolutely paradoxical, we can't understand it, and we don't know what it means, but we have proved it, and therefore we know it must be the truth.' "[83]

In a paper Peirce gave before the American Academy of Arts and Sciences five years after he distributed his *Linear Associative Algebra*, he at once expressed his awe for the square root of minus one and his enthusiasm for Hamilton and quaternions: "The square root of minus one . . . may be more definitely distinguished as the symbol of *semi-inversion*. This symbol is restricted to a precise signification as the representative of perpendicularity in quaternions, and this wonderful algebra of space is intimately dependent upon the special use of the symbol for its symmetry, elegance, and power. But the strongest use of the symbol is to be found in its magical power of doubling the actual universe, and placing by its side an ideal universe, its exact counterpart, with which it can be compared and . . . from which modern analysis has developed her surpassing geometry."[84]

Despite his son's objections to Peirce's use of complex numbers, his doing so was one of the factors that brought boldness to his work. It was the complex coefficients that introduced divisors of zero into algebras. If he had used real coefficients, he would have avoided them. When the British mathematician, William Kingdon Clifford, published his work on quaternions in 1873, he was familiar with Peirce's work, and concurred with him on the use of complex numbers.[85]

$$\Phi^n$$

While Peirce was working on linear associative algebra, he presented several papers on his research to the National Academy of Sciences.[86] Although the National Academy was comprised of the best scientists in the country, few had mathematical expertise, and Peirce's presentations were far beyond them:

> Professor Peirce at one of the sessions of the Academy had just finished, with the inspired eye and prophetic manner[,] delivering to a confounded audience one of the most abstruse portions of his researches, and a respectful silence had ensued when Agassiz rose and said:
>
> > I have listened to my friend with great attention and have failed to comprehend a single word of what he has said. If I did not know him to be a man of great mind, if I had not had frequent occasion to feel his power, to admire his judgment and discrimination on ground where our several lines of study touch, in organic morphology, in physics of the globe, I could have imagined that I was listening to the vagaries of a madman, or at best to empty and baseless literal dialectics. But knowing my friend to be not only profound but fruitful in all matters on which I have any judgment, I am forced to the conclusion that there are modes of thought familiar to him, which are inaccessible to me, and I accept in faith not only the logical truth of his investigations, but also their value as means of opening to our comprehension the laws of the universe.
>
> Upon this epilogue there was what the French call a sensation in the hall, and the audience recovered their cheerfulness.[87]

The National Academy of Sciences intended to publish Peirce's *Linear Associative Algebra* as part of its *Memoirs,* but was never able to do so. Presumably Peirce could have attempted to have his work published in Europe, but he did not.[88] Hilgard, Peirce's assistant in the Coast Survey, anticipating a long delay in the *Algebra's* publication, suggested running off one hundred lithograph copies. The work was done by members of the Coast Survey in their spare time.[89]

Most of the hundred lithographed copies were distributed to Peirce's colleagues and friends in the United States, but a few were sent to foreign societies. Peirce's son Charles presented a copy to De Morgan when he went to Europe to view the total eclipse of the sun in 1870.[90] After viewing the same eclipse, Benjamin Peirce went to England, where he lectured before the London Mathematical Society on his *Linear Associative Algebra,* and personally presented the society with a copy of his treatise.[91] His work was well received there. In 1872, William Spottiswoode discussed Peirce's *Linear Associative Algebra* extensively, summarizing most of Peirce's results, in his address to the London Mathematical Society upon his vacating the office of president of the society.[92] The British mathematician, Arthur Cayley, also praised Peirce's work in 1883, after it had appeared in the *American Journal of Mathematics.*[93]

Unfortunately Peirce's *Linear Associative Algebra* had a disappointing reception in Germany. Peirce was particularly anxious to have his work known there, since Germany was then the world's leading country in mathematics. He sent two copies to his friend and former employer, George Bancroft, who was then the United States minister to Prussia. One of these copies was a gift for Bancroft. Peirce asked his old friend to give the other copy to the Prussian Academy, but no notice was taken of the work.[94] Later, after Peirce's *Linear Associative Algebra* was published, there was a notice in the *Jahrbuch über die Fortschritte der Mathematik* promising a review of the work in a later issue of the journal, but it never appeared.[95]

The extremely limited circulation of only one hundred lithographed copies in 1870 coupled with the fact that few people in the United States had the mathematical expertise to appreciate Peirce's work caused a poor initial reception. In 1881, Peirce's son Charles edited his father's 1870 work for publication in the *American Journal of Mathematics;* Charles added footnotes, a related paper Benjamin had given in 1875, and additions of his own. These additions were important results and one of the few immediate influences of his father's *Linear Associative Algebra.*[96] In addition, Peirce's *Algebra* was pertinent to Charles's theory of relatives, although Charles used different notation than his father. Many of Charles's footnotes give equivalent expressions in his notation for relatives. Others make simple clarifications or corrections to the original text. A few are lengthy commentaries or replacements for some of Peirce's laconic proofs.

Charles had great hopes for his father's work. In an introductory note, he wrote "This publication . . . may almost be entitled to take rank as the *Principia* of the philosophical study of the laws of algebraical operation."[97] James J. Sylvester, the editor of the *American Journal of Mathematics,* was also anxious to have his friend's work published in the journal.[98] Even after its publication, few continental mathematicians noticed the work, and it continued to have little impact on other mathematical work.

One reason for the work's poor reception was simply that at that time, Europeans tended to overlook American scientific accomplishments.[99] Another reason was its novelty. Although it fits easily within the modern concept of mathematics, Peirce's contemporaries felt that it was more philosophy than mathematics. The British mathematician, Arthur Cayley, was impressed with Peirce's work on algebra, but he felt it was "outside of ordinary mathematics," and was not part of "ordinary algebra."[100]

The American mathematician, H. A. Newton, felt that Peirce's work was valuable, but feared that his colleagues might feel that it was mere philosophical speculation: "To some mathematicians, and other men of science, it may yet be a question if the time has not come for them to say with entire certainty whether this work is to share the fate of Plato's barren speculations about numbers, or to become the solid basis of a wide extension of the laws of our thinking. Those who have thought the most on the course which contemporary mathematical science is taking will probably agree that the new ground thus broken can hardly fail to bring forth precious fruit in the future by adding to the powers of mathematics as an instrument."[101]

It was not until Herbert Hawkes wrote a paper for the *American Journal of Mathematics,* "Estimate of Peirce's 'Linear Associative Algebra' " in 1902, that Peirce's *Algebra* received significant notice. In the meantime much of Peirce's work had been duplicated by the German mathematicians, Eduard Study and George Sheffers. Hawkes pointed out Peirce's priority over Study and Sheffers. He also suggested that continental mathematicians overlooked Peirce's work because of his unfortunate choice of definitions and his sometimes faulty proofs.[102] At about the same time Hawkes wrote an additional paper based on Peirce's work, in which he placed "Peirce's work on a clear and rigorous basis."[103]

After the appearance of this paper a number of American mathematicians, including Hawkes, did work in the spirit of Peirce's *Algebra.* Although they did not generally cite him, they were almost surely aware of his work.[104] Peirce's *Algebra* is also cited in Alfred North Whitehead's important work on symbolic algebra, *A Treatise on Universal Algebra.*[105] Perhaps the work's most important influence was on Joseph Wedderburn's seminal structure theorem. Wedderburn combined Peirce's "decomposition of an algebra relative to an idempotent with a theory of complexes to prove the famous structure theorem."[106] In this case, Wedderburn was aware of Peirce's work and cited it in his paper.[107]

Peirce's *Algebra* is highly regarded today; it marks the beginning of the investigation of abstract algebraic structures, using the modern method of postulates. The work makes Peirce at once a pioneer in American mathematics and in modern algebra.[108]

$$\Phi^n$$

The first university in the United States that emphasized research and graduate work was Johns Hopkins. In many ways it was the realization of the university that Peirce and his Lazzaroni cronies had hoped to found in Albany back in 1851.[109] When Daniel Coit Gilman became its first president in 1875, he began searching for outstanding faculty. One man Gilman was considering to head the university's mathematical program was Peirce's old friend, James J. Sylvester.

As a young man, Sylvester had met with strong anti-Semitism in the United Kingdom. He was denied a degree from Cambridge because he was Jewish, and, after finishing his education there, opportunities for employment were severely limited. University College, London, was essentially the only university in England that permitted Jewish faculty. Since the professorship in mathematics there was held by Augustus De Morgan, and thus unavailable, Sylvester obtained the professorship in natural philosophy there. Despite the similarities in the fields, Sylvester found himself unsuited for the position, and resigned after three years to take the professorship in mathematics at the University of Virginia.[110] After Sylvester's brief and disappointing sojourn in the United States, he, of course, found no better academic opportunities when he returned to England in 1843.[111] Sylvester searched for a job for over a year, and finally found a position as an actuary.[112] He subsequently studied the law and became a lawyer, although despite his legal work, he still found time to do a prodigious amount of mathematical research. It was not until 1855 that he finally obtained an academic position at the Royal Military Academy at Woolwich.[113]

Woolwich had a distinguished mathematical tradition; two notable English mathematicians, Thomas Simpson and Robert Hutton, had taught there. But the school remained mathematically conservative; it had been reluctant to discard fluxions in favor of Continental mathematics, and it generally lagged far behind its French counterpart, the École Polytechnique, in mathematics. Moreover, Sylvester's position there, like almost all other academic positions in the United

Kingdom at the time, stressed teaching and offered little reward for success in research. Sylvester's priorities were quite the opposite, and he had a stormy fifteen years there on the faculty.[114]

In 1870, when he was only fifty-six, he lost this professorship at Woolwich; the War Office enacted regulations forcing all of the academic staff at Woolwich over fifty-five to retire.[115] Sylvester found no other employment and struggled, with only partial success, to obtain a pension.[116] Yet, despite the lack of attractive academic opportunities, Sylvester's mathematical work was outstanding. He had established himself as one of the world's leading mathematicians.

When Sylvester learned that Gilman was considering him as a professor for the new university, he wrote to Peirce.[117] Peirce lost no time in responding, urging Gilman to hire Sylvester:

> Hearing that you are in England I take the liberty to write you concerning an appointment in your new university, which I think it would be greatly to the benefit of our country and American science if you could make it. It is that of one of the two greatest geometers of England, J. J. Sylvester. If you inquire about him you will hear his genius universally recognized, but his power of teaching will probably be said to be quite deficient. Now there is no man living who is more luminous in his language, to those who have the capacity to comprehend him, than Sylvester, provided the hearer is in a lucid interval. But as the barn fowl cannot understand the flight of the eagle, so it is the eaglet only who will be nourished by his instruction. But as the greatness of a university must depend upon its few able scholars, you cannot have a great university without such great men as Sylvester in your corps of teachers. Among your pupils, sooner or later, there must be one who has a genius for geometry—He will be Sylvester's special pupil, the one pupil who will derive from his master, knowledge and enthusiasm—and that one pupil will give more reputation to your institution than the ten thousand who will complain of the obscurity of Sylvester, and for whom you will provide another class of teachers. . . . I hope you will find it in your heart to do for Sylvester what his own country has failed to do—place him where he belongs—and the time will come when all the world will applaud the wisdom of your selection.[118]

Although Sylvester had also been recommended by Joseph Henry and the distinguished English botanist, Joseph Hooker, Gilman was cautious about hiring him. He feared that at sixty-one years of age, Sylvester was too old, and his productive years may well be behind him. He also remained skeptical about Sylvester's academic abilities, despite Peirce's glowing recommendation. Later, after Sylvester had been hired, Gilman complained that many mathematicians had not heard of Sylvester. Peirce reassured him by listing Sylvester's honors, and then adding, "Not to know Sylvester is the surest evidence that a mathematical professor is himself unknown."[119]

Even though Sylvester had been unemployed for over four years when he was offered the position at Johns Hopkins, he was reluctant to accept it. He refused an offer of $5,000, the highest salary for a professor in the United States at the time. The highest salary at Harvard was $4,000; Yale's top salary was $3,500, and these were atypically high.[120] When Gilman upped the offer to $6,000, Sylvester accepted.

Sylvester's qualms about taking the position at Johns Hopkins were not limited to salary, however. He feared he could never "regard America as a home."[121] And he also had misgivings about his ability to found an outstanding graduate program in mathematics in Baltimore.[122] After accepting the job, Sylvester went to Cambridge for an extended visit with the Peirces.[123] Peirce's son James wrote his brother, "We are all quite enjoying Sylvester's visit. He is truly a delightful old bore & certainly one of the kindest & best of men. Aunt Lizzie is reminded of Moses & thinks his poetry far more interesting than David's. . . . He has taken rooms . . . & means to stay all summer." James added that "Father has not seemed very well, & has appeared to be growing old, but Sylvester has brightened him up amazingly."[124]

Once in Cambridge, Sylvester sought Peirce's advice about an assistant professor for his department.[125] Both Benjamin and James recommended two of Peirce's former students, William E. Story and William Elwood Byerly. Byerly had remained at Harvard to do his graduate work under Peirce. Although Story was one of the first students to graduate with honors from Harvard and had taken extensive course work in mathematics there,[126] his most impressive accomplishment was that he had studied in Europe and earned a PhD at Leipzig. Although Byerly showed excellent promise as a teacher, Story's potential for research was judged to be superior, and Sylvester offered him the position.[127]

Story, presently a tutor at Harvard, also sought Peirce's advice about accepting the job at Hopkins. He wrote to Peirce's son James, "I feel that it is desirable to obtain a situation which will make me independent and where I shall have a reasonable prospect of further advancement. I should wish to devote my leisure time to original work as a mathematician, not merely as a student."[128] Story hoped to obtain a better position at Harvard than that of a tutor; when that failed, he took the position at Johns Hopkins. After thirteen years at Johns Hopkins, Story had failed to become either a full professor or editor of the *American Journal of Mathematics;* he then left Hopkins to develop the influential graduate program in mathematics at Clark University.[129]

Φ^n

At Hopkins, Sylvester was successful in creating a research level school of mathematics.[130] His doing so gave an impetus to the development of graduate programs at established universities, like Harvard and Yale, and was followed by new programs at such schools as Clark and the University of Chicago.[131] The creation of a graduate program at Johns Hopkins was one of the most important events for the advancement of mathematics in the United States during Peirce's lifetime.

One of the important contributions of the Johns Hopkins mathematical program to the development of American mathematics was the founding of the *American Journal of Mathematics*. This was the first mathematical journal in the United States solely devoted to mathematical research. The need for such a journal is seen in the lack of a place for Peirce to publish his *Linear Associative Algebra*. The impetus for this journal came not from Sylvester, but independently from both Story and Gilman.[132] Indeed, Sylvester initially resisted launching the journal, fearing that there wasn't enough good mathematics in the country to fill it:

> You have spoken about our *Mathematical Journal*. Who is the founder? Mr. Gilman is continually telling people that I founded it. That is one of my claims to recognition which I strongly deny. I assert that he is the founder. Almost the first day that I landed in Baltimore, when I dined with him in the presence of Reverend Johnson and Judge Brown, I think, from the first moment he began to plague me to found a *Mathematical Journal* on this side of the water something similar to the *Quarterly Journal of Pure and Applied Mathematics* with which my name was connected as a nominal editor. I said it was useless, there were no materials for it. Again and again he returned to the charge, and again and again I threw all the cold water I could on the scheme, and nothing but the most obstinate persistence and perseverance brought his views to prevail. To him and to him alone, therefore, is really due whatever importance attaches to the foundation of the *American Journal of Mathematics*.[133]

Predictably, Peirce was happy to hear that such a journal was to commence publication. He wrote Sylvester, "I shall be glad to do anything I can to advance the mathematical journal. It is a nice thing for Gilman to take it up—and a wise thing for the Johns Hopkins University."[134] Despite Sylvester's initial reticence about the journal, he soon became optimistic. He wrote to Peirce, "I hope we shall have your promised article for our first number of the American Journal of Mathematics. With that and a very pretty paper of Newcomb's, and

I am told a very valuable one of Hill's, one of my own and one that Story hopes to give . . . we shall I think make a very respectable debut."[135] Story also urged Peirce to submit his paper on mechanics and quaternions.[136]

When the time neared for the publication of the first issue, Sylvester wrote to Peirce, inviting him to come to Baltimore to help celebrate the birth of the new journal:

> Mr. Gilman tells me that he has asked you to name some day when you can if convenient combine a visit to Washington with one to us at the University to dine with us and some of the supporters of the Mathematical Journal to celebrate its birth which is now daily expected and which you have done so much to promote. . . .
>
> We have over a hundred annual subscriptions actually paid up so that I think it will not perish for want of external support—Story is a most careful managing editor and a most valuable man [in the] university in all respects and an honor to the university and its teaching from whence he received his initiation.
>
> We have some very promising mathematicians in our small troop and two or three of them will I believe make their mark in the world of science and do something to be remembered by—I hope it will not be long before you fulfill your promise of sending us a contribution to the Journal—it will be very welcome—we have some good articles in hand for the second number.[137]

Peirce's paper on mechanics and quaternions did not appear in the first issue of the journal, nor, despite additional requests,[138] in any subsequent issue. Peirce remained interested in scientific questions, but as a productive scientist, the torch had been passed to his son Charles whose work appeared frequently in the new journal.[139]

15

My Best Love

With the exception of his brother-in-law, Charles Mills, all of Peirce's closest friends were scientists: Alexander Dallas Bache, Louis Agassiz, John Le Conte, James J. Sylvester, and Charles Henry Davis.[1] The Baches visited the Peirces in Cambridge twice a year, the Le Contes made many long visits, and James J. Sylvester made extended visits to the Peirce household separated by over thirty years. Other scientific visitors included Elias Loomis and Joseph Henry. Many of Peirce's students, and former students, visited the Peirce home regularly, including Benjamin Gould, Thomas Hill, James E. Oliver, John D. Runkle, Chauncey Wright, Simon Newcomb, and George W. Hill.

But Peirce had broad interests and many friends outside the sciences. He was one of the original eleven members of the famous Saturday Club, which met once a month for dinner in Boston.[2] In addition to Peirce and Agassiz, its membership included such notable Americans as the transcendentalist Ralph Waldo Emerson, the poets Henry Wadsworth Longfellow and John Greenleaf Whittier, and the author Nathaniel Hawthorne. Another member, Charles E. Norton, was a professor of art at Harvard and significantly influenced American fiction as an editor. With James Russell Lowell, he edited the *North American Review* and founded the *Nation*. James Russell Lowell, also Peirce's friend and a member of the Saturday Club, began his career as a lawyer, but had a distinguished literary career. He was an effective satirist and the editor of the *Atlantic Monthly*. Another member of the Saturday Club, the Brahmin physician Oliver Wendell Holmes, taught at the Harvard Medical School, but is perhaps best known for his essays, including *The Autocrat of the Breakfast Table*. Two additional members were Charles Sumner and Ebenezer Rockwood Hoar. Sumner, a noted abolitionist, served in the United States Senate. Also an abolitionist, the witty lawyer, Hoar, was a member of Congress and the United States attorney general.

The Peirces were active in Cambridge society. Many of the leading literary figures in the United States visited the Peirce home. One was

Francis Bowen, who edited the *North American Review*. Another, the feminist Margaret Fuller, was the first woman allowed to use the Harvard Library, for her research on western expansion. She also became the first woman to work for a newspaper, when Horace Greeley made her literary editor of the *New York Daily Tribune*. John Bartlett, who also visited the Peirce home, became well known in Cambridge for his extraordinary ability to recall general information. He is best remembered today for *Bartlett's Familiar Quotations*.

Peirce was a friend of the sculptors Horatio Greenough, who designed Bunker Hill Monument, and William W. Story, the innovative neoclassicist.[3]

Mrs. Harrison Gray Otis was at one time a next door neighbor of the Peirces. Her husband made a fortune in real estate and was mayor of Boston, a congressman, and senator. Other politicians who visited the Peirce home were Rufus Choate, an exceptional courtroom lawyer, who served in the United States House of Representatives and Senate; Daniel Webster who replaced Sarah's father in the Senate when he retired and was a family friend; and George Bancroft, the historian and politician, who was not only Peirce's former employer at the Round Hill School, but also an intimate friend of Sarah's family in Northampton.

The Peirces were friends with several notable ministers, including Fredrick Dan Huntington, an Episcopalian, who knew the Peirces before they married. Andrew Preston Peabody was a Unitarian minister who tutored mathematics with Peirce at Harvard. The Unitarian minister James Freeman Clarke, Peirce's classmate at Harvard, promoted liberal religious ideas through his *Western Messenger,* and founded the Church of the Disciples in Boston.[4]

The Peirces felt quite at ease with distinguished visitors from abroad. When Lady Napier, the pretty, youthful wife of the British minister to the United States, unexpectedly appeared at Harvard's commencement, Sarah and Benjamin had no trouble entertaining her. Peirce's sister, Elizabeth, wrote:

Last Commencement day, shortly before dinner Ben went home & said to Sarah "Have you got anything for dinner?" "Why yes," said she, "something, but not much." "Well" said Ben, "Just now Mr. Winthrop came to me & said here is Lady Napier, come to Commencement! Cannot you invite her to dine at your house?" Mr. Winthrop & indeed all the gentlemen were to dine together at the Hall. So Ben asked Lady Napier to dine with Sarah. She said "She would." Sarah received her as she does all her guests with great politeness, & Ben dined at home, & invited Agassiz also to dine with him. Everything passed off charmingly—Lady Napier made herself very pleasant, was simple, easy, witty, and entertaining—the dinner was just a turtle bean soup & stewed lobster; Lady N. ate some of each.

Peirce home at 9 Oxford St., Cambridge. By permission of the Harvard University Archives, call # HUV 138 (1–3).

Then roast lamb & vegetables—then custard & pies—and then raspberries. So Lady N. had a chance to see a simple family dinner in this country, which I venture to say, pleased her more than all the splendid entertainments which have been given to her. Certainly nowhere could she have had the honor and pleasure of being entertained & talked to by both Ben & Agassiz.[5]

Φ^n

Peirce was particularly close to his sister, Charlotte Elizabeth, or Lizzie. Elizabeth and Benjamin's mother lived with his mother's sister and her husband, the wealthy Charles Saunders. Peirce frequently walked over to their house for breakfast, often accompanied by one or more of his children. There he entertained his sister with his wit. Elizabeth was particularly fond of Benjamin, and one wonders if one of the reasons she never married was that she never found a man who was, in her eyes, the equal of her brother. She expresses her feelings toward Benjamin in a letter to her Aunt Nichols: "Dear Ben how good

& delightful he is! There are not many such pearls as he . . . —so warm
hearted, so genial, so child-like! . . . I always have a feeling of inex-
pressible tenderness come over me when I think of Ben—& I verily
believe, that the angels love <u>him</u>."[6]

Elizabeth was a polyglot, an extraordinary letter writer, and an
enthusiastic gardener. "My Sweet peas are indeed the pride of Cambridge,"[7] she declared.

Living with her rich uncle was not always easy. When her uncle
died, Elizabeth wrote: "No sensible wife could be <u>pleased</u> to find that
her husband had deceived her with regard to his property or indeed
in anything. Then came the attempt by Uncle's family to break the
will. & all the consequent trouble & annoyance to Aunt & her final
disappointment. I alone knew how much she felt. She was greatly displeased with the offered compromise which her wisest & best friends
told her she must not accept if only from self-respect (A devoted wife
accepting a second place to that claimed by the nephews & nieces &
sister of the husband!)"

After noting that Aunt Saunders told Elizabeth she would be her
sole heir, she explained:

My mother & I have done more than I can tell to contribute to Uncle's
happiness in many ways for many years. . . .

He paid us $850 dollars a year—I was obliged to give lessons in German,
French & History to meet the family expenses—I also translated for pay.
At this time my brother was so much dissatisfied that we should be thus as
it were sacrificing ourselves that he very seriously thought on insisting on
our breaking up the establishment & going to live at his house, but he
found that Mother would be so unhappy to leave Aunt that he decided
not to interfere.[8]

Φⁿ

Peirce "was subject at times to violent outbreaks of impatience, and
temper,"[9] but he was devoted to his wife, children, and family, and it
is doubtful that his bad humor unduly spoiled a congenial domestic
environment. Henry Cabot Lodge, the son of Sarah's cousin, gave a
warm description of Peirce:

I cannot recall that time when Benjamin Peirce . . . was not at once
familiar and impressive to me. . . . the "Professor" was constantly at our
house. . . . he made a profound impression on my imagination. I heard
him spoken of always with admiration, and I gathered that he was a man
of vast and mysterious knowledge, not understood by most people, which
was true enough, but the effect on my mind was to make me regard him

as a species of necromancer or magician. His appearance fostered the idea. He wore his black hair very long, after the fashion of his youth. He had a noble leonine head and dark, deep-set eyes. His voice possessed a peculiar quality. It was without any metallic or ringing note, but as if slightly veiled, and very attractive for some reason which I have never clearly defined. Altogether he had a fascination which even a child felt, and all the more because he was full of humor, with an abounding love of nonsense, one of the best human possessions in this vale of tears. I know that I was delighted to see him, because he was so gentle, so kind, so full of jokes with me, and "so funny." As time went on, I came as a man to know him well and to value him more justly, but the love of the child, and the sense of fascination which the child felt, only grew with the years.[10]

Peirce loved children. He found it surprising that the horrible doctrine of infant damnation could have grown out of a religion whose cornerstone was love.[11] But he did not only love children in the abstract; he enjoyed the company of his own children: "Before breakfast he always went to walk with his younger children, now a delightful memory to them. This man, who could divine and see remotest suns in space, amused his little ones by allowing no pin to hide from his eyes in the dust of the sidewalk;—'although he never seemed to be looking for them, he would suddenly stoop to pick up a pin. He had various 'pincushions'; one was the trunk of an elm tree near our gate, others on Harvard and Brattle Streets. Those on Quincy and Kirkland Streets are still standing.' At home, his daughter says, 'He was such a great, big ray of Light and Goodness, always so simple, cheerful and showing more than amiability, that his great power did not seem to assert itself.' "[12]

Peirce took to children generally. His grandchildren were no exception: "Benjamin Peirce, mathematician and mystic, was not always on the heights. Calling at his house one day to consult him on some abstruse problem I found him on all fours in the parlor playing bear with one of his grandchildren, and I was invited to take part in the game."[13]

$$\Phi^n$$

The principal responsibility for managing domestic affairs, of course, fell to Sarah. It was not always a light burden. She wrote to her son Benjamin, who was studying in Paris: "I have just now my hands full for we in America are going through a crisis in Domestic affairs. No servants in the true sense of the word to be had except at fabulous prices & such 'girls to help' as come within our means just good for nothing. So we work ourselves. How long this state of things will last I do not know, but it is not pleasant. I have really now great hesitation in asking any friends to the house, and I cannot receive or entertain them as I have been accustomed to . . . & must do so much myself that I am unfit to act as hostess. I hope before you come home affairs will be on a pleasant footing else you will wish yourself back."[14]

Fortunately there were more happy moments in Sarah's house-keeping. Sarah exclaimed that their new cook "makes <u>such</u> an apple <u>Charlotte</u>!! Such an one as you never ate & of which you can form no conception! Color crimson—substance, solid jelly flavour, the fruit of the tree of paradise compounded with a bare soupcon of ever balmy & spicy delight."[15]

Beyond being a diligent homemaker, Sarah was a virtuous and pious woman. A bit too much so for some of her family at times. Elizabeth remarked, "How thankful I am that I did not live in those stormy times with Philip the second for my King—I felt as I read of all his wickedness, & depravity, as if I would have liked myself to scratch out his eyes <u>myself</u> & verily believe that I should have been tempted to make the attempt—which Sarah would have considered very naughty in me."[16] On one occasion when Elizabeth heard a long, dry sermon, she remarked, "I got rather fidgety towards the last—but I suppose Sarah would have enjoyed every word."[17] Sarah's brother-in-law, Charles Henry Davis, feared Sarah's righteous indignation. When he wanted Benjamin to help him with some geological investigations, he realized "I know that Sarah will regard me in the light of wicked sinner who is attempting to seduce her husband into the paths of destruction."[18]

Yet, if Sarah was too religious for some members of her family at times, she was neither narrow-minded nor a fundamentalist. She wrote to Elizabeth:

> I hope you will adhere to your independence in your Bible discussions —or if you must rest upon human Authority or take anybody's views why are not those of a devout & learned & profound mind like Bunsen's as much to be relied on at least as Mr. McKensie's or even his—"Murphy"? Whose book he gave you to read? You say <u>he</u> (Murphy) does "not pretend to go behind the scenes but simply explains & moralizes—But he cannot explain the sun standing still & to moralize about it can be of but little use any way. . . . & for my part I am quite content to let such statements alone & all discussions about them. It is very easy however to see that astounding events like these may have been merely delusions—easier than to think that all the laws which govern the Universe should have been set aside for the convenience of General Joshua & his army. Now if this sounds profane to you, you must forget that I have said it—I said it with all reverence & you asked me what I thought.[19]

Sarah Peirce. By permission of the Houghton Library, Harvard University, MS CSP 1643.

Sarah's willingness to make religion conform to the laws of science no doubt made her religious orientation amenable to her husband. At any rate, Benjamin never complained of his wife's excessive piety. His letters to his wife expressed not only the deepest love to Sarah, but to his children as well. The following passage is typical: "I closed my letter in such haste that I did not send any of the expressions of endearment to the children with which my heart was full charged, and more. . . . Give my best love and a thousand kisses to the children and be sure that I shall not forget you unless in regard to them. Kiss darling little Bertie, Dear sweet gentle Lellie, warm hearted, inquisitive Benjie, and bright soled fervently loving Charlie. And when Jem gets home (the gem of my life) tell him that he has grown every day dearer to me, and that I am every day prouder of such a son. God bless you my own true loving wife—dearer to me than ever. How dearly I love you! God grant that when I return, I shall find you with a diminished cough. . . . Dearest Sarah, kiss my mother for me and Elizabeth, and give my best love to Aunt Saunders, to your mother.[20]

A letter from Benjamin's friend, Josephine Le Conte, gives a charming description of the Peirces' domestic life. "What have you been amusing yourselves with since our departure? I fancy Mrs. Peirce's little figure busily occupied at the sewing machine and you as indefatigable as ever in your exertions to find out some theory of curves. . . . Am I not right?"[21]

The sewing machine was a big hit in the Peirce household; Benjamin's sister Elizabeth wrote: "I believe I never told you about Sarah's sewing machine, given to her by Charles Mills. It works admirably. Ben, however, sews with it better than any of the family. I hear of his hemming towels, making shirts, skirts & c. Most famously it is a grand thing to be able to employ the gentlemen so usefully. Charley & James also are very expert in the use of it."[22]

Φⁿ

If there was a woman who competed with Sarah for Peirce's affections, it was Josephine Le Conte. Peirce's correspondence with her sometimes seems to express more than friendship, for example:

My Dearest, my darling friend—
   "My friend, Mine. Are you not mine? How can I tell you with what deep heart feeling I received those dear flowers—which had held on your bedside watching over you for so many days of sadness and sickness! I could not permit another to intrude upon such delicious joy, and their coming is not known to others. When shall I see you, my dear Josephine, this winter?"[23]

Another example: "My own dear friend! How is your presence missed! You have left your shadow every where—but I find you nowhere! In this chair—In the corner of the parlour sofa! Now and then and you are not there. I can do no work. All I can do is to think of you. I went to the library and walked all over it, and the Queen was everywhere—looking at the prints—ascending the stairs—entering the alcoves—standing with me in my own alcove—and yet she was not there. . . . Always when I came from town—I trace my torturous way over the shadows, which I crossed with the Queen when she said 'You had me whithersoever you will.' The shadow of the Queen was over all the way—the grass, the shadows of the trees—and upon the heart of the loving functionary. But the Queen was gone. I can do no work."[24]

Despite effusive protestations of affection, most if not all of Peirce's letters can be attributed to his almost pathological sensibility. For one thing, there is no evidence that Josephine ever aroused Sarah's jealousy, although Sarah did become jealous on at least one other occasion. In this instance, Benjamin was quick to apologize and promised to change his ways:

But my dear Sarah, your letter is cold and I cannot understand why except it be that you feel annoyed at my saying nothing about a young lady's going back with me to Cambridge. I did write that May Lovering was going with me—but I presumed that Alice Goodnick's talk about going was nothing but a joke and unless you have heard otherwise, I am sure that it will be proven to be the case and my dear wife, let me say to you myself that as to any feeling or interest for so young a girl—I am wide awake to the ridiculous absurdity of such a fancy—and on my honor as a gentleman, by all that is in nature, I never have felt the least temptation to be anything but amused by her sprightliness. But I will go farther and tell you that I think she is not a good looking girl and that your opinion of her exhibition of herself at Fanny Huntington's corresponds to my real opinion of her. But never it trouble you, dearest wife, I will take more pains not to annoy you by even the appearance of attention to her during my absence from you, my dearest. I have thought much—very much of all your love for me and your image shines most brightly in my heart. Your letter makes me sad—O! Dearest Sarah! I do love you earnestly—faithfully—and sincerely; and I feel in all my heart the faithfulness of your generous love of me and I assure you that there is not the least shadow of untrue attraction for me in this girl. I have not taken much pains to read her—but I am inclined to think from what I saw in N. Y. that there is a private understanding between her and young Fessenden—and that her ostentatious demonstration towards myself and Felton and other men is only to screen herself. . . . and if you read me well you know that I rather enjoy this being viewed as a lover. It is . . . because I see that there is no danger that I pretend to carry on a flirtation. The fun of the thing to me, indeed, is that while I look as if I were flirting, there is not the least value of the thing in fact. But if it has troubled you, I have done wrong, and I won't do it any more. Oh! My dear Sarah! I wish that I could just press you to my heart for one half instant, and in one instant, I would drive away from your spirit all doubt upon this matter.[25]

There was also a competition between Peirce, Agassiz, Henry, and Bache for Josephine's attention, but no one seemed to take such antics seriously.[26] Moreover Sarah, Josephine, Benjamin, and John all had strong feelings of friendship toward each other. John wrote Benjamin, "A few moments ago, our hearts were gladdened by the receipt of your letter to me, dated June 27th. Josie came bouncing to me like a mountain deer when she saw it. By the way, she brought your photograph with her, and it is now hung up on the wall in our room."[27] Peirce described the relationship of the four friends: "We have never talked so much and with such earnestness of love of you and Josie as since the Montreal Meeting, I hardly know which half of this pair is most in love with which half of that pair—but to spare the special blushes of the pair whom I now specially address, the warmest affection of my wife for yours is of a major kind of which we should both be jealous if we were not the best and most generous of husbands."[28]

$$\Phi^n$$

Peirce's devotion to Sarah was not limited to the piffle in his letters. He showed great concern for her well-being and was extremely devoted to her when she was ill. She was ill, for instance, when it appeared that Henry Darwin Rogers would be appointed to the Rumford chair in 1846. Despite Peirce's conviction of Rogers's incompetence and the importance of the appointment, he temporarily ignored the situation in order to nurse his wife.[29]

Sarah was also seriously ill in the spring of 1860. Peirce's sister, Elizabeth, wrote:

> Sarah has been so very ill that I have not written lately. I had as much as I could possibly do to be at Ben's & keep things going on right at home. Mrs. Mills is at Sarah's this winter, and when Sarah was first taken sick her mother was able to be with her a great part of the time, but the fatigue & anxiety proved to be too much for her, & she became ill herself. So that the whole care of Sarah & the family seemed to rest on Ben & myself. Ben fortunately excels in nursing almost if not quite as much as in mathematics & always seemed to do just the right thing. . . .

> I suffered for myself—for Sarah—for the children, but I think most of all for dear Ben. I wish you could have seen him throughout Sarah's illness.

As he is my brother I don't like to tell even you what I think of him—you his dear Aunt & Uncle & Cousins.[30]

Among her illnesses, Sarah had a difficult delivery of their fourth child, Helen, and Benjamin nursed her personally. He wrote to his friend Elias Loomis: "My wife has given me her first daughter after three sons. She did very finely until a fortnight ago, when she was attacked by pleurisy under which she is still suffering."[31]

$$\Phi^n$$

Helen, the Peirces' only girl, was named after Sarah's sister, Helen Mills Huntington, who died shortly before the Peirces' daughter's birth in 1845.[32] The birth was a difficult one not only for mother, but for child. Helen, called Lellie or Nellie, was a sickly baby, suffering from bouts of digestive disturbances and fevers. Fearing her maladies might prove fatal, Sarah took her to Swampscott, a seaside community just south of Salem, where she responded favorably to the healthy air and a wet nurse.

When she was some years older, her Aunt Elizabeth described her as a "good quiet little girl."[33] Her mother wrote to Helen's brother, Benjamin, "What a pity that a young girl's life should be so made up of follies as it almost necessarily is in the present state of Society. I do not wish to complain of Nellie for she is quite helpful at home & is becoming expert with her needle & capable in many ways—but when Mannie is here their days & nights are frittered away in nothings."[34]

Everyone liked Helen. She had many girlfriends and plenty of beaux. She had a long courtship with one of them, William Ellis: "Helen & Will Ellis are going on swimmingly & of course are very happy—I feel better about it than I expected—I am sure however that they have no reason to complain of the opposition which they have encountered. It would have been the most absurd impudence to have allowed an engagement before he had even graduated—He has now obtained a situation which promises something for him in the course of years—& I am satisfied that they should be engaged—your father has always thought & felt with me about it altho' all the opposition apparently came from me—He feels reconciled to it now & Will seems already like one of the family— . . . I hope you will like him— He is a man of few words, He is good looking & a gentleman in appearance & in fact—of most unexceptional moral character & in this respect a very safe match. As for marriage it is of course in the very distant future! & this is an objectionable feature of it."[35]

Helen Peirce. By permission of the Houghton Library, Harvard University, MS CSP 1643.

When wedding presents started to arrive, Sarah observed: "Helen says I must tell you of her present from Miss Greenough—a lovely little tête a tête tea set—on a waiter—plates but of elegant forms & altogether beautiful—She & Will are delighted with it & already imagine themselves using it in 'their' rooms with all their other pretty things about them—in a perpetual Honey Moon."[36]

At this time Sarah warned her husband that "you have never married a daughter & don't know what it is—it takes up money too, let me assure you, so that we shall feel a little poor for some time."[37] The Peirces survived the wedding, and acquired a congenial son-in-law. Helen found joy in the domestic life. Peirce wrote, "Dear Helen is just as much of an angel as ever is pure & good & lovable & it makes me happy 2 witness her happiness in her husband & children & how Will seems to dote on her—& how tender he is of her."[38]

$$\Phi^n$$

Despite Helen's difficult birth, Sarah and Benjamin had a fifth and final child, Herbert Henry Davis Peirce, or Bertie.[39] Herbert enjoyed putting on amateur theatricals, as did all the family, Benjamin included. On one occasion Sarah wrote: "Berts is crazying me about a dress for some theatrical they are getting up. . . . First dress rehearsal on Saturday & nothing done about his dress yet. He speaks altogether too much on these things & allows them I fear to interfere with his studies—He is doing better at school I think . . . but he will never be studious I fear—he has all the time some great business of this kind to absorb him."[40]

As a student at Cambridge High School, Herbert's lack of industry increasingly concerned his parents, as college approached. Sarah wrote, "Berts . . . does not it must be admitted love study—& . . . therefore if I had my way I would not send [him] to college at all—altho' he wants to go. I fear he will not study much anywhere—He learns easily & is satisfied with a superficial knowledge of his lessons."[41]

Much as his mother expected, Herbert failed his entrance exams: "Bertie I am sorry to say failed at his examinations—but means to try if he can to be examined again at the end of vacation—or if this is not allowed, to study up to enter Sophomore next year. I had no expectations that he would get in & therefore did not feel disappointed altho' he is very much so."[42]

Herbert then attended Exeter Academy for a year and did enter Harvard as a sophomore. He failed to pass his exams that year, however, and was permanently separated from his college class.[43] His parents thought that he might benefit from living with his brother Ben-

Herbert Henry D. Peirce, ca. 1871. By permission of the Harvard University Archives, call # HUP [1].

jamin in Marquette, Michigan. Herbert stayed with him for a while, but that too did not work out. Herbert lost some money there, and his father wrote, "I feel that Bertie is too great a load of anxiety to you to be weighing upon you, and so I have sent for him to come home."[44] Upon returning home, Herbert entered the Lawrence Scientific School, though his mother's prognosis was as pessimistic as it was correct: "Bertie has entered the engineering department of the scientific school—& seems somewhat interested in the drawing he has to do although the mathematics seems a bugbear to him. He does not & never will I fear study much."[45]

Herbert dropped out of school once again, soon after he began his engineering studies, and began a career in business. He was employed for a while at the iron works in Allentown, Pennsylvania, and later had business dealings in New York and Boston. He also worked for his father for a time on the Coast Survey.

Despite Herbert's inability to succeed academically or with his brother Benjamin, his mother did see potential in him, "I have faith to believe that he will come out right after all if he can be put to any regular & definite active work."[46] Herbert certainly exhibited his ability to act well in an emergency when he was accompanying his parents to Washington:

> In getting on to the ferry boat at N Y just at dusk with no lights properly placed to guide passengers your father fell between the boat & the pier, but caught by his arms to the boat & found himself with the boat rapidly nearing up at the corner by the chain you know without the least power to help himself owing to the small space for his legs & body—in another instant he would have been crushed—but Bert with admirable presence of mind put his arms around his body & drew him up just in time. . . .
> Your father scraped one of his shins badly—but would not do anything about it while we were gone—& is consequently having a little trouble with it now & keeps his foot up altho' he is not wholly confined to the house.[47]

Sarah's faith in her son's abilities eventually came to fruition. Herbert began a distinguished career in the Foreign Service in 1894, when President Cleveland appointed him secretary of the legation at St. Petersburg, Russia. When the legation was raised to an embassy in 1898, he was made its first secretary. He remained in this position for three years and gained renown as an authority on Russian affairs. Then, in 1891, President Roosevelt appointed him third assistant secretary of state. He served in this position for five years; during this time he oversaw several significant visits to Washington, including that of Prince Henry of Prussia, the Crown Prince of Siam, Prince

Fushimi of Japan, and Prince Tsu of China. He also played an important role in the peace negotiations at Portsmouth, New Hampshire, between Russia and Japan, which ended the Russo-Japanese War in 1905. The next year he was appointed envoy extraordinary and minister plenipotentiary to Norway, the highest post in the United States Embassy.

Herbert married Helen Nelson, the daughter of the wealthy Horatio Hose of Portland, Maine; Herbert and Helen went through her sizable fortune by lavish living. He was Peirce's only son who had children.[48]

Φⁿ

The Peirces' oldest child, James Mills Peirce,[49] called Jem, was born in Cambridge on May 1, 1834. Benjamin's mother approved of the manner in which Sarah and her son were raising their first child: "As to little Jemmy, Ma was saying only yesterday that we would admire to have him with us all the time only that you & Ben will make him a better man than anyone else could. Ma thinks you manage him just right—which is a great thing for her to say considering her estimate of her own power and capacity in the 'bringing up of children.' "[50]

James graduated from Harvard in 1853, a classmate of Charles Eliot, with whom he maintained a close association throughout his life. James then studied law for a year, but abandoned the law to become a tutor in mathematics at his alma mater. (Eliot was also a tutor in mathematics at this time.)[51] Then in 1857, James entered the Harvard Divinity School, although he retained his position as tutor until 1858. Upon graduating from Divinity School in 1859, he spent two years preaching at Unitarian churches in and around Boston, in New Bedford, Massachusetts, and Charlestown, South Carolina. His father wrote the Le Contes, "James has preached the two last Sundays in my own native town of Salem, and as far as I can learn, not unacceptably."[52] Peirce's sister, Elizabeth, was happy that he was not a disciple of liberal Unitarians like Peirce's friend Theodore Parker. She wrote, "I heard James preach in Cambridge a few Sundays ago & liked him very much. He is not in the least a Parkerite—He wants to preach the Bible."[53]

When James had an opportunity to preach in the South, his father wrote Josephine Le Conte, "But James, lucky dog, is going to South Carolina to preach in Charleston. He has received so very handsome an invitation, that he could not decline it and he will arrive in Columbia on his way to Charleston, in about a fortnight or three weeks, if I understand his arrangements and will pass a day or two with you."[54]

James Mills Peirce. By permission of the Harvard University Archives, call # HUP [6b].

James saw the Le Contes several times. He first saw Josephine, while she was in Charleston for race week. "Notwithstanding all the discomforts incident to the immense throng which congregated in the city at that time [Race week at Charleston, S. C.], she enjoyed the visit very much. She met your Son, James, at the Mills House (the Hotel at which both of them stayed), and she reports, that they had a 'power of fun'! We expect James Peirce here on next Friday or Saturday; and we hope he will enjoy the extreme quiet of his short sojourn here, as contrasted with the bustle and confusion of <u>race week</u> in Charleston."[55]

Despite James's success in the ministry, when James's father went to Europe in 1860, James returned to Harvard and took over his classes, his father paying him directly.[56] James apparently found being a professor more appealing than the ministry and stayed at Harvard, being appointed assistant professor of mathematics in 1861. James was tempted to accept a position at the University of California in 1869, but a generous salary and a salubrious climate were not enough to tempt him to leave Cambridge.[57] His mother wrote, "Jem . . . has pretty much decided to remain in Cambridge. The Corporation have offered him a full Professorship & full salary, with a place in the Scientific Faculty—& have been hearty and unanimous in their action so that he feels as if he would be able to endure a life which in his former position was a mere <u>grind</u>."[58] In 1885 James was made Perkins Professor of Mathematics and Astronomy, the same chair that his father had occupied for so many years.[59]

From the time James and Eliot started out as tutors of mathematics, they pushed for academic reform at Harvard. Among other things, they instituted written exams and fought for electives. When Eliot became president of Harvard, James found him to be an excellent administrator and a congenial ally for academic reform.[60] Knowing his father's antipathy towards Eliot, he wrote to him defending the new president and his plans for Harvard.[61]

During Eliot's administration of Harvard, James became the first dean of the graduate school and worked to develop graduate education at Harvard. He later served as Dean of the Faculty of Arts and Sciences. Although a sound administrator and an excellent teacher, James lacked the mathematical ability of either his father or his brother Charles.[62]

James was a discreet, but confirmed, homosexual, and never married. Perhaps his father suspected his sexual orientation when he complained to Bache "only one of them [my sons] is married, I am sorry to say, and I fear that the oldest is destined to be in the perpetual enjoyment of single cursedness."[63]

Φⁿ

The Peirces' third son, Benjamin Mills Peirce, or Benjie, was born in 1844. Young Benjamin exhibited genius, but lacked prudence and self-discipline. If the frequency that his father referred to him in his correspondence is significant, he was one of Peirce's favorite children. His father writes of building his son a hennery,[64] collecting rocks for his aquarium,[65] and notes that "Benjie has purchased a nice little microscope with three lenses with which he is preparing to make some mighty philosophical discovery."[66] But he was most enthusiastic about his son's acting ability. He wrote his friend, Bache: "Benjie has come out an actor. He played the part of Caliban the other night to everybody's astonishment. His low growls, his curses of Prospero, and his crouching under Prospero's wand were quite inestimable; and it was done with an unconsciousness which added greatly to its effect.—So that everybody who was present has spoken about it; as professor Parsons said that he pitied me—for 'you have a genius for a son.' "[67]

Peirce recorded one delightful narrative of his son's activities when he was a boy of about fourteen:

> Let me tell you a story of my son Benjie, which I have just heard and which is unjustly thought by some to betray his birth. At a party at Mrs. Eustis's, the other day, given to the children—a cake was divided and Benjie, much to his annoyance, obtained the ring. Under the awful necessity of choosing to whom he should give it, and with the fate of Paris hanging over his head, he contrived that the fifteen candidates should form a ring about him and that he should try the ring on their various and successive fingers—And my informants said that the way in which he tantalized the trembling hopes of each and every candidate was cruel in the extreme—and he prolonged their misery in his innocent desire to gratify them all by retaining the ring the longest possible time upon the finger to which he applied but which it did not fit. "But,["] says Benjie, ["]they all insisted that it did fit." "Was it a pretty girl that it actually fitted" asked the father. "Of course[,"] replied the ennoble son. ["]She was the prettiest of them all, and everybody admitted that it was a perfect fit. When I came to an ugly girl—I made her show her thumb—and the ring was always too small for them. One little bit of a child begged me to give it to her, but I tried it on her little finger, and it was too large. I was horribly afraid that Katy Toffy [a little beauty, and the most terrible little coquette (this papa's remark)] would get it and so I made her show her thumb and it did not fit"[68]

When her nephew was eleven, Aunt Elizabeth reported that "Benjie goes to the town-school & takes great interest in his studies."[69] Benjamin graduated from Cambridge High School, and entered Harvard, where his high spirits got him into a bit of trouble. After a number of offenses, ranging from cheating on a mathematics test to being absent from prayers, Benjamin was rusticated, in his sophomore year, to Westford, Massachusetts. There he could continue his studies, contemplate his conduct, and regain his bearings. Rustication was a common means of disciplining students at Harvard. The student was sent away from the college, for about six weeks, usually under a minister's supervision.[70]

Benjamin's father, relieved that his son had received no worse punishment than rustication, wrote to him while he was being punished, giving him encouragement:

> I am sorry to deny you anything and will not deny anything which my duty & love can consent to give you. But I feel that it would be unwise in me to permit you to return to the lazy influences of a college room. I shall be greatly disappointed if you do not come back to college with a firm determination to take hold as a scholar. If this is not your intention, it is better that you should leave college altogether. Your future life on this earth will be hard and bitter, if you do not use your present opportunities for acquiring the foundation of a shining and inquisitive character. You complain too much of your present lot, it is not manly to suffer yourself in such despondency. Were it punishment, you have brought it upon yourself and have no right to complain. But it was on the contrary the opportunity which was offered you to repair your own errors, and you ought to be thankful that it was granted you. It has been a very difficult effort upon the part of your parents to enable you to have this opportunity. It is a pecuniary expense, which it is hard for us to undergo and we think reasonable to expect that you will do your part to make it serviceable to you. . . . God bless and preserve you.[71]

When the sophomore rustication proved to be ineffective in reforming Benjamin, he was rusticated a second time near the end of his junior year.[72]

After Benjamin graduated from Harvard, he went to France to study at the prestigious École Polytechnique.[73] He was the only Peirce child to study abroad.[74] Aunt Elizabeth was excited about his being in Paris, wanting to know all about French fashions. "In January, if the Philadelphia & Wilmington Railroad does its duty, & pays a good dividend, I shall be able I trust to treat myself to a good dish of fashions from you."[75]

Although brilliant, Benjamin, like his father, felt things deeply. He was passionately involved in French politics, when his cousin Henry Cabot Lodge visited him in Paris:

> Ben. Peirce, the son of Professor Peirce, the eminent mathematician, who had graduated from Harvard the year before [1865], was completing his

Benjamin Mills Peirce. By permission of the Harvard University Archives, call # HUP [1a].

education at the École Polytechnic and lived in two small rooms in the Latin Quarter. . . . Ben. Peirce was a man of really brilliant talent, but was wearing himself out by a reckless disregard of health and of all the necessary limitations of human existence. He understood almost everything except self-control. He worked very hard and distinguished himself in his studies; he also played very hard, and in short burned the candle not only at both ends, but at every other point on its surface. Not content with all the amusements affected by the students of Paris, he flung himself violently into French politics. He was one of the drollest human beings I ever knew, as well as one of the most hard-working, and he had a wild humor which would carry him into all sorts of excesses, and very dangerous when applied to French politics, which at that particular time were not of the safest. Haunting the cafés frequented by students, and speaking French with the utmost fluency, although with a strange accent, he became a violent republican and an ardent foe of the empire.

He was wont to discourse about the infamy of established government, half seriously and half humorously, but with a violence and an eloquence which used to startle my youthful mind. Unfortunately he would not always stop there. One night, returning from dinner with his cousins, he insisted upon climbing up onto the high fence of the Tuileries, and from that point of vantage shouting: "Vive la République!" "A bas l'Empereur!" . . . The natural result was the appearance of the sergeant de ville and his immediate arrest. He was rescued with difficulty by his cousins, who explained that he was an American, and that he was only joking. . . . To me Ben. Peirce seemed then, as he does now in memory, one of the most fascinating beings I had ever seen. His fun and humor were unbounded, but he was equally interested in serious matters, and if he did not always think soundly, he rarely failed in originality. He graduated at the Polytechnique with distinction, came home and entered at once upon a career which was full of promise. But the candle had been burned too freely and in too many places.[76]

After returning from France, Benjamin compiled a report for the United States State Department on the condition and resources of Greenland and Iceland. In the summer of 1868, he obtained a position as a mining engineer at Lake Superior in the iron mines near Marquette, Michigan.[77] His father regretted his being so far away: "I cannot help feeling sad at having you go off again although I presume it to be for the best. I wished that something might turn up to keep you here—but if you should ever need any more assistance, my dear boy, to get along—remember that the best part of my life is to keep my children alive—and that I will rejoice to assist you."[78]

Later, Peirce wrote a note to Benjamin urging moderation, "Don't undertake too many things at once. But remember that repose at heart is the best attitude for success. Your success is certain, if you will not be impatient. . . . I know the rocks upon which most men of great power are apt to be wrecked and I would be the better to warn you."[79] His mother was also concerned that young Benjamin was working too hard.[80]

Benjamin Mills had protracted problems with his health, having acquired a chronic cough sometime before September 1868. Both of his parents were concerned and sent him advice about his health.[81] By January 1870, his condition appeared serious, and his mother pleaded with him to take care of himself.[82] Two months later she said, resignedly, "Well—I commend you to the Good God & your kind Doctor to whom I feel very grateful."[83] Peirce was concerned as well as Sarah; he wrote:

> My Dear darling son,
>   We cannot help feeling quite troubled about you, and wish that you were here with us. Your friends are evidently most kind to you and are plainly doing all they can to relieve your illness; but they are not father and mother. Can you not come to us by land without any very great expense. Do come if you possibly can.
> Your Loving Papa[84]

Young Benjamin rallied in the middle of March. When his mother received the good news, she wrote: "You laugh at & abuse me for my anxiety about you but I must have been more or less than mother to have heard of your being confined with sore throat and cough so long with the suggestion of night perspirations & no appetite & then horror—among them, nothing to eat & not be anxious. Well thank God that you are better. Your father has been persecuting me to go to Washington with him this week, but I told him I could not go unless I heard first from you, so he sent the telegram."[85]

Despite the temporary recovery, young Benjamin was dead before the end of April, at the age of twenty-six, probably from syphilis.[86] His father, accompanied by Benjamin's brother Charles, went to Michigan to bring the body back to Cambridge. Peirce telegraphed his son James: "Reached Albany Depot Wednesday, eleven A. M. Arrange services for same day. Propose Cambridge Cemetery. Decide with Mother. Body invisible—death calm—are well."[87]

$$\Phi^n$$

In a letter to his mother, Peirce whimsically predicted the genius of his second child, Charles Saunders, who, in fact, has been credited with the most brilliant mind the Americas have yet produced:[88]

Good news! Dearest mother, Sarah is fine and so is the, yet anonymous boy. Sarah sends her very best love. She has had the least bad time she ever had, the least exhausting. She suffered for about two hours and at 12 was confined. The boy weighs 8 3/4 pounds, and is as hearty as possible. He offered to go out and throw the vase with me today; a certain sign that he will either make an eminent lawyer or a thief. As soon as he publishes his "Celestial Mechanics" I will send you a copy, and I have no doubt he would be glad to correspond with you about your last mathematical researches. He has two splendid optical instruments each of a single achromatic lens which is capable of adjustment!—as so wonderful is the contrivance for adjustment that, by a mere act of the will, he is able to adapt it to any distance at pleasure. But one fault has been found with these instruments and that is, the images are inverted; our new born philosopher, however, our male Minerva, contends that this is a great advantage in the present topsy-turvy state of the world. By the way, he calls himself Minervus. The first proof of his genius which he exhibited to the world consisted in sounding most lustily, a wonderful acoustical instrument whose tones, in noise and discordancy, were not unlike those of fame's fish-horn. Is not this a singular coincidence? a sure omen of his coming, almost come, celebrity?[89]

In an equally jocular tone, Peirce announced Charles's birth to his brother and sister. His reference to suicide eerily foreshadows his son's self-destructive traits that plagued his tragic life: "Today at 12 o'clock was born a boy. Its mother and it are both doing 'finely' and send you their very best love. The boy would have written, but is prevented by circumstances over which he has no control. He does not like this blue ink he says. He hung himself this afternoon to a pair of steelyards; but postponed the further execution of his wicked designs upon himself, because he found he wanted one quarter of a pound of nine pounds which he regards as the minimum of genteel and fashionable suicide. At 8 1/2 this morning his mother—if she can be called his mother before he was—his future mother, or more transcendentally, the mother of this child of futurity was well—or nearly so, the shadows of coming events having but slightly obscured the brightness of her countenance."[90]

Though brilliant, Charles didn't do particularly well in school. He attended Cambridge High School, where he was suspended several times. After he graduated, his parents then sent him to E. S. Dixwell's school for a semester to prepare him for Harvard. Mr. Dixwell was not impressed with young Charles's ability.[91] Charles's father retained more faith in his son than in Mr. Dixwell. He wrote, "Charlie appears very well, and I have great hopes that he is coming out bright, Dixwell to the contrary not withstanding."[92]

Charles Saunders Peirce. By permission of the Houghton Library, Harvard University, MS CSP 1643.

At Harvard, Charles remained a mediocre student. Not until after graduating, when he studied chemistry at the newly opened Lawrence Scientific School, did he gain academic distinction. The prime source of his education, however, was neither from the Cambridge High School, nor Harvard College, but his father. His brother Herbert and sister, Helen, wrote: "As a child he was forever digging into encyclopedias and other books in search of knowledge upon abstruse subjects, which discussions with his learned father upon profound questions of science, especially higher mathematics and philosophy, were common matters of astonishment, not only to his brothers and sister, but to his parents as well."[93]

Charles put it simply, "I was educated by my Father ... & if I do anything it will be his work."[94] Charles described the intellectual training he received from his father: "He very seldom could be entrapped into disclosing to me any theorem or rule for arithmetic. He would give an example; but the rest I must think out for myself. He gave me at a tender age a table of logarithms, with an example of the way of using it to find the logarithm of a number, and another to illustrate a simple multiplication by logarithms; but beyond that he would give me nothing. He took great pains to teach me concentration of mind to keep my attention upon the strain for a long time. From time to time, he would put me to the test by keeping me playing rapid games of double-dummy from ten in the evening until sunrise, and sharply criticizing every error."[95]

Charles began to study chemistry at an early age. "Although I was not a precocious child, at the age of 8 I took up of my own accord the study of chemistry, to which the following year I added natural philosophy; so that by the time I went to college, I was already a fairly expert analyst."[96] His uncle, Charles Henry Peirce, who was Eben Norton Horsford's assistant in the Lawrence School, and his Aunt Elizabeth helped him set up a laboratory at home. There he made his way through Liebig's program of quantitative analysis.[97]

Although Charles acquired sound intellectual training from his father, he often complained about the lack of moral guidance he received: "But as to moral self-control he [my father] unfortunately presumed that I would have inherited his own nobility of character, which was so far from being the case that for long years I suffered unspeakably, being an excessively emotional fellow, from ignorance of how to go to work to acquire sovereignty over myself."[98]

$$\Phi^n$$

The Peirces were delighted when Charles became engaged to Harriet Melusina Fay, or Zina, who became a noted feminist and social reformer. Peirce wrote Bache, "Charlie is getting married to the daughter of a old friend and chum in College."[99] Aunt Elizabeth observed, "Charles Peirce's engagement is a delightful one. Miss Fay is a very intelligent & highly educated & amiable girl. . . . her mother was a daughter of Bishop Hopkiss."[100] Under Zina's influence, Charles became an Episcopalian. As in other things, the family felt that Zina was exerting a positive influence on Charles's religious life. Far from being upset by Charles's conversion to Episcopalianism, Peirce was pleased by Charles's being more religious.[101]

Since Charles had little income, he and Zina lived with Benjamin and Sarah for two years and took their meals with them for an additional six. Peirce was thoroughly delighted with his daughter-in-law, "Mrs. Charles S. Peirce is come to stay with us and she is the most charming acquisition to the family circle."[102] The marriage proved to be an unhappy one. Among other problems, Zina suspected that Charles was unfaithful, and left him after fourteen years.[103]

$$\Phi^n$$

Charles first worked for the Coast Survey the summer before he graduated from Harvard. Despite his brilliance, and such influential family and friends as his father, Bache, William James, and his first cousin, Henry Cabot Lodge, Charles found no permanent employment outside of the Coast Survey.[104] He did acquire a part-time position at the Harvard Observatory doing photometric research, under his father's protégé, Joseph Winlock, and delivered several courses of lectures at Harvard. After Winlock died, Charles became embroiled in an unfortunate dispute with Harvard's president, Charles Eliot, that precluded his ever obtaining a professorship at Harvard.[105]

After Charles's father became head of the Coast Survey, he furthered Charles's career in several instances, and also did not hold him to the same degree of accountability that he did other Survey employees. Although there was no question as to Charles's brilliance or ability, these actions smacked so blatantly of nepotism, that they eventually had serious consequences for Charles.[106]

$$\Phi^n$$

Early in 1878, Daniel Coit Gilman, the president of the newly founded Johns Hopkins, began negotiating with Charles about a half-time lectureship in logic. By June 1879, Charles accepted the post.[107]

He retained his position with the Coast Survey, although he was finding his work there increasingly distasteful. Johns Hopkins was the first great research university in America. By 1880, four years after the founding of Johns Hopkins, the university's faculty had produced almost as much published research as that from all other universities in the United States for the previous twenty years.[108] Among Charles's most notable students at Johns Hopkins were John Dewey, Fabian Franklin, Benjamin Ives Gilman, Joseph Jastrow, Christine Ladd (Franklin), Allan Marquand, Oscar Howard Mitchell, and Thorstein Veblen.[109]

After spending several years as a lecturer at Johns Hopkins, the university's president, Daniel C. Gilman, assured Charles that he would be offered a permanent position at Hopkins.[110] All during this time, Charles had been stretching himself almost to the breaking point by performing the duties of both his academic job and his position with the Coast Survey. Consequently his health began to fail. Not only was he having trouble with his physical health, but with his mental health as well. After consulting doctors in New York, he wrote to Gilman, "My physician . . . informed me that he considered the state of my brain rather alarming. Not that he particularly feared regular insanity, but he did fear something of the sort."[111]

One of Charles's physical problems was trigeminal neuralgia.[112] The disease causes excruciating facial pain. Attacks may be brought on by such a simple thing as touching one's cheek or by a slight breeze hitting one's face. They often occur when someone suffering from the disease is under stress, but attacks may also occur from no apparent cause. The illness is not fatal, but it is extraordinarily painful. The only way to treat the disease at that time was with ether or laudanum (tincture of opium). Since Charles had a severe case of the disease, he was almost surely addicted to opium. Succumbing to the pressure of managing both of his jobs, he probably began using cocaine to give him sufficient energy. Cocaine, like opium, was widely available at this time and used by many notable people.[113]

Whether from the heavy workload, from his own arrogance and tendency towards procrastination, or because of the nature of the work, much of his work on pendulums for the Coast Survey had not been written up. Charles also occasionally missed classes at Johns Hopkins, offering various excuses to Gilman. Nevertheless, by 1883, the last year he spent at Johns Hopkins, Charles had produced an extraordinary series of publications, many of them seminal contributions to philosophy and logic. Such a record of scholarly production would normally win a man an academic post, even if he had many other shortcomings.[114]

Φ[n]

In 1876 Charles met Juliette Annette Froissy, a woman shrouded in mystery. Although she claimed to be a widow, she was only twelve years of age. She also claimed to be a Hapsburg princess, who as a child had played with the boy who became Kaiser Wilhelm II. More likely she was a Spanish gypsy, whose true name was Fabiola de Lopez. She came to this country from France, apparently without a passport or visa, and she wished to keep her identity unknown.[115] Charles's Aunt Elizabeth did not care for women with a mysterious past. She wrote, "I do wish Charley would come out openly & tell me her true history—yet I am afraid if he did, I should not like her any better than I do now."[116]

In the years between 1876 and 1883, Charles had an open love affair with Juliette.[117] He finally obtained a divorce from Zina, his estranged first wife, and married Juliette only a few days after the divorce became final. Juliette was nineteen; Charles was forty-four.[118] American mores during the second half of the nineteenth century could countenance a discreet affair, but a man who openly lived with a woman and then had the effrontery to marry her was thought to be making a mockery of the institution of marriage and committing an unforgivable act. Charles's own family disapproved of the marriage, as did Boston and Cambridge society.[119] After they married, they went to Cambridge to visit the family. Charles's mother, Sarah, wrote to her daughter, Helen. "I dread the troubles which are sure to be in store for them & thro' them for me—in the family and in society. Poor Charlie is like a child about conventionalities & has no idea that anything stands in the way of her being received everywhere!!"[120] Charles's Aunt Elizabeth was outraged by both the match and the visit. She wrote: "I hope and trust they will go this week and never return. They could not have a worse place for them than Cambridge —where all the White Mountain scandal [Charles and Juliette had indiscreetly conducted their affair there three years previously] & others of a similar nature is known. However it is not worthwhile to worry. No doubt of her badness & I fear of his also—I have sufficient that is personal against them to prevent my having *any* intercourse with them."[121]

Aunt Elizabeth had not softened by the next spring; she wrote: "I do pity Charley, still it was his own choice to marry J. [uliette] contrary to the advice or wishes of all his friends. I never could imagine what he could discover fascinating in or about her. It was not her beauty—since she *had none.*—& never could have had much if any of that article. The fact is she bewitched him, I suppose. If she had lived in Salem during witch time she wd certainly have been tried as a witch & been sentenced accordingly."[122]

Johns Hopkins was no happier with Charles's marriage than was his family. When the knowledge of his relationship with Juliette was made known to the trustees of Johns Hopkins, his lectureship was not renewed. Charles was so enraged and devastated about losing his job that he delayed telling his family in Cambridge about it for as long as possible.[123]

Two years after Charles's marriage, in July 1885, the Coast Survey came under severe censure and investigation from Congress for waste. Hilgard, the current head of the Coast Survey, had problems with alcohol that dated back at least to the time when Peirce was superintendent of the Coast Survey. He was accused of dereliction of duty and replaced. Charles came under heavy condemnation as well, and was also accused of dereliction of duty. Although he was somewhat exonerated, and his resignation from the Coast Survey was not accepted, the combination of his troubles with the Coast Survey, his relationship with Juliette, and his dismissal from Johns Hopkins gave him an unsavory reputation, sufficiently so that he never obtained the academic position that he hoped for, and that his ability certainly justified him for.[124]

In response to Hilgard's public drunkenness and poor management of the Coast Survey, the Allison Commission from Congress made a thorough investigation. As a result, the Survey became a bureaucracy that was interested in producing only utilitarian results. Gone were the halcyon days under Alexander Bache and Benjamin Peirce, when the Survey was an institution that sponsored basic scientific research. These changes made an already unattractive position in the Coast Survey, for Charles, now close to intolerable.[125]

$$\Phi^n$$

Although Charles had many professional spats, most people who had disagreements with him were still sympathetic to him. Sylvester, for example, had a protracted and heated dispute with Charles over priority in a publication on logic. When Sylvester heard of his dismissal from Johns Hopkins, however, he wrote to Gilman, expressing regret about this action as well as the spat he had had with Charles. Sylvester assured Gilman that he, himself, was largely to blame for the disagreement.[126] Similarly, in most cases men who denied him opportunities did so both reluctantly and without vindictiveness, recognizing Peirce's genius and charm. Not so with Simon Newcomb, one of his father's students, and Charles's supposed friend. Newcomb took an active part in Charles's downfall.

Newcomb became the most honored and recognized American scientist of his time.[127] Yet Newcomb's background was vastly different from that of the son of a leading and influential American scientist, and whose maternal grandfather had been a United States senator. Newcomb, the son of an itinerant schoolteacher in Nova Scotia, had very little formal education, receiving only a rudimentary education from his parents. He was apprenticed to a doctor, who turned out to be an unprincipled and abusive quack. He worked Newcomb hard, but failed to provide the promised instruction. After two years as an apprentice, Newcomb feared that his formative years were being wasted and fled to the United States. He eventually joined his father, who was now living in Maryland. Newcomb found work there as a schoolteacher and tutor.

He made numerous trips to the Smithsonian to use its library in order to pursue his scientific interests. There he met Joseph Henry, who, impressed with his intelligence and diligence, suggested that he might obtain a position with the Coast Survey. Unfortunately no positions were available, so Henry suggested Newcomb might find employment as a computer with the *Nautical Almanac*.[128] Newcomb was reluctant to take such a position, hoping to be something more than a "computational drudge."[129]

The exact circumstances that led Newcomb to the *Nautical Almanac* are vague, as is the part Benjamin Peirce played in them.[130] In the summer of 1855, Newcomb's father visited Harvard University and told Peirce about a paper his son had written on the motion of the moon. Peirce requested Newcomb to send him a copy of his paper, which he did.[131] The contact with Peirce may well have been the crucial factor in drawing Newcomb to Cambridge, which Newcomb later described as "a world of sweetness and light."[132] Newcomb said, "Life in the new atmosphere was in such pleasant and striking contrast to that of my former world that I intensely enjoyed it."[133] Certainly coming to Cambridge was the key event that changed Newcomb's life and gave him the opportunities for scientific training and employment that he so deeply desired and so well deserved.

If Peirce was not the principal figure in opening up this opportunity to Newcomb, he was a crucial one. The duties of the *Nautical Almanac* office were light, so Newcomb had ample time to pursue mathematical and astronomical studies under Peirce. Peirce became Newcomb's mentor, encouraging him in his work. He invited Newcomb to social activities at his home, and arranged for him to attend meetings of the American Academy of Arts and Sciences. It was almost surely Peirce's influence that enabled Newcomb to borrow books from the American Academy's library.[134]

Newcomb did not appreciate what Benjamin Peirce had done for him. He later wrote despairingly of his experience at the Lawrence

School, studying with Peirce. "As a student of mathematics it could hardly be said that anything was required of me either in the way of attendance on lectures or examinations until I came up for the degree of Bachelor of Science. I was supposed, however, to pursue my studies under the direction of Professor Peirce."[135]

Winlock, who was then head of the *Almanac* office, immediately recognized Newcomb's ability, gave him more interesting work, and singled him out for advancement. Yet despite favorable treatment from both Winlock and Peirce, Newcomb may never have really felt comfortable in Cambridge. The Peirces were not wealthy, but they were comfortably in the upper middle classes; furthermore, the Peirces moved easily among the social and intellectual elite of Cambridge and Boston. Probably more significantly, despite Newcomb's unquestionable scientific ability, he tended to be a conventional and conservative thinker.[136] This may explain why he was never brought into the small circle of exceptional young men, including Charles Peirce, Chauncey Wright, and William James, which eventually became the Metaphysical Club.[137]

Newcomb continued to correspond with Benjamin Peirce about mutual scientific interests after he left Cambridge, but his letters lack the warmth, not to mention the adoration, that is typical of letters from many of Peirce's former students and protégés. If Newcomb did have any feelings of gratitude toward Peirce, they did not extend to his son Charles. Perhaps Newcomb resented Charles's being born to privilege and found his arrogance unbearable. Perhaps he deplored Charles's disregard of current moral values. Whatever his motivation, he consistently and determinedly undermined Charles's career.

It was Newcomb's conversation and follow-up letter to Gilman that made the Hopkins trustees aware of Charles relationship with Juliette and subsequently led to his dismissal.[138] In another instance, Sylvester had published the first part of Charles's "Algebra of Logic" in the *American Journal of Mathematics* when he was editor. When Newcomb succeeded him as editor, he refused to publish the second part of Charles's "Algebra of Logic," claiming that its subject was not mathematics.[139]

Charles's work on pendulums was chronically and consistently late. When criticized for his tardiness by Superintendent Thomas Corwin Mendenhall, Charles replied, "Now anybody who had ever done such work in such a way—ask such men as Langley or Newcomb—will tell you that it is impossible to make any reliable estimate of the time it will take."[140] In 1890, Charles submitted part of his long overdue report on pendulums. Mendenhall, apparently regarding his own expertise insufficient to judge the report, gave it to

Simon Newcomb. Newcomb thought it substandard, but suggested having it reviewed by other scientists. Mendenhall took Newcomb's advice and referred the report to a three-member panel, which included Newcomb.

Charles was the only American with an international reputation in geodesy, and the report was, in fact, brilliant and innovative. It has been called "the best work of its kind in the nineteenth century."[141] Nevertheless, the report was judged unsatisfactory by a two-to-one vote. Neither Newcomb nor Mendenhall had a sufficient knowledge of geodesy and mathematics to appreciate Charles's work. Nevertheless, Mendenhall generously gave Charles nearly a year to finish his work on pendulums. When it did not appear, Mendenhall asked for Charles's resignation.[142] Charles never had regular employment again. With no pension plan yet instated for government workers, after thirty years of service in the Coast Survey, Charles was left without any income and little prospect of employment.[143]

Perhaps Charles's best opportunity for employment after being asked to resign from the Coast Survey was to head the newly formed Department of Weights and Measures, since he had been in charge of weights and measures while employed with the Coast Survey. When Charles sought this job in 1899, Newcomb refused to support, and probably acted against Charles's application. In 1902, Charles applied to the Carnegie Institution for a grant to complete his seminal work on logic. Once again, Newcomb's negative report on the importance of Charles's work prevented him from receiving the grant.[144]

Charles's arrogance, his procrastination, his disregard for the moral code of his country, his frequent spats with colleagues, and his extravagance were sufficient reasons to prevent his having a successful career, but Newcomb was certainly there at every turn to put a nail in his coffin.

$$\Phi^n$$

Two years before Charles was ousted from the Coast Survey, he bought a farm outside of Milford, Pennsylvania. Some of the money for the purchase came from a recent inheritance from his Aunt Charlotte Elizabeth. One of Charles's many fanciful schemes to obtain wealth was to found a school for logic in Milford. As with Charles's other such schemes, this failed, but he spent the rest of his life in this house in Milford, which he named Arisbe, after the ancient Greek colony on Miletus, where western philosophy, science, and mathematics were born.[145]

Now devoid of income, except for occasional money from his writing and his wife's annuity, Charles was literally dying of malnutrition when his brother James came to his aid, and his friend William James gave him a modest stipend for the remainder of his life. He received no aid from his other siblings, including his sister, who was quite wealthy at this time. When he died, the only family member to attend his funeral was a nephew. Although he had been denied a permanent faculty position at Harvard, the Harvard Department of Philosophy lost no time in sending a team out to Milford to collect his papers.[146]

Φ n

Benjamin Peirce's health began to decline early in 1879, about the same time he was asked to give the Lowell lectures on science and religion. His health was no better when he went to Baltimore to repeat them the next year. At Johns Hopkins, Peirce was happy to see Charles and other friends, including Gilman, and was also gratified with the reception of his lectures. "My last night's lecture seems to have been a success. As great as any which I have had. . . . I heard Charlie lecture this morning and was delighted with him. I saw President Gilman and I believe that Charlie will be permanently connected with the university."[147] Nevertheless Peirce immediately became lonesome and despondent. The next day he wrote to Sarah:

> Oh my dearest Sarah! No tongue can begin to tell how weary I am of being here! It is paying vastly too dear for the article. I was the obedient poor Christian—to have come for these lectures. . . . And I want to be at home with my dearest loving girl. . . . I hear no whisper from Patterson. This seems to me most probable that he looks upon his consulting geometer as a useless exercise, from which he would readily be delivered. He is evidently not coming near the lectures, and making arrangements with President Gilman about which he does not consult me. The world has had enough of me altogether and it is time to get under the sod.
>
> Your dearest lovely, loving letter is just come, and it is . . . a most delightful restorative . . . without you life is nothing.

Now a happier Peirce continued: "My lecture was a thorough success—of that I am certain—and everybody says how well my voice filled the house. . . . And even Charlie spoke of my having taken exactly the right key. . . . My health is greatly improved since I left home—that is—I am stronger than I have been for years. Gilman gives me a reception tomorrow evening and I am not afraid that I shall find it too much for me. God bless you Dearest and Best."[148]

In May 1880, Peirce became seriously ill. Peirce's old friend, Thomas Hill, wrote Sarah, concerned about Peirce's illness. He reminisced about the effect Peirce had had on his life: "Once in the Connecticut Valley away up near Wells River, I said something which was in bad taste, and which an enemy might have interpreted to show that there was something much more than bad taste in me. Prof. Peirce said not a word, but he gave me a glance of astonished puzzled inquiry, inquiry appended to himself, whether he had mistaken me; that glance did more for me than a dozen sermons on the government of the tongue could have done. I cannot think of it even today without a mingled sense of shame for myself & gratitude to him. Yet I doubt if he did it consciously."[149]

Aunt Elizabeth wrote to her nephew Charles bemoaning her expected loss:

> Your father has slept great part of day—& has not taken any morphine since night before last—The Doctor said—this sleeping is a phase of the disease. I will write again tomorrow—if only a postal—unless Jem or your mother writes. I feel so dull and forlorn today! Dear Charlie what will become of me—all alone—as I shall feel myself to be? Jem came this morning—but he hardly speaks to me—& therefore for all the good his presence does me he might just as well be in China. But I must get used to the condition of things & the sooner the better. I can see now that I shall feel your father's loss more & more—nobody to love me as he did—& he was always so kind and good—It is only to you that I can bear to speak of him & express my misery at the thought of living without him. One comfort I have is that I have always done what I could for him—in little and great things—for him, the darling of my life. I do not think that I have ever <u>realized</u> till today that he is soon to leave us—all along—it has seemed to me more as if he was going on a journey—well I shall soon follow him and when he is in Heaven & I am still on earth we shall be really together if I can only manage to be good enough to keep the Father with me—for in whose happy presence dear Ben will be—so if the Father is with me—& Ben is happy with the Father, shall we not be together?[150]

Peirce's condition gradually worsened, and after a painful illness, he died on October 6, 1880. His funeral services took place in Appleton Chapel three days later. The service was conducted by two of Peirce's lifelong friends, Andrew Peabody and James Freeman Clarke. His pallbearers were comprised of a combination of Boston Brahmins and the nation's scientific elite: Oliver Wendell Holmes, Charles Eliot, Thomas Hill, Simon Newcomb, Joseph Lovering, Jonathan Ingersoll Bowditch, James J. Sylvester, and Captain C. P. Patterson.[151]

Appendixes

I
Presidents of Harvard during Peirce's Lifetime

1806–10	Samuel Webber
1810–28	John Thornton Kirkland
1829–45	Josiah Quincy
1846–49	Edward Everett
1849–53	Jared Sparks
1853–60	James Walker
1860–62	Cornelius Conway Felton
1862–68	Thomas Hill
1868–69	Andrew Preston Peabody, Acting President
1869–1909	Charles William Eliot

II
Chronology

1778		Benjamin Peirce (Peirce's father) is born.
1781	January 3	Lydia Ropes Nichols (Peirce's mother) is born.
1809	April 4	Peirce is born in Salem, Massachusetts.
1823		The Round Hill School is founded.
1825		Peirce enters Harvard.
		Peirce publishes his first mathematical work (the solution to a problem) in the *Mathematical Diary*.
1826		Peirce's grandfather's and father's business fails.
1829	May 5	Sarah Mill's father, Elijah Hunt Mills, dies.
	Spring	Peirce graduates from Harvard.
	Spring	Ward graduates from the Round Hill School.
	Fall	Peirce begins teaching at the Round Hill School.
1830	May 18	Peirce meets Sarah Mills.
1831	July 26	Peirce's father dies at age fifty-three.
		Peirce is appointed a tutor in mathematics at Harvard College.

1832		Peirce's paper on perfect numbers is published in the *Mathematical Diary*.
		The *Mathematical Diary* ceases publication.
1833		Peirce becomes University Professor of Mathematics and Natural Philosophy at Harvard College.
	July 23	Peirce marries Sarah Mills.
1834	May 1	James Mills Peirce (Jem) is born.
1835		Peirce's first textbook, *An Elementary Treatise on Plane Trigonometry*, is published.
1836		The *Mathematical Miscellany* begins publication.
1839	September 10	Charles Saunders Peirce is born.
		The *Mathematical Miscellany* ceases publication.
		A committee, headed by John Q. Adams, is formed to organize the Harvard Observatory.
		William Cranch Bond comes to the Harvard Observatory.
1841	November	J. J. Sylvester comes to Virginia.
1842		Peirce is appointed Perkins Professor of Mathematics and Astronomy.
	April	Peirce revives the *Mathematical Miscellany*, as the *Cambridge Miscellany*. The publication lasts only one year, or four issues.
		Peirce is elected a member of the American Philosophical Society.
		Charles Henry Davis marries Harriette Blake Mills, the sister of Sarah Mills Peirce.
1843	March 22	Peirce lectures in Boston on the great comet; the lecture stimulates support for the installation of a telescope at the Harvard Observatory.
		Peirce meets Alexander Dallas Bache.
	June	Sylvester leaves America.
	December	Bache becomes superintendent of the U. S. Coast Survey.
1844	March 19	Benjamin Mills Peirce is born.
1845		Helen Huntington Peirce is born.
1846	January	The Rumford Professorship becomes vacant.
	October	Louis Agassiz comes to America.
	December 3	Joseph Henry becomes the first secretary of the Smithsonian.
	December 23	Using Leverrier's computations, Galle discovers the planet Neptune.
1847		Abbot Lawrence donates $50,000 to Harvard University for the Lawrence Scientific School.
	March	Peirce serves on a three-member panel to investigate a controversy between A. D. Bache and James Ferguson.

	March	The American Association of Geologists and Naturalists resolves to become the American Association for the Advancement of Science.
		Peirce, Agassiz, and Henry D. Rogers draft a constitution for the new organization.
	March 16	Peirce announces that Leverrier's discovery of Neptune was a happy accident.
		A telescope is installed in the Harvard Observatory.
		Peirce receives an LLD from University of North Carolina.
		Peirce is appointed by the American Academy of Arts and Sciences to a five-member committee to draw up a program for organization of the Smithsonian.
1848		Peirce's former student, B. A. Gould returns to the United States with a German PhD.
		Peirce publishes his work on the perturbations of Uranus and Neptune.
1849		American Nautical Almanac is established at Cambridge, and Peirce becomes the consulting astronomer. Peirce serves in this capacity until 1867.
	April 11	Herbert Henry Davis Peirce (Bertie) is born.
1850		Peirce is elected an associate of the Royal Astronomical Society, London.
1851–52		Civic leaders, scientists, and scholars attempt to found a National University at Albany.
1852		Peirce publishes his "Criterion for the Rejection of Doubtful Observations."
		Peirce becomes director of the longitude determinations of the United States Coast Survey. He retains this position until 1867.
	May 16	Peirce first mentions "Florentine Academy" in a letter to Bache.
	November 26	Peirce is notified that he has been elected a foreign member of the Royal Society of London.
1853		Peirce is president of American Association for the Advancement of Science.
	August 30	Peirce first uses the word "Lazzaroni" to refer to the group of scientists led by Bache.
1854		Peirce gives his address upon retiring as the president of the American Association for the Advancement of Science.
1855		Peirce's *Analytical Mechanics* is published.
1855–58		Peirce is a member of the Scientific Council (with Bache and Henry) of the Dudley Observatory, Albany.

1857	January & February	Peirce delivers lectures at the Smithsonian in Washington.
		Warner-Winslow affair begins.
1857–59		Dudley Observatory controversy.
1858	Spring	Peirce resigns from the American Academy of Arts and Sciences. When Peirce reapplies, his application is denied. Eventually he is reinstated.
1860		Peirce is elected an Honorary Fellow of the University of St. Vladimir, at Kiev, Russia.
	May 26	Peirce leaves for his first trip to Europe.
	December 20	South Carolina cedes from the Union.
1861	April 12	The Confederate States fire on Fort Sumter. The Civil War begins.
		Peirce is elected a corresponding member of the British Association for the Advancement of Sciences.
1862	September	Lincoln issues the Emancipation Proclamation.
1863		Wolcott Gibbs is appointed to the Rumford Chair. Charles Eliot leaves Harvard.
1863	March 3	The National Academy of Sciences is founded.
1865	April 9	Lee surrenders to Grant.
	April 14	Lincoln is assassinated.
	May 26	Last Confederate troops surrender.
1867	February 17	Bache dies.
		Peirce testifies in the Hetty Robinson trial.
		Peirce is elected an honorary fellow of the Royal Society of Edinburgh.
		Peirce is elected a correspondent in the mathe- matical class of the Royal Society of Sciences at Göttingen.
		Peirce receives an LLD from Harvard.
	February 26	Peirce becomes superintendent of the Coast Survey.
1869	August	Peirce sends five parties to various locations in the United States to observe a solar eclipse.
		Charles Eliot becomes president of Harvard.
1870	April 22	Benjamin Mills Peirce dies.
		Peirce's *Linear Associative Algebra* is published.
	December	Peirce is the leader of the United States' expedition to Sicily and Spain to observe an eclipse of the sun.
1871	March 3	The scope of the Coast Survey is extended by a geodetic connection between the Atlantic and Pacific coasts.
1874	February 16	Peirce resigns as superintendent of the Coast Survey.
1876	Summer	J. J. Sylvester comes to Johns Hopkins University.

1878		The *American Journal of Mathematics* is founded at Johns Hopkins University.
1879	February	Peirce begins a series of lectures at the Lowell Institute. (These lectures are posthumously published as *Ideality in the Physical Sciences*.)
1880	January	Peirce begins a series of lectures at the Peabody Institute.
1880	October 6	Peirce dies at Cambridge, Massachusetts.

Notes

LRP   = Lydia Ropes Peirce
MHSL = Massachusetts Historical Society Library
NACP = National Archives at College Park, Maryland
NFB   = Nancy Fowler Bache
PCHH = Benjamin Peirce Correspondence, MS Am 2368, by permission
        of the Houghton Library, Harvard University
PPHH = Benjamin Peirce Papers, by permission of the Houghton
        Library, Harvard University
SA   = Smithsonian Archives
SCL   = South Carolina Library, University of South Carolina
SMP   = Sarah Mills Peirce
YB   = Beinecke Rare Book and Manuscript Library, Yale University
YUL   = Manuscripts and Archives, Yale University Library
ZP   = Harriet Melusina Fay Peirce

PREFACE

1. Fisch, "Introductory Note," in Sebeok, *The Play of Musement*, 17.
2. BP to JoLC, January 18, 1852, LFPB.
3. BP to ADB, March 12,1864, ADB file, PCHH.
4. BP to SMP October 4, 1853, SMP file, PCHH.

CHAPTER 1. HURRAH FOR YOUNG AMERICA

1. I describe Peirce here as he appeared during most of his adult life, not nec-essarily as he appeared in 1847. The description is based on an article in the *New Orleans Delta*, December 21, 1856 (transcription of this article is in the American Association for the Advancement of Science file, PCHH); Hoar, *Autobiography*, vol. 1, 99; Lodge, *Early Memories*, 55; Sargent, *Sketches*, 331–32; and Archibald et al., "Ben-jamin Peirce," 5 and 7. When Peirce was older, he added a moustache. The earliest known portrait of Peirce, a daguerreotype taken about 1845, shows Peirce with long but not shoulder-length hair, and extensive sideburns, but no beard; see Archibald et al., *Benjamin Peirce 1809–1880*.

2. BAG to BP, August 6, 1847, BAG Jr. file, PCHH; Hoar *Autobiography*, vol. 1, 100. Gauss reaffirmed his position to George Bond, when Bond visited him in 1851; see Diary of George Bond, September 4, 1851, in Holden, *Memorials*, 109.

3. Peirce, "Two Hundred and Ninety-third Meeting," 65. The American Acad-emy of Arts and Sciences was founded in Boston in 1780, and is the country's second oldest learned society, after the American Philosophical Society, founded in Philadelphia in 1743 by Benjamin Franklin.

4. Grosser, *The Discovery of Neptune*, 116–18.

5. Also written Le Verrier.

6. Clerke, *Popular History*, 78–82; Hanson, "Leverrier," 359–65.

7. BP to Loomis, April 21, 1847, YB; *Boston Courier*, October 23, 1846, 2; Hubbell and Smith, "Neptune in America," 263. This excellent article gives more technical details about the discovery of Neptune than I do in this chapter.

8. [Mitchel], "Le Verrier's Planet," 44.

9. "To the Editor of the Courier," *Boston Courier,* October 23, 1846, 2.

10. "The New Planet," The *Daily Union,* October 26, 1846.

11. See, for example, "The New Planet," *National Intelligencer,* October 27, 1846, 2; and Hubbell and Smith, "Neptune in America," 263.

12. Hubbell and Smith, "Neptune in America," 263.

13. Walker, "Researches," 10.

14. *ANB;* BP to the corporation, July 29, 1842, Harvard College Papers, 2nd series, UA I.5.131.10, HUA.

15. AU are astronomical units, one AU being the mean distance from the earth to the sun, or about 93 million miles.

16. The validity of the law had already been discredited by Gauss, however. See Gould, *Report,* 30.

17. Although eye trouble prevented Herrick from attending college, his work in astronomy and entomology won him an international reputation as a scientist. See Elliott, *Biographical Dictionary,* 123–24, and DAB.

18. Walker, "Researches," 10; Peirce's activities are reported in Peirce, "Two Hundred and Eighty-eighth Meeting," 41–42.

19. BP to Herrick, December 10, 1846, Edward Claudius Herrick Papers, YUL.

20. Ibid. Peirce reported that Wartmann's Star wasn't Neptune in Peirce, "Two Hundred and Eighty-eighth Meeting," 41–42; see also Peirce, "Two Hundred and Ninety-third Meeting," 57.

21. BP to Herrick, January 7, 1847, Edward Claudius Herrick Papers, YUL.

22. Peirce, "Two Hundred and Ninety-third Meeting," 57, 62–64.

23. S. C. Walker to BP, February 9, 1847, S. C. Walker file, PCHH.

24. Walker, "Researches," 10–12.

25. *ANB.*

26. BP to Herrick, April 14, 1847, Edward Claudius Herrick Papers, YUL. Since the Naval Observatory was the only observatory supported by the federal government, it was referred to as the national observatory.

27. Hubbell and Smith, "Neptune in America," 266.

28. Walker, "Researches," 10–13.

29. Peirce, "Two Hundred and Ninety-third Meeting," 65 (emphasis in the original). Geometer was often used as a synonym for mathematician in the nineteenth century.

30. Peirce, "Two Hundred and Ninety-third Meeting," 66.

31. Ibid.

32. Walker, "Astronomy," 134. Leverrier referred to Uranus by the name of its discoverer, William Herschel, hoping that the new planet would be named after its discoverer as well.

33. Struik, *Yankee Science in the Making,* 416.

34. Peirce, "On the New Planet."

35. S. C. Walker to BP, March 29, 1847, S. C. Walker file, PCHH.

36. JH to ADB, March 31, 1847, Henry, *Papers,* vol. 7, 74.

37. Sparks to BP, March 19, 1847, Sparks file, PCHH.

38. "Le Verrier's Planet," *Sidereal Messenger* 1 (1847), 85–86 (emphasis in the original), on p. 86; and also, for example, "Le Verrier's Planet," *National Intelligencer* March 26, 1847, 3.

39. Mitchel to BP March 29, 1847, O. M. Mitchel file, PCHH.

40. Peirce, "Two Hundred and Ninety-third Meeting," 66.

41. BP to J. P. Nichol, [1847], John Pringle Nichol file, PCHH. This is probably the same Nichol who wrote, *Contemplations on the Solar System.* Peirce wrote an

anonymous review of this book; see Peirce *"Contemplations."* For another example of Peirce's assurance of his admiration for Leverrier, see BP to Ormsby Mitchel, May 9, 1847, Mss VF 4049, Cincinnati Museum Center.

42. Peirce, "The Century's Great Men in Science," 694–95.

43. Letter to the *National Intelligencer,* dated Cambridge, March 13, 1848; clipping in the *National Intelligencer* file, PCHH.

44. Gould, *Report,* 55.

45. Peirce, "Two Hundred and Ninety-fifth Meeting," 144.

46. Ibid. 332.

47. Gould, *Report,* 44.

48. U. J. J. Leverrier to Matthew Fontaine Maury, February 9, 1848, from the *National Intelligencer,* March 10, 1848, in Reingold, *Science in Nineteenth-Century America,* 140–41. Leverrier's criticism is quoted below.

49. Letter from BP, March 13, 1848, *National Intelligencer,* March 23, 1848, 1; clipping in the *National Intelligencer* file, PCHH.

50. Peirce, "Two Hundred and Ninety-fifth Meeting," 144.

51. BP to [O. M. Mitchel] (draft), April 7, 1848, O. M. Mitchel file, PCHH. Peirce, "Report," 338, made essentially the same statement in his report to the American Academy on April 4, 1848.

52. U. J. J. Leverrier to Matthew Fontaine Maury, February 9, 1848, from the *National Intelligencer* March 10, 1848, in Reingold, *Science in Nineteenth-Century America,* 140–41. The letter was also printed in the *Sidereal Messenger* 2 (1848), 68–70. Leverrier also published refutations of Peirce's position in the *American Journal of Science;* see Leverrier, "Le Verrier's Remarks" and Leverrier, "Le Verrier's Further Vindication."

53. BP to O. M. Mitchel, March 13, 1848, in the *Sidereal Messenger* 2 (1848), 70–71 (emphasis in the original); a clipping of the letter, with *National Intelligencer* written on it, is in the *National Intelligencer* file, PCHH.

54. From the Diaries of George Bond, in Holden, *Memorials,* 91–92.

55. Dana to Haldeman, March 14, 1848, in the Haldeman Papers, Academy of Natural Sciences of Philadelphia. Quoted by Daniels, *Science in American Society,* 144.

56. O. M. Mitchel to BP, March 22, 1848, O. M. Mitchel file, PCHH; BP to Herrick, April 14, 1847, Edward Claudius Herrick Papers, YUL.

57. Reingold, *Science in Nineteenth-Century America,* 145–46.

58. Jones, "Diary of the Two Bonds," 185–86.

59. [O. M. Mitchel], "The New Planet," *Sidereal Messenger* 1 (1847), 107, as quoted by Hubbell and Smith, "Neptune in America," 277–78.

60. Mitchel to BP, September 15, 1847, O. M. Mitchel file, PCHH.

61. Mitchel, *The Orbs of Heaven,* 140.

62. Ibid., 146.

63. Asa Gray to BP, March 26, 1848, Gray file, PCHH (Gray's emphasis).

64. C. S. Walker to BP, May 30, 1847, C. S. Walker file, PCHH.

65. ADB to BP October 25, 1847, ADB file, PCHH.

66. BP to N. Bowditch, June 26, [1830], BPL.

67. Loomis 1848. BP to Loomis, April 21, 1847, YB; Loomis to BP, January 1, 1848, Loomis file, PCHH; BP to Loomis, January 13, 1848, YB; and BP to Loomis, April 7, 1848, YB. Loomis, "Historical Notice of the Discovery of Neptune."

68. BP to Loomis, April 7, 1848, YB.

69. Loomis, *Recent Progress of Astronomy.*

70. JH to Elias Loomis, October 30, 1849, Loomis Papers, YB, in Henry, *Papers,* vol. 7, 611.

71. Gould, *Report*.

72. Ibid., 44.

73. So who was right? Peirce or Leverrier? As Hubbell and Smith (Neptune in America," 282) point out "That at least two interpretations of the discovery were current suggests that the 'data' did not (and do not) by their very nature compel adherence to any one viewpoint." Ignoring the complex mathematics involved in the argument, it seems highly unlikely that a planet would be found where Leverrier said it would be, purely by chance. After all, it is a very big universe. On the other hand, Neptune's orbit, and mass, turned out to be considerably different than what Leverrier had predicted. Differences of opinion between scientists and historians of science remain. For example, Pannekoek, "Discovery of Neptune," argues convincingly that Peirce was right, whereas Rawlins, "Some Simple Results," concludes that Peirce was wrong. For other opinions see Reingold, *Science in Nineteenth-Century America*, 136; and Struik, *Yankee Science in the Making*, 416–17.

74. See for example, Emerson, *Early Years of the Saturday Club*, 100; Hill, "Benjamin Peirce," also reprinted in King, *Benjamin Peirce*, 8.
Newspaper article from the *New Orleans Delta*, December 21, 1856. A transcription of this article is in the American Association for the Advancement of Science file, PCHH.

75. Newspaper article from the *New Orleans Delta*, December 21, 1856. A transcription of this article is in the American Association for the Advancement of Science file, PCHH.

76. Hubbell and Smith, "Neptune in America," 279.

77. Walker, "Researches," 9.

78. Archibald et al., "Benjamin Peirce," 14; Walsh, "Relations between Logic and Mathematics," 17; Peterson, "Benjamin Peirce," 92.

79. W. W. Smyth to BP, November 26, 1852, Royal Society of London file, PCHH.

80. Henry, *Papers*, vol. 5, 142–44; and vol. 6, 265–66. Peirce was regarded as a leading American mathematician well before the Leverrier controversy, and received widespread recognition as a scientist throughout his life. He became a member of the American Philosophical Society in 1842, and was made a Fellow of the American Academy of Arts and Sciences in 1858. He received an honorary doctorate from the University of North Carolina in 1847 and one from Harvard in 1867. He was one of the original members of the Royal Societies of Edinburgh and Göttingen, and an Honorary Fellow of the Imperial University of St. Vladimir, at Kiev. See Matz, "Benjamin Peirce," 175, and Archibald et al., "Benjamin Peirce," 10–11.

CHAPTER 2. THE FATHER OF AMERICAN GEOMETRY

1. On Salem see Phillips, *When Salem Sailed the Seven Seas;* Phillips, *Salem and the Indies;* Marvin, *American Merchant Marine;* and Morison, *Maritime History of Massachusetts*. On the library see Phillips, *Salem and the Indies*, 192–93, and Rantoul, "Memoir of Benjamin Peirce," 163.

2. Pulsifer, *Witch's Breed*, 47, 52, 58–59, 202. On Peirce's family background and genealogy see Pulsifer, *Witch's Breed;* Pulsifer, *Supplement to Witch's Breed;* and Rantoul, "Memoir of Benjamin Peirce."

3. When it is questionable whether I am referring to father or son, I will refer to Peirce's father as Peirce Senior, or Benjamin Senior. I will refer to Peirce as simply Peirce or Benjamin unless I feel the identification is particularly confusing, in which

case I will refer to the younger Peirce as Benjamin Junior, young Peirce, or young Benjamin.

4. Hannah Sawyer to LRP, June 18, 1798, L 700, CSPP (emphasis in the original).

5. BPSr. to LRP, August 31, 1799, L 660, CSSP (emphasis in the original). Pulsifer, *Witch's Breed*, 188 gives 1802 as Benjamin Sr.'s graduation date, but Peabody, *Harvard Reminiscences*, 68 gives the date of 1801, as does Rantoul, "Memoir of Benjamin Peirce," 162. LRP to BPSr, May 1801, L 684, CSPP, and the HUA substantiates this date.

6. LRP to BPSr, September 18, 1797, L 684, CSPP.

7. BPSr to his sister, Sarah, April 7, 1798, L 660, CSPP.

8. LRP to BPSr, June 28, 1799, L 684, CSPP (emphasis in the original).

9. Pulsifer, *Witch's Breed*, 188.

10. LRP to Joseph Story, [no date, but before 1800], L 684, CSPP. Story was called Henry.

11. LRP to Joseph Story, [no date, but before 1800], L 684, CSPP.

12. Pulsifer, *Witch's Breed*, 179, 189. See also Hannah Sawyer to LRP, September 1800, L 700, CSPP.

13. Ichabod Nichols to LRP, September 6, 1800, in Pulsifer, *Witch's Breed*, 189–90.

14. LRP to Ichabod Nichols, [no date], L 684, CSPP.

15. Ichabod Nichols to Joseph Story, March 27, 1801, L 649, CSPP.

16. Ichabod Nichols to BPSr, April 20,1801, in Pulsifer, *Witch's Breed*, 192.

17. BPSr to Ichabod Nichols, April 22, 1801, L 660, CSPP.

18. BPSr to Jerathmiel Peirce, May 1801, L 660, CSPP; *ANB*.

19. LRP to BPSr, May 1801, L 684, CSPP.

20. BPSr to LRP, August 8, 1801, Benjamin Peirce Senior Correspondence 1796–1826, HUG 1680.10 VT, HUA.

21. LRP to BPSr, [August 10, 1801], L 684, CSPP (emphasis in the original).

22. BPSr to LRP, August 12, 1801, Benjamin Peirce Senior Correspondence 1796–1826, HUG 1680.10 VT, HUA.

23. LRP to BPSr, [August 12, 1801], L 684, CSPP (emphasis in the original).

24. Joseph Story to LRP, Salem, May 19, 1802, in Pulsifer, *Witch's Breed*, 195–96.

25. BPSr to Ichabod Nichols, October 29, 1803, L 660, CSPP.

26. Ichabod Nichols to BPSr, November 29, 1803, in Pulsifer, *Witch's Breed*, 196–97.

27. Newmyer, *Supreme Court Justice Joseph Story;* Dunne, *Justice Joseph Story;* and *ANB*.

28. Pulsifer, *Witch's Breed*, 198.

29. Newspaper clipping, dated July 7, 1812, L 663, CSSP. The talk was published; see Peirce, *An Oration*. The War of 1812 was extremely unpopular in New England.

30. Pulsifer, *Witch's Breed*, 197.

31. BPSr to Sarah Peirce, September 1798, as quoted by Ketner, *His Glassy Essence*, 69.

32. Elizabeth Peirce to BPSr, 1796, L 677, CSPP.

33. Ibid., 1797, L 677, CSPP.

34. We also get an indication of Lydia's feminist viewpoints from her letters from Hannah Sawyer. See, for example, Hannah Sawyer to LRP, June 18, 1798, February 27, [1801], and July 15, 1798, L 700, CSPP.

35. LRP to BPSr, September 18, 1797, L 684, CSSP.

36. BPSr. to LRP, September 26, 1797, L 660, CSPP (emphasis in the original).

37. LRP to BPSr, September 2, 1798, L 684, CSPP.

38. BPSr to Sarah Peirce, February 26, 1798, L 660, CSPP (emphasis in the original).

39. Ibid.

40. Ibid., April 7, 1798, L 660, CSPP (emphasis in the original).

41. There is an extensive literature both by and about Mary Wollstonecraft; see for example, Sunstein, *A Different Face;* James, *Mary Wollstonecraft, a Sketch;* Johnson, *Mary Wollstonecraft;* and Wollstonecraft, *A Vindication.* Rendall, *The Origins of Modern Feminism,* 55–72, discusses the place of Wollstonecraft in the eighteenth-century feminist movement. On the influence of Wollstonecraft on Lydia see LRP to BPSr, February 28, 1799, and Dudley L. Pickman to LRP, [February 1801], L 690, CSPP.

42. Wollstonecraft's second child.

43. *Columbia Centinel:* "The Latitudinarian," January 24 1801; "To the Author of the Latitudinarian, signed 'Sarah Frankly,' " January 31, 1801; "The Latitudinarian," February 7, 1801; "The Latitudinarian," February 14, 1801; "To the Author of the Latitudinarian, signed 'One of Many' " February 18, 1801; "Latitudinarian," February 25, 1801; "To the Author of One of Many," February 28, 1801; "Letter to the Author by Joseph Story," March 7, 1801; "To Joseph Story," March 11, 1801; "To the Author of the Latitudinarian," March 21, 1801.

44. Archaic spelling for "style."

45. A friend to her sex to the *Columbia Centinel,* February 14, 1801 (emphasis in the original).

46. BPSr to LRP, February 23, 1801, L 660, CSSP (emphasis in the original).

47. One of many to the *Columbia Centinel,* February 18, 1801 (emphasis in the original).

48. *Columbia Centinel,* February 28, 1801 (emphasis in the original). See also *Columbia Centinel,* March 21, 1801; Hannah Sawyer to LRP, February 27, 1801; and March 23, 1801, L 700, CSPP.

49. *Columbia Centinel,* February 25, 1801.

50. Ibid., February 28, 1801.

51. LRP to BPSr, February 28, [1801?], L 684, CSPP (emphasis in the original).

52. Ibid., March 7, 1801, L 684, CSPP (emphasis in the original).

53. Ibid., February 28, 1799, L 684, CSPP.

54. "Solutions of Algebraic Problems," April 14, 1801, Harvard University Mathematical Theses 1782–1837, HUC 872.514 PF, HUA. This is creditable work in mathematics for an American college senior in 1801, but it falls far short of his son's thesis. The difference in the two theses is probably more attributable to the progress that the United States made in mathematics in a generation than the difference in abilities between father and son.

55. Pulsifer, *Witch's Breed,* 160; Rantoul, "Memoir of Benjamin Peirce," 162. The attainments necessary to qualify as an eminent mathematician at the beginning of the eighteenth century were modest. Nichols did not contribute to any of this country's early mathematical periodicals, which had many amateur contributors.

56. In addition to Charlotte Elizabeth Peirce and Charles Henry Peirce, who are mentioned elsewhere, the Peirces had a son, John Nichols Peirce, who died as a child. See Pulsifer, *Witch's Breed,* 197 and 203.

57. Rantoul, "Memoir of Benjamin Peirce," 164. Michael Walsh, *A New System of Mercantile Arithmetic: Adapted to the Commerce of the United States, in its Domestic and Foreign Relations; with forms of accounts and other writings usually occurring in trade,* Newburyport, Mass., printed by Edmond M. Blunt, 1801. This book went through twenty-four editions. Walsh also wrote a work on mensuration, which probably was not published. See Karpinski, *Bibliography,* 137–40, 551, and 556.

58. The other was Robert Adrain; see Hogan, "Robert Adrain."

59. Newton, "Benjamin Peirce," 167; Matz, "Benjamin Peirce," 174; Anon., "Benjamin Peirce," 443.

60. Peirce, *A System of Analytical Mechanics*, vi. Peirce also dedicated his senior thesis (HUC 8782.514, in HUA) to Bowditch and wrote an article about Bowditch's mathematical work for the *North American Review*. See Peirce and Palfrey, "*Mécanique Céleste.*"

61. Matz, "Benjamin Peirce," 174.

62. Lydia Peirce to Hannah Lee, June 23, 1823, in Pulsifer, *Witch's Breed*, 200.

63. Diary of LRP, entries for September 7, and September 10, 1824, L 685, CSPP. Bailey, *Historical Sketches of Andover*, 543–49. See also Rantoul, "Memoir of Benjamin Peirce," 165.

64. BPSr to BP, January 11, 1825, L 660, CSPP (emphasis in the original). " 'Mens sasa in corpes sano.' Verbem sat." translates: " 'a sound mind in a sound body,' as the saying goes."

65. BPSr to LRP, July 24, 1825, L 660, CSPP.

66. Peabody, *Harvard Reminiscences*, 181–82.

67. Clarke, *Autobiography*, 34.

68. Emerson, *Reminiscences*, 32–34. See also Smith and Ginsburg, *Mathematics in America*, 96–97. On Farrar see Ackerberg-Hastings, "Mathematics is a Gentlemen's Art;" Cajori, *Teaching and History of Mathematics*, 128–33; Peabody, *Harvard Reminiscences*, 70–73; and Palfrey, *Professor Farrar.*

69. Emerson, *Reminiscences*, 32. For most of his career, Emerson was principal of the Boston English Classical School. His primary scientific interests were in botany and ecology. See Elliott, *Biographical Dictionary*, 86; and *ANB*.

70. Cajori, *Teaching and History of Mathematics*, 128.

71. Karpinski, *Bibliography*, 215–16, 218–20, 228–29, 233–34, 246, 256, 263–64, and 325. For contemporary evaluations of Farrar's textbooks, see [Walker], "Farrar's Mathematics;" Anon., "Review of the Cambridge Course;" Anon., "Cambridge Course of Mathematics;" Emerson, "Course of Mathematics;" and Anon., "Cambridge Course of Mathematics. See also Pycior "Styles of Algebra," 129–37; Ackerberg-Hastings, "Mathematics is a Gentlemen's Art," 166–69; and Rosenstein, "The Best Method," 78–79.

72. I have no hard evidence of this, but what little research Farrar did was in meteorology and astronomy, not mathematics. Furthermore, he contributed to none of the nation's nascent mathematical periodicals. All of this implies that his interest in mathematics did not go beyond teaching the limited Harvard curriculum.

73. On Bowditch see, [Bowditch], "Memoir of Nathaniel Bowditch;" Pickering, "Eulogy;" Reingold, Science *in Nineteenth-Century America*, 11–14; Struik, *Yankee Science in the Making*, 108–14, 227–32; Berry, *Yankee Stargazer;* and *ANB*.

74. Peabody, *Harvard Reminiscences*, 181; Matz,"Benjamin Peirce," 173.

75. Morison, *Three Centuries*, 264, reports that Nathaniel Bowditch said this, and it is likely to be true. Morison may, however, have confused Farrar with the tutor, James Hayward, whose unfavorable comparison with the undergraduate Peirce is well documented in Morison, "The Great Rebellion," 107.

76. Ackerberg-Hastings, "Mathematics is a Gentlemen's Art," 213–14, agrees with my assessment. She says, "The only time Peirce mentioned Farrar graciously in public was at the annual meeting of the American Academy of Arts and Sciences after the older man's death, where he 'ascribed to [Farrar], more than to any other man, the adoption of the present admirable system of instruction in the mathematical sciences.' " The quotation is from Lovering, "Communication," 39.

77. BP to President Josiah Quincy, February 14, 1845, L 664, CSPP. Part of the letter appears to be in Sarah Peirce's hand, so it is likely a draft. The mechanics text was almost surely John Farrar's *An Elementary Treatise on Mechanics*, Cambridge:

Hilliard and Metcalf (1825). The instructor may well have been Farrar. If either is true, it may explain Peirce's lack of praise for Farrar. The diagram accompanying Euclid's proof that base angles of an isosceles triangle are equal resembles the trusses on a bridge. Students who understood the proof of this theorem were said to have passed over the Pons Assinorum, or Asses Bridge. Such students usually had little trouble mastering the rest of Euclid's *Elements*.

78. The Dutch lawyer, Hugo Grotius (1583–1645), is considered to be the founder of international law.

79. BP to LRP, October 22, 1825, LRP file, PCHH.

80. The following description is taken from Peabody, *Harvard Reminiscences*, 196–200. Peabody was a senior when Peirce was a freshman.

81. BP to LRP, April 21, [1826], LRP file, PCHH.

82. Peabody, *Harvard Reminiscences*, 200.

83. See Morison, "The Great Rebellion" and Bailyn, "Why Kirkland Failed," 27–30.

84. BP to JoLC, April 4, 1858, LFPB.

85. Morison, *Three Centuries of Harvard*, 210.

86. Bailyn, "Why Kirkland Failed," 41. Harvard is governed by two bodies, the corporation, consisting of the president, treasurer, and five fellows, and the overseers. See Morison, *Three Centuries of Harvard*, 17–19 and 211–13.

87. Farrar's health had been bad for some time, and the added load caused a further decline, see Palfrey, *Professor Farrar*, 13.

88. As quoted by Morison, "The Great Rebellion," 107. On the troubles between Kirkland and Bowditch, see Morison, "The Great Rebellion," 102–212; Morison, *Three Centuries of Harvard*, 200; Bailyn, "Why Kirkland Failed," 41; and Berry, *Yankee Stargazer*, 206–10.

89. Bowditch's sons denied that such an outburst by their father had precipitated Kirkland's resignation. See Henry Wadsworth Longfellow to Stephen Longfellow, October 11, 1840, and accompanying notes, in Hilen *The Letters of Henry Wadsworth Longfellow*, vol. 2, 253–55.

90. The first editor of the *Diary* was Robert Adrain; see Hogan, "Robert Adrain."

91. Hogan, "Robert Adrain."

92. *Mathematical Dairy*, vol. 1, iii. Similar statements of purpose were in other early American mathematical periodicals. Peirce, himself, extolled the educational value of such journals in [Peirce] 1837b. A letter from BP to Gill, June 6, 1836, Charles Gill Papers, #1444. Divison of Rare and Manuscript Collections, Cornell University Library, identifies Peirce as the author of this article.

93. *Mathematical Diary* vol. 1, 277; the other two solutions are on pp. 281 and 286.

94. Peirce's thesis, *Solutions of Questions . . . from the Mathematical Diary, &c.*, HUC 8782.514, is in HUA.

95. *Mathematical Diary* vol. 2, 116, 118, 128, 129, 131, and 134.

96. See BP to BPSr, February 20, 1830, PCHH and BP to BPSr, August 1, 1830, BPSr file, PCHH.

CHAPTER 3. THE FINEST LADY IN NORTHAMPTON

1. Pulsifer, *Witch's Breed*, 403–4. The exact reasons for the reversal in the business are unknown. Pulsifer blames the failure on The Embargo and Non-Intercourse Acts, and on the crash of 1826. But The Embargo and Non-Intercourse Acts occurred

about two decades earlier. There was an economic depression in 1826 [Fite and Reese, *Economic History*, 134]. In addition, Pulsifer notes that by 1826 the importance of Salem as a seaport was declining. Salem harbor was gradually filling with silt, preventing large ships from coming up to the wharves.

2. Since Charlotte Elizabeth Peirce's family called her Elizabeth or Lizzie, I will refer to her as Elizabeth rather than Charlotte.

3. Pulsifer, *Witch's Breed*, 198.

4. BPSr. to Pres. John Thornton Kirkland, August 15, 1826, L 660, CSSP.

5. Day, *Biography of a Church*, 44–45.

6. CEP to [Betsy Peirce], [1825], L 676, CSSP.

7. LRP to BPSr, August 3, 1826, L 684, CSSP.

8. See the Mill Dam Papers, L 662, CSPP.

9. BPSr to LRP, August 16, 1826, L 660, CSPP. BPSr to LRP, July 24, 1825, L 660, CSPP, mentions that the family was involved in manufacturing. I have found nothing on how the move affected Benjamin Peirce Jr.'s life. Apparently it affected him little; he remained a student at Harvard.

10. LRP to BPSr, August 19, 1826, L 684, CSPP (emphasis in the original).

11. Timothy Walker to BPSr, October 11, 1830, L 709, CSPP. Lydia Peirce wrote about an early religious experience that solidified her faith in Christianity; see L 685, CSPP.

12. CEP to JEM, September 12, 1882, L 676, CSPP.

13. BPSr to LRP, August 20, 1826, 1826, L 660, CSPP.

14. BP to LRP, May 18, 1830, LRP file, PCHH.

15. BPSr to N. Bowditch, September 24, and December 4, 1827. I have been unable to determine if Benjamin was successful.

16. BP to Rev. Dr. Pierce, March 9, 1829, Ferdinand Julius Dreer Collection, case 6, box 18, Historical Society of Pennsylvania.

17. Matz, "Benjamin Peirce," 173, notes that, due to Bowditch's preparing Peirce well beyond the limits of the under-graduate course in mathematics, he was able to benefit from the lectures on mathematics given by Francis Grund. Grund later became known as an author; he was born in Germany and educated at the Vienna Polyteknik. He taught at the military school in Rio Janeiro, Brazil, in 1825, before coming to Philadelphia. There he did journalistic work. On Grund see the *National Cyclopaedia of American Biography*, vol. 23, 131–32. I have found no other mention of these lectures or their value to Peirce.

18. BP to Nathaniel Bowditch, November 22, 1829, Miscellaneous Bound Manuscripts, MHS.

19. Archibald et al., "Benjamin Peirce," 9.

20. Peirce replaced Timothy Walker, mentioned above, who returned to Harvard Law School and boarded with Peirce's parents.

21. Ticknor was an outstanding American scholar. His most notable accomplishments are his *History of Spanish Literature*, and his part in the founding of the BPL.

22. On the Round Hill School, see Bassett, *"The Round Hill School"* and Bellows, *"The Round Hill School."*

23. BP to LRP, May 18, 1830, LRP file, PCHH

24. BP to BPSr, November 29, 1829, BPSr file, PCHH.

25. BP to Nathaniel Bowditch, November 22, 1829, Miscellaneous Bound Manuscripts, MHS.

26. Ibid.

27. BP to BPSr, November 29, 1829, BPSr file, PCHH.

28. Ibid.

29. Ibid., February 20, 1830, BPSr file, PCHH.

30. Ibid.

31. Ticknor, *Life of Joseph Green Cogswell,* 17–19.

32. Bassett, *"The Round Hill School,"* 26–27.

33. BP to BPSr, February 21, 1830, BPSr file, PCHH.

34. BP to LRP, May 18, 1830, LRP file, PCHH.

35. A triangular playing field just west of Memorial Hall and separated from the yard by Cambridge Street; Morison, *Three Centuries of Harvard,* 206–07.

36. Emerson, *Early Years of the Saturday Club,* 99.

37. Ibid., 100–101. Emerson identifies the teller of the story only as one who attended the concert. The incident was also reported in the *Boston Daily Globe,* August 27, 1856, Benjamin Peirce file, HUG 300, HUA.

38. BP to N. Bowditch, June 26, [1830], BPL.

39. BP to Charles Henry Peirce, December 1829, Charles Henry Peirce file, PCHH.

40. BP to BPSr, August 1, 1830, BPSr file, PCHH.

41. BP to LRP, May 18, 1830, LRP file, PCHH.

42. BP to BPSr, undated, BPSr file, PCHH.

43. On Eijah Hunt Mills, see *ANB;* some information about the Mills family is in Baker et al., *History and Genealogy,* 254.

44. BP to N. Bowditch, June 26, [1830], BPL.

45. BP to Charles Henry Peirce, August 3, 1831, L 664, CSPP.

46. Ibid.

47. Presumably Bowditch, which displays Peirce's gross provinciality, extraordinary patriotism, or commendable loyalty. Bowditch was not in the same league as the leading European mathematicians, but he was widely recognized in his day as the leading American mathematician; Adrain was not considered his equal until near the end of the nineteenth century.

48. BP to SMP, August 1831, SMP file, PCHH. This letter is in the folder with other letters from Peirce to his wife and was presumably written and sent to her. There is no salutation, however (perhaps part of the letter is missing). Since there is neither salutation nor signature, perhaps Peirce wrote this letter, but never mailed it, or it was a rough draft of a letter he did send. But these details do not detract from the overwhelming ambition Peirce expresses in it. "Amore duce, nil desperandum" translates "Sweet love, without desperation."

49. Diary of Charles Phelps Huntington, April 17, 1832, as quoted by Ketner, *His Glassy Essence,* 80–81.

50. BP to Mrs. C. P. Huntington, December 27, 1832 [typescript copy], L 664, CSPP (Peirce's emphasis).

51. LRP to BP, Rhinebeck, January 19, [1834], L 684, CSPP.

52. Corporation Records, vol. 7, May 16, 1833, 1827–36, UAI 530.2, HUA.

53. Diary of Charles Phelps Huntington, December 29, 1833, as cited by Ketner, *His Glassy Essence,* 81.

CHAPTER 4. THE VOICE OF GOD

1. Pulsifer, *Witch's Breed,* 197; Parkman 1830.

2. Peirce's father died on July 26, 1831.

3. BP to CHP, August 3, 1831, L 660, CSPP.

4. BP to Dr. George C. Shattuck, July 27, 1831, George Cheyne Shattuck papers, MHS.

5. BP to Benjamin R. Nichols, Benjamin R. Nichols file, PCHH.

6. CEP to her Aunt, March 9, [1830], L 676, CSPP.

7. CEP to LRP, [no date], L 676, CSSP (emphasis in the original). Smith's arithmetic was probably: Roswell C. Smith's *Practical and Mental Arithmetic,* Providence: Microcosm Office (1823), a popular arithmetic text that went through many editions; see Karpinski, *Bibliography,* 278–83. Warren Colburn's *An Arithmetic on the Plan of Pestalozzi,* Boston: Cummings and Hilliard (1821), was one of the most innovative and important arithmetic textbooks published in the United States during the first half of the nineteenth century. It took an inductive approach to arithmetic and differed markedly from contemporary textbooks. It met with unprecedented popularity and went through many editions; see Karpinski, *Bibliography,* 236–39; Cajori, *Teaching and History of Mathematics,* 106–8.

8. LRP to BP and CHP, May 6, 1832, L 684, CSPP. See also LRP to CHP and BP, April 29, 1832, L 684, CSSP.

9. Pulsifer, *Witch's Breed,* 197. The lawyer and philologist, John Pickering, was one of the most highly respected Bostonians of his day. Among other things, he was the leading authority of his time on the languages of North American Indians. See *DAB.*

10. Peirce Sr., *History of Harvard.*

11. LRP to BP, November 6, 1832, L 684, CSSP.

12. LRP to Sarah P. Nichols, March 15, 1833, L 684, CSSP.

13. LRP to Elizabeth Peirce (her sister-in-law), March 15, 1833, in Pulsifer, *Witch's Breed,* 204–6.

14. LRP to her mother and father, April 19, [1833], L 684, CSPP.

15. CEP to her grandmother, May 11, 1833, L 676, CSPP.

16. CEP to her grandmother and grandfather, December 29, 1833, L 676, CSPP.

17. Judge Samuel Jones to LRP, February 8, 1834, L 632, CSPP.

18. Ichabod Nichols (Peirce's grandfather) to Henry Nichols (Ichabod's son), May 21, 1837, and February 20, 1837, in Pulsifer, *Witch's Breed,* 127 and143.

19. Pulsifer, *Witch's Breed,* 208. Lydia and Elizabeth may not have moved in with her brother-in-law immediately. Elizabeth mentioned that in 1847 they were living in Roxbury; see CEP to Aunt Nichols, August 28, 1857, L 676, CSPP.

20. Corporation Records, vol. 7, 1827–36, May 19, 1831, UAI 530.2, HUA.

21. Eliza Farrar (wife of Prof. Farrar) to Mrs. Charles Saunders, June 16, 1831, L 609, CSPP; Corporation Records, vol. 7, 1827–36, October 5, 1831, June 21, 1832, UAI 530.2, HUA; Palfrey, *Notice,* 13–14; Ackerberg-Hastings, "Mathematics is a Gentlemen's Art," 177–80.

22. Matz, "Benjamin Peirce," 174.

23. Corporation Records, vol. 7, 1827–36, May 16, 1833, UAI 530.2, HUA.

24. Corporation Records, vol. 8, 1836–47, February 26, 1842, UAI 530.2, HUA.

25. Morison, *Three Centuries of Harvard,* 34–35.

26. BP to Edward Everett, September 28, 1846, Harvard College Papers, 2nd series, UA I.5.131.10, HUA.

27. Peirce's classmate, James Freeman Clarke (Clarke, *Autobiobraphy,* 34) said that the majority of the students were this age when Peirce entered Harvard.

28. Hofstadter and Smith, *American Higher Education,* 272–73.

29. Peabody, *Harvard Reminiscences,* 201. Peabody says that most of the instruction was by lecture, but this differs from other accounts. Clarke, *Autobiobraphy,* 40–41, for example, states that in the freshman and sophomore years three hours a day were spent in recitations.

30. Peabody, *Harvard Reminiscences,* 201–2.

31. Bailyn, "Why Kirkland Failed," 24.

32. Anon., "Introduction to Algebra," 246. There are many similar descriptions of teaching mathematics; for example see, Anon., "Review," 310; Emerson, "Course of Mathematics," 369, and Cajori, *Teaching and History of Mathematics*, 49, 56, and 64.

33. Morison, *Three Centuries of Harvard*, 224–38; Bailyn, "Why Kirkland Failed," 33–34.

34. McCaughey, *Josiah Quincy*, 139.

35. Josiah Quincy to BP, June 22, 1832, Corporation Papers, 2nd series, box 6, UA 1.5.130, HUA. I have not found Peirce's reply to this request.

36. BP to Josiah Quincy, October 23, 1832, Harvard College Papers, 2nd series, vol. 5: 144, 220, UA I.5.131.10, HUA.

37. Corporation Records, vol. 7, 1827–36, October 25, 1832, UAI 530.2, HUA. Peirce's plane trigonometry text was accepted as a textbook by the corporation for use in the mathematics department on May 21, 1835, see Corporation Records, vol. 7, 1827–36, UAI 530.2, HUA.

38. Cajori, *Teaching and History of Mathematics*, 70, agrees with this assessment.

39. Karpinski, *Bibliography*, lists the textbooks with their editions published before 1850. Archibald et al., "Benjamin Peirce," gives an excellent bibliography of Peirce's published works, including his textbooks. See also: Peirce, *First Part;* Peirce, *Elementary Treatise on Sound;* Peirce, *Elementary Treatise on Algebra;* Peirce, *Plane and Solid Geometry;* Peirce, *Elementary Treatise on Plane and Spherical Trigonometry;* Peirce, *Elementary Treatise on Curves;* and Peirce, *A Sysytem of Analytical Mechanics.*

40. Peirce, *First Part*, iii-iv.

41. Hassler was a Swiss immigrant, who became the first head of the Coast Survey. See Hassler, *Elements of Analytical Trigonometry.* Bonnycastle was an English-born mathematician, who taught at the University of Virginia; he used the modern definitions for trigonometry in his *Inductive Geometry.* See Bonnycastle, *Inductive Geometry.*

42. A negative reaction to Hassler's approach, the same as used by Peirce, is seen in Anon., "Geometry and Calculus."

43. Cajori, *Teaching and History of Mathematics*, 135, 241.

44. Cajori, "Attempts," 200; Cajori, *Teaching and History of Mathematics*, 134–35.

45. Hill, "Benjamin Peirce," 91.

46. Peirce received letters from professors praising his textbooks; see for example, I. Smith Fowler to BP, March 31, 1848, I. Smith Fowler file, PCHH; and H. B. Lane to BP, December 28, 1840, Lane file, PCHH. Anon., "Day's Algebra," a contemporary review of Day's *Algebra*, found Peirce's textbooks the best available for those with a "decided taste for mathematics."

47. Ketner, *His Glassy Essence*, 83; some of them still survive in libraries. This is one clue to the correct pronunciation of the family name. See also Peterson, "Benjamin Peirce," 97.

48. Cajori, *Teaching and History of Mathematics*, 140–41.

49. Quoted by Cajori, *Teaching and History of Mathematics*, 141.

50. BP to Edward Everett, September 11, 1848, Harvard College Papers, 2nd series, vol. 16, 144, UA I.5.131.10, HUA. See also BP to Edward Everett, February 8, 1848, Corporation Papers, 2nd series, 1848, box 6, UA 1.5.130, HUA. Pycior, "Styles of Algebra," 143–45 discusses Peirce's *Algebra*. Rosenstein, "The Best Method," 84–85, describes Peirce's calculus textbook as, "mathematically intriguing, but pedagogically painful."

51. Hutton, *Course of Mathematics.*

52. Cajori, *Teaching and History of Mathematics*, 141.

53. Matz, "Benjamin Peirce," 174.

54. McCaughey, *Josiah Quincy*, 147. Day and Kingsley, "Original Papers" is usually called the Yale Report of 1828. For an alternative view of the Yale Report, see Guralnick, *Science*, 28–32.

55. Day and Kingsley, "Original Papers," 300 (emphasis in the original).

56. Peirce had several students who achieved prominence as mathematicians or astronomers. They include two of his sons, James Mills Peirce and Charles Saunders Peirce; T. H. Safford, the mathematical prodigy; G. P. Bond, who became director of the Harvard Observatory; Benjamin A. Gould, who received a PhD in astronomy from Göttingen; Asaph Hall, the astronomer who discovered the moons of Mars; Simon Newcomb, the mathematical astronomer; G. W. Hill, also a mathematical astronomer; Chauncey Wright, the mathematician and philosopher; Calvin M. Woodward, the educational reformer; J. D. Runkle, professor at the Massachusetts Institute of Technology; W. Watson, also professor at the Massachusetts Institute of Technology; J. E. Oliver, professor at Cornell University; and William Byerly, professor at Cornell and Harvard.

In addition, three of Peirce's students with strong mathematical interests were subsequently presidents of Harvard. Charles William Eliot was a tutor in mathematics at Harvard College from 1854–58, and assistant professor of mathematics from 1858–61. He was also assistant professor of chemistry from 1858–63. He served as president of Harvard from 1869–1909. A. Lawrence Lowell was president of Harvard from 1909–33, and Rev. Thomas Hill was president of Harvard from 1862–68. In addition, Peirce's student, Arnold B. Chace, was chancellor of Brown University.

57. See for example, BP to Edward Everett, September 18, 1846, Harvard College Papers, 2nd series, HA UA I.5.131.10, HUA. See also Pycior, "Styles of Algebra," 144, and Peirce, "Intelligent Organization."

58. McCaughey, *Josiah Quincy*, 167–68.

59. Josiah Quincy to BP, March 11, 1835, UAI.15.882, HUA.

60. BP to Josiah Quincy, March 15, 1835, Josiah Quincy Papers, (Sophomore, Freshman, and Mathematics are capitalized in the original), UAI.15.882, HUA.

61. Ibid.

62. BP to Josiah Quincy, May 21, 1835, Harvard College Papers, 2nd series, HA UA I.5.131.10, HUA.

63. Ibid., October 14, 1835, Corporation Papers, 2nd series, 1835–44, Box 3, UA I.5.130, HUA.

64. Ibid.

65. Ibid.

66. Ibid., October 21, 1835, Corporation Papers, 2nd series, 1835–44, Box 3, UA I.5.130, HUA.

67. McCaughey, *Josiah Quincy*, 172, 174–75.

68. Ibid., 170.

69. BP to Josiah Quincy, January 21, 1838, Harvard College Papers, 2nd series, HA UA I.5.131.10, HUA.

70. Ibid., May 15, 1838, and BP to Committee on Mathematical Reform, Josiah Quincy, Chairman, May 26, 1838, Harvard College Papers, 2nd series, HA UA I.5.131.10, HUA.

71. *A Catalogue of the Officers and Students of Harvard University for the Academical Year 1839–1840.* HU 20.41 Harvard College Catalogue 1819–54/5, reel 2, HUA. See also *Regulations* [1838], 1, HUC 8838.175, Harvard University Archives.

72. BP to Josiah Quincy, August 1839, Harvard College Papers, 2nd series, 291, UA I.5.131.10, HUA.

73. Ibid.

74. Ibid (emphasis in the original).

75. Cajori, *Teaching and History of Mathematics,* 137.

76. Ibid., 135; Morison, *Three Centuries of Harvard,* 234–37; Hebrew was made an elective as early as 1755 (Morison, *Three Centuries of Harvard,* 58).

77. Beck to Josiah Quincy, June 16, 1838, Harvard College Papers, 2nd series, vols. 9–10, UA I.5.131.10, HUA.

78. Corporation Records, vol. 8, May 26, and June 2, 1838, UAI 5 30.2, HUA.

79. Felton to Josiah Quincy, August 16, 1839, Josiah Quincy Papers, HUA, as quoted by McCaughey, *Josiah Quincy,* 177.

80. Parsons, "Report," 2.

81. BP to Edward Everett, September 18, 1846, Harvard College Papers, 2nd series, HA UA I.5.131.10, HUA.

82. Ibid.

83. McCaughey, *Josiah Quincy,* 178.

84. BP to Josiah Quincy, May 15, 1838, Harvard College Papers, 2nd series, HA UA I.5.131.10, HUA.

85. Ibid., August 15, 1838, Harvard College Papers, 2nd series, vol. 9, 66, UA I.5.131.10, HUA.

86. Corporation Records, vol. 8, 1836–47, August 16, 1838, UAI 530.2, HUA. Judge Story was Joseph Story, the man who courted Peirce's mother. Judge Lemuel Shaw was Chief Justice of the Massachusetts Supreme Judicial Court.

87. BP to Josiah Quincy, July 8, 1839, Harvard College Papers, 2nd series, vol. 9, 264, UA I.5.131.10, HUA.

88. Ibid., April 10, 1840, Harvard College Papers, 2nd series, HA UA I.5.131.10, HUA.

89. School of Civil Engineers, Harvard College Papers, April 10, 1840, 2nd Series, UA I.5.131.10, HUA.

90. Corporation Records, vol. 8, 1836–47, April 25, 1840, UAI 530.2, HUA.

91. Daniel Treadwell, an inventor, was Rumford Professor on the Application of Science to the Useful Arts; Webster was the infamous professor of chemistry and mineralogy (he appears in a later chapter); Lovering was professor of natural philosophy, or physics.

92. Rossiter, "Emergence," 68.

93. Dupree, *Science,* 106–12; Bruce, *Launching of Modern American Science,* 40.

94. Asa Gray to Charles Darwin, June [1863], as quoted by Dupree, *Science,* 290–91. See Peirce, "Mathematical Investigations," and Gray "Composition."

95. See for example, BP to ADB, March 27, 1863, ADB file, PCHH; BP to ADB, May 8, 1854, ADB file, PCHH; and BP to ADB, June 25, 1857, ADB file, PCHH.

96. Struik, *Yankee Science,* 86–91.

97. Ibid., 90.

98. Henry, *Papers,* vol. 6, xxiv; and Gray to JH, January 12, 1846, Henry, *Papers,* vol. 6, 368–69.

99. BP to ADB, January 29, 1846, ADB file, PCHH (emphasis in the original).

100. Rossiter, "Emergence," 70.

101. BP to ADB, January 29, 1846, ADB file, PCHH; Bruce, *Launching of Modern American Science,* 25.

102. BP to ADB, 29 January 1846, ADB file, PCHH.

103. Rossiter, "Emergence," 71.

104. Gray to JH, January 12, 1846, Mary Henry Copy, Henry Papers, Smithsonian

Archives, in Henry, *Papers,* vol. 6, 368–69. See also JH to James Henry, January 31, 1846, Henry, *Papers,* vol. 6, 371–74.

105. Gray to JH, November 28, 1846, Henry, *Papers,* vol. 6, 541.

106. Frothingham, *Edward Everett,* 61–68; Rossiter, "Emergence," 70.

107. Frothingham, *Edward Everett,* 81.

108. Edward Everett to Peter C. Brooks, July 18, 1845, in Frothingham, *Edward Everett,* 267.

109. Frothingham, *Edward Everett,* 302–72; Rossiter, "Emergence," 70.

110. Henry, *Papers,* vol. 6, 368.

111. See BP to Edward Everett and the Honourable, the fellows of the Corporation of Harvard University, February 27, 1846, Harvard College Papers, 2nd series, HA UA I.5.131.10, HUA.

112. Ibid.; and Benjamin Peirce, "Plan of a School of Practical and Theoretical Science," February 27, 1846, HUF 1680.100, HUA.

113. BP to Edward Everett and the Honourable, the fellows of the Corporation of Harvard University, February 27, 1846, Harvard College Papers, 2nd series, HA UA I.5.131.10, HUA.

114. Everett, *Orations,* 2:497.

115. Rossiter, "Emergence," 73.

116. Abbott Lawrence to Samuel Eliot, treasurer of Harvard College, June 7, 1847, Harvard College Papers, 2nd series, HA UA I.5.131.10, HUA.

117. Bruce, *Launching of Modern American Science,* 163–64.

118. Abbott Lawrence to Samuel Eliot, treasurer of Harvard College, June 7, 1847, Harvard College Papers, 2nd series, HA UA I.5.131.10, HUA (emphasis in the original).

119. Rossiter, "Emergence," 75–84; Bruce, *Launching of Modern American Science,* 24.

120. On Agassiz, see Lurie, *Louis Agassiz.* See also Bruce, *Launching of Modern American Science,* chapter 4; Menand, *Metaphysical Club,* chapter 5; and Struik, *Yankee Science,* 347–61.

121. Lurie, *Louis Agassiz,* 138–40.

122. Charles Saunders Peirce; "Studies in Meaning," ms 619, CSPP.

123. Bruce, *Launching of Modern American Science,* 27–28, 232–34, 288.

124. *James,* "Engineering an Environment," 72.

125. Cajori, *Teaching and History of Mathematics,* 137.

126. BP to Josiah Quincy, February 14, 1845, (draft), L 664, CSPP.

127. Peirce et al., "Remarks on a National University," 89.

128. In addition to those cited in this chapter, see Cajori, *Teaching and History of Mathematics* 139–40; Archibald et al., "Benjamin Peirce," Hale, *James Russell Lowell,* 23–24, 128; and Emerson, *Early Years of the Saturday Club,* 97–99.

129. Lillie Greenough to her mother, 1856, in De Hegermann-Lindencrone, *Courts of Memory,* 1–5. Actually it was Agassiz's insatiable desire for specimens for his museum, not their house, that prompted Agassiz's wife, Elizabeth Cabot Cary Agassiz, to open her school (Lurie, *Louis Agassiz,* 200–201).

130. CEP to Aunt, Uncle, and Cousins, April 2, [1859?], L 676, CSPP.

131. Lillie Greenough to her mother, 1856, in De Hegermann-Lindencrone, *Courts of Memory,* 4–5.

132. Ketner, *His Glassy Essence,* 199.

133. See De Hegermann-Lindencrone, *Courts of Memory;* De Hegermann-Lindencrone, *Sunny Side;* and Ketner, *His Glassy Essence,* 198–201.

134. Lillie Greenough Moulton, to Zina Peirce, November 10, 1870, Lillie Greenough Moulton file, PCHH.

135. It is not my purpose to defend Peirce as a teacher, but some of the negative assessments of his teaching are questionable. For example, Cajori, *Teaching and History of Mathematics,* 140, quotes a passage from a letter written by William F. Allen (Harvard, class of 1851): "I am no mathematician, but that I am so little of one is due to the wretched instruction at Harvard. Professor Peirce was admirable for students with mathematical minds, but had no capacity with others. He took only elective classes, and of course I didn't elect." All that Allen says may be true, but he never took a course from Peirce; he had no firsthand experience.

136. Archibald et al., "Benjamin Peirce," 2.

137. Ibid., 4.

138. Ibid.

139. Morison, *Three Centuries of Harvard,* 260.

140. Ibid., 251.

141. Ibid., 252–54; McCaughey, *Josiah Quincy,* 152,161.

142. Dupree, *Science,* 123–25.

143. Emerson, *Early Years of the Saturday Club,* 99; Peabody, *Harvard Reminiscences,* 182–86.

144. Quoted by Emerson, *Early Years of the Saturday Club,* 97–98.

145. Archibald et al., "Benjamin Peirce," 3.

146. Hale, "T Is Sixty Years Since," 497–98. See also Hale, *New England Boyhood,* and Hale, "College Days."

Chapter 5. The Feeling of Mutual Goodwill

1. See, for instance, Josiah Quincy to Thomas H. Perkins, June 4, 1842, Josiah Quincy file, PCHH; and Gill to BP, June 19, 1843, Gill file, PCHH.

2. The *Analyst* was published until Hendricks's health failed in 1883, and then continued as the *Annals of Mathematics;* see Cajori, *Teaching and History of Mathematics,* 280 and 284. On Hendricks see Colaw, "Dr. Joel Hendricks."

3. Cajori, *Teaching and History of Mathematics,* 94–97, 277–86; Smith and Ginsburg, *Mathematics in America,* 85–91; and Karpinski, *Bibliography.*

4. BAG to BP, November 28, 1845, BAG Jr. file, PCHH. The *Mathematical Miscellany,* to which Gould refers was published soon after the demise of the *Diary,* and is discussed in this chapter.

5. Ward to his father, November 22, 1831, Samuel Ward Papers, Manuscripts and Archives Division, The New York Public Library, Aster, Lenox and Tilden Foundations; Ward to BP, January 8, 1832, Samuel Ward file, PCHH. On Ward see Elliott, *Uncle Sam Ward;* Tharp, *Three Saints;* Thomas, *Sam Ward;* and Hogan, "A Proper Spirit."

6. Ward to BP, December 13, 183[1] (the letter is dated 1832, but appears to have been written in 1831), Samuel Ward file, PCHH; Ward to BP, January 8, 1832; February 14, 1832, Samuel Ward file, PCHH.

7. Ward to BP, January 8, 1832, Samuel Ward file, PCHH (emphasis in the original). Ward is referring to Peirce's solution of the problem, "In a given ellipse, it is required to inscribe the greatest possible equilateral triangle," *Mathematical Diary,* vol. 2, 221–22.

8. Ward to BP, February 14, 1832, Samuel Ward file, PCHH.

9. Peirce, "On Perfect Numbers."

10. The other two are 496 and 8128. As of July 2006, there are thirty-nine known perfect numbers. See Devlin, *New Golden Age,* 20.

11. In his article, Peirce incorrectly credited Euler with Euclid's result. Peirce, "On Perfect Numbers," 267.

12. Stillwell, *Mathematics*, 38–41.

13. *Mathematical Diary*, vol. 1, 135.

14. Ibid., 314.

15. *Mathematical Diary*, vol. 2, 211–12.

16. Dickson, *History*, does not mention Peirce's proof, but does note that Lebesgue had obtained the same result, some years later, in 1844 [Dickson, *History*, vol. 1, 120–21.] See also Archibald et al., "Benjamin Peirce," 20.

17. Ward to BP, February 14, 1832, Samuel Ward file, PCHH.

18. Ward to BP, January 8, 1832, Samuel Ward file, PCHH; Hogan, "A Proper Spirit," 162; and Cajori, *Teaching and History of Mathematics*, 95–96.

19. Elliott, *Uncle Sam Ward*, 155.

20. Ward to BP, January 8, 1832, Samuel Ward file, PCHH. Diophantine analysis is another name for number theory; Diophantus (ca. 250) was the great number theorist from antiquity.

21. Hogan, "Mathematical Miscellany."

22. Gill to BP, January 5, 1836, Gill file, PCHH.

23. BP to Gill, January 18, 1836, Charles Gill Papers, #1444, Division of Rare and Manuscript Collections, Cornell University Library.

24. Gill to BP, August 1, 1836, Gill File, PCHH.

25. BP to Gill, February 16, 1838, Charles Gill Papers, #1444, Division of Rare and Manuscript Collections, Cornell University Library.

26. [Peirce], "Mathematical Miscellany." The notice is anonymous, but Peirce is identified as the author in a letter from Benjamin Silliman, the journal's editor, to BP, July 20, 1836, Benjamin Silliman file, PCHH.

27. [Peirce], "Mathematical Miscellany," 185.

28. BP to Gill, May 9, 1849, in McClintock, "Charles Gill," vol. 15, 19.

29. John Lee to BP, June 12, 1837, Hillhouse Family Papers, YUL (emphasis in the original).

30. Kohlstedt, *Formation of the American Scientific Community*, lists a John C. Lee of Salem as a member of the AAAS.

31. See Hogan, "Mathematical Miscellany," 250–53.

32. Gill to BP, August 1, 1836, Gill file, PCHH.

33. Some of this material has been preserved in the PCHH. Peirce also states that he took over the editorship of the *Miscellany*, BP to Gill, May 9, 1849, in McClintock, "Charles Gill," vol. 14, 19. Much of the material in the first issue of the *Cambridge Miscellany* is identified as coming from Gill's *Miscellany*.

34. BP to President and Corporation of Harvard University, February 8, 1842, Harvard College Papers, HA UA I.5.131.10, HUA.

35. Details of how Peirce and Lovering interacted as co-editors is unknown. On Lovering see B. O. Peirce, "Joseph Lovering;" Anon., "Sketch;" and *DAB*.

36. See, for example, Lane to BP, February 21, 1842, H. B. Lane file, PCHH.

37. BP to Gill, June 21, 1838, Charles Gill Papers, #1444, Division of Rare and Manuscript Collections, Cornell University Library.

38. [Peirce and Palfrey], "Bowditch's *Translation*." Other articles by Peirce in the *North American Review* are [Peirce], "American Almanac," "Norton's Elementary Treatise," "Bartlett's Elementary Treatise," "Worcester's Geography," "Magnetic Observations," "Meteors," and "Nichol's *Contemplations*." Archibald et al., "Benjamin Peirce," identifies Peirce as the author of some of these articles. The *North American*

Review Papers, bms Am 1704.10 (70), (259–60), (328), and (474), Houghton Library, Harvard University, identify Peirce as the author of the others.

39. [Peirce and Palfrey], "Bowditch's *Translation,*" 144–45.

40. Ibid., 175. The Baconian philosophy was popular in antebellum America. It was based on the Scottish philosophers, and stressed empiricism. See Daniels, *American Science,* chapter 3.

41. [Peirce and Palfrey], "Bowditch's *Translation,*" 176–77.

42. On Sylvester and the University of Virginia, see Feuer, "First Jewish Professor;" Archibald, "Unpublished Letters;" Yates 1937; and Hogan, "A Proper Spirit." These views of what transpired during Sylvester's stay in the United States during the early 1840s differ sharply. I find Feuer's account by far the most persuasive.

43. Feuer, "First Jewish Professor," 155–58.

44. Ibid., 155–69, Yates 1937, 199.

45. Sylvester to BP, September 5, 1842, James Joseph Sylvester Papers, Brown University Library (emphasis in the original). See also Archibald, "Unpublished Letters," 116.

46. Ibid., (emphasis in the original). See also Archibald, "Unpublished Letters," 117.

47. *DSB.*

48. Sylvester to BP, [May 22, 1843], Sylvester file, PCHH (emphasis in the original). The letter is also published in Archibald, "Unpublished Letters," 121–23.

49. *DSB.*

50. ADB to BP, August 22, 1842, ADB file, PCHH. See also C. H. Davis to ADB, May 23, 1842, BPLC.

51. ADB and John Ludlow to BP, May 26, 1843, ADB file, PCHH; Ketner, *His Glassy Essence,* 90.

52. Bruce, *Launching,* 17.

53. Slotten, *Patronage,* 82.

54. Odgers, *Alexander Dallas Bache,* 33–72.

55. CEP to her Aunt, August 10, [1856], L 676, CSPP.

56. Cajori, *Chequered Career,* 15–36.

57. Ibid., 37–70.

58. Ibid., 59–70, 83.

59. Ibid., 83–87, 92.

60. Ibid., 87, 118; Slotten, *Patronage,* 52–53.

61. Cajori, *Chequered Career,* 119, 123–24.

62. Ibid., 161, 191, 202.

63. Ibid., 62, 78, 163.

64. Ibid., 180; Slotten, *Patronage,* 58.

65. BP to ADB, December 8, 1843, ADB file, PCHH.

66. Ibid., November 28, 1843, as quoted by Ketner, *His Glassy Essence,* 91.

67. Ibid., December 8, 1843, ADB file, PCHH.

68. Ibid., November 28, 1843, as quoted by Ketner, *His Glassy Essence,* 91.

69. BP to ADB, February 1, 1845, HL.

70. Ibid., April 7, 1850, ADB file, PCHH.

71. Reingold, *Science in Nineteenth-Century America,* 152.

72. ADB to BP, April 21, 1845, ADB file, PCHH. The computers that Bache refers to are human computers, that made arithmetical computations.

73. BP to ADB, April 25, 1845, HL. See also BP to ADB, January 29, 1846, ADB file, PCHH, and *Report of the Secretary of the Treasury communicating a report of the Superintendent of the Coast Survey, showing the progress of the survey during the year ending November 1846,* 34–35, 29th Congress, 2nd session.

74. BP to ADB, May 16, 1852, ADB file, PCHH.

75. Ibid., May 22, 1852, ADB file, PCHH; ADB to BP, May 25, 1852, ADB file, PCHH; see also ADB to BP, April 24, 1852, ADB file, PCHH; and ADB to BP, September 25, 1852, ADB file, PCHH.

76. BP to ADB, May 22, 1852, ADB file, PCHH.

77. Slotten, *Patronage,* 86.

78. ADB to BP, January 9, 1847, ADB file, PCHH; ADB to BP, January 21, 1847, ADB file, PCHH; and ADB to BP, February 10, 1847, ADB file, PCHH; and Slotten, *Patronage,* 87.

79. Wright, *Horatio Greenough.*

80. American Jupiter.

81. BP to SMP, February 20, 1847, SMP file, PCHH.

82. See *ANB.*

83. BP to JnLC, April 8, 1859, LFPB.

84. BP to ADB, May 8, 1854, ADB file, PCHH.

85. Shaler, *Autobiography,* 188–91,

86. Williams, *Matthew Fontaine Maury* is a sympathetic biography of Maury, as is Lewis, *Matthew Fontaine Maury.* Jahns, *Maury & Henry,* is more critical. See also Bruce, *Launching,* 176–77.

87. Jahns, *Maury & Henry,* 11–25

88. *The American Journal of Science and Arts,* often called Silliman's Journal, because Benjamin Silliman Sr. was the founder and editor.

89. Williams, *Matthew Fontaine Maury,* 97–98; 105–11; Jahns, *Maury & Henry,* 41–61.

90. Miller et al., *The Lazzaroni,* 99.

91. The United States South Seas Exploring Expedition (1838–42) was under the command of the navy's Captain Charles Wilkes.

92. Bruce, *Launching of Modern American Science,* 177. On the Wilkes's expedition see Reingold, *Science in Nineteenth-Century America,* 108–10; on Gilliss see *ANB.*

93. BP to George Bancroft, May 14, 1845, George Bancroft papers, MHS.

94. Williams, *Matthew Fontaine Maury,* 148–57.

95. Stevens, "Sketch," 24; Miller et al., *The Lazzaroni,* 97–103.

96. Jahns, *Maury & Henry,* 106–9; Williams, *Matthew Fontaine Maury,* 178–95.

97. Maury to BP, September 9, 1845, Maury file, PCHH.

98. BP to Ormsby Mitchel, March 22, 1848, Department of Special Collections, Kelvin Smith Library, Case Western Reserve University.

99. Williams, *Matthew Fontaine Maury,* 167–68.

100. Miller et al., *The Lazzaroni,* 102.

101. Ibid.

102. [Le Conte], "Physical Geography of the Sea."

103. Bruce, *Launching,* 185–86

104. Genuth, "Heaven's Alarm," 41.

105. Morison, *Three Centuries of Harvard,* 30, 264, and *passim.* In chapter 1, Jones and Boyd, *Harvard College Observatory,* give an excellent account of astronomy at Harvard before the building of the observatory.

106. Loomis, *Recent Progress,* 160–65.

107. Jones and Boyd, *Harvard College Observatory,* 38. Only eleven astronomical observatories were founded in the United States before 1840, but after that time many were built. See Herrmann, "Gould and His Astronomical Journal," 98; Loomis, *Recent Progress,* 160–202.

108. Jones and Boyd, *Harvard College Observatory,* 41.

109. Ibid., 45.

110. [Peirce], "American Astronomical and Magnetic Observations," 26.

111. BP to Corporation of Harvard University, July 29, 1842, Harvard College Papers, 2nd series, UA I.5.131.10, HUA.

112. [BP] to [Joseph Cranch], August 14, 1842, Harvard Observatory Papers, microfilm roll # 3, HUA. The Dorpat telescope was soon bettered by a fifteen-inch instrument at Pulkovo, which became the model for Harvard's. See Jones and Boyd, *Harvard College Observatory*, 454–55, fn 5.

113. BP and William Cranch Bond to the president and Corporation of Harvard University, September 24, 1842, Harvard College Papers, 2nd series, vols. 11–12, UA I.5.131.10, HUA.

114. Hindle, *Pursuit of Science*, 96. The comet of 1758 was named after Halley.

115. Jones and Boyd, *Harvard College Observatory*, 48. The Millerites were followers of William Miller, the founder of the Adventist movement. Miller predicted that the second coming of Christ would occur in 1843. After being proved wrong, he gave a new date of October 22, 1844.

116. Undated clipping, Professor Peirce's Lecture on the Comet, *Astronomical and Meteorological Scrapbook, 1843–51*, Harvard University Archives. Peirce's colleague, Henry Wadsworth Longfellow, reported that Peirce's lecture was "very good." See H. W. Longfellow to Stephen Longfellow, April 2, 1843, in Hilen, *Letters of Longfellow*, vol. 2, 524–25.

117. Jones and Boyd, *Harvard College Observatory*, 49.

118. Rothenberg, "Patronage," 44–45.

119. Jones and Boyd, *Harvard College Observatory*, 49–50.

120. Ibid., 51.

121. Jones, "Diary of the Two Bonds," 50.

122. BP to Loomis, July 14, 1847, YB.

123. Ibid., April 2, 1845, YB.

124. Eliot to BP, December 6, 1844, Samuel Eliot file, PCHH (emphasis in the original). Eliot led a distinguished life dedicated to public service. He was a mayor of Boston and a member of the United States Congress. See *DAB*.

125. See, for example, BP to Loomis, April 21, 1847, YB; and Jones and Boyd, *Harvard College Observatory*, 53.

126. Stephens, "Astronomy as Public Utility," 24.

127. BP to ADB, April 25, 1845, HL.

128. Stephens, "Astronomy as Public Utility," 22, 29.

129. Groves, "Unsung Sailor," 28.

130. Ibid.

131. Davis, *Life of Charles Henry Davis*, 73.

132. Ibid., 75.

133. Ibid.

134. Ibid., 80–81.

135. Newcomb, *Reminiscences*, 63. Davis said the *Almanac* was located in Cambridge "for the advantages which could only be derived from the proximity of the University," (Davis, *Life of Charles Henry Davis*, 89.)

136. Newcomb, *Reminiscences*, 62. On Davis and the *Nautical Almanac*, see also Waff, "Charles Henry Davis."

137. Davis, *Life of Charles Henry Davis*, 88; Tyler, "John Daniel Runkle," 183.

138. CEP to Aunt Nichols, Thanksgiving Day, 1856, L 676, CSPP.

139. Ms 619, CSPP.

140. Davis, *Life of Charles Henry Davis*, 91.

141. BP to ADB, May 8, 1854, ADB file, PCHH.

142. Ibid., (emphasis in the original).

143. BP to ADB, May 23, 1854, ADB file, PCHH.

Chapter 6. A Prince of the Humbugs

1. Winslow to Warner, July 7, 1860, John Warner Papers, APSL.

2. Thomas Hill was a minister who had been Peirce's student. He later became president of Harvard.

3. Winslow to Warner, September 20, 1857, John Warner Papers, APSL.

4. Ibid., October 2, 1857, John Warner Papers, APSL.

5. Daniels, "Process of Professionalization," 151; *American Science,* chapter 2.

6. ADB to Felton, December 19, 1850, Bache Papers, Smithsonian Institution Archives, Washington, D. C.; BP to Frazer, November 2, 1858, and Gould to Frazer, November 1856, Frazer Papers, APSL; Gould to Dana, November 22, 1856, Dana Papers, YB; all as cited by Beach, "Scientific Lazzaroni," 118–19.

7. Miller et al., *The Lazzaroni,* 4.

8. BP to SMP, February 14, 1858, SMP file, PCHH.

9. I have used this symbol to separate sections within chapters.

10. Beach, "Scientific Lazzaroni," 119.

11. James, *Dudley Observatory,* 419, fn. 62.

12. BP to ADB, May 8, 1854, ADB file, PCHH.

13. Ibid., November 1, 1863, ADB file, PCHH.

14. ADB to BP, October 11, 1860, ADB file, PCHH.

15. On the Lazzaroni see: Beach, "Scientific Lazzaroni;" Miller et al., *The Lazzaroni;* Bruce, *Launching of Modern American Science,* 217–24; Kohlstedt, *Formation of the American Scientific Community,* 156–59, 164–68.

16. Peirce, "Opening Address," xvii–xix.

17. Ibid., xviii.

18. Peirce, "Presidential Address," 15.

19. Henry, *Papers,* vol. 6, 463.

20. ADB to JH, December 4, 1846, General Manuscripts Collection, Department of Special Collections, Van Pelt Library, University of Pennsylvania, in Henry, *Papers,* vol. 6, 564–67 (emphasis in original).

21. Miller et al., *The Lazzaroni,* 31; Coulson, *Joseph Henry;* and *ANB.*

22. JH to Asa Gray, January 10, 1848, Loomis Papers, YUL, in Reingold, *Science in Nineteenth-Century America,* 158.

23. Bruce, *Launching of Modern American Science,* 188–90.

24. Ibid., 191.

25. BP to ADB, May 8, 1854, ADB file, PCHH.

26. Ibid., July 3, 1854, HL.

27. See BP to ADB, January 17, 1855 and BP to ADB, May 23, 1854, both ADB file, PCHH.

28. BP to Banks, January 7, 1855, HL.

29. Bruce, *Launching of Modern American Science,* 191.

30. Ibid., 252.

31. Ibid., 254.

32. JH to ADB, August 9, 1838, in Reingold, *Science in Nineteenth-Century America,* 81–90; see especially 87–88.

33. The United States South Seas Exploring Expedition (1838–42) was under the command of the navy's Captain Charles Wilkes.

34. Kohlstedt, *Formation of the American Scientific Community,* 54–57; and Bruce, *Launching of Modern American Science,* 252.

35. Kohlstedt, *Formation of the American Scientific Community,* 79–80.

36. Fairchild, "American Association," 366; Kohlstedt, *Formation of the American Scientific Community,* 80–86; Bruce, *Launching of Modern American Science,* 254.

37. Miller et al., *The Lazzaroni,* 7.

38. Anon., *Spiritualism.* For a different view of the proceedings see Nichols, *Brothers Davenport,* 83–91; and Redman, *Mystic Hours,* 307–17. Redman was another medium that took part in the demonstration.

39. JH to Anonymous, December 5, 1853, Henry Papers, Smithsonian Archives, in Henry, *Papers,* vol. 8, 498–500. The *Albany State Register* had reported that Henry was a believer in spiritualism. He wrote the *Register* denying this allegation, as he did most vehemently in this letter.

40. JnLC to BP, July 4, 1857 (emphasis in the original); see also, JnLC to BP, January 9, 1859, both LFPB.

41. This is verified in the *Boston Daily Courier,* August 19, 1857. I have been unable to find Winslow's letter to Hill.

42. CEP to Aunt, August 28, 1857. L 676, CSPP (emphasis in the original).

43. Winslow, *Cosmography.* Winslow stated that he gave the book to Peirce in *The Boston Daily Journal,* October 26, 1857; also see the *Boston Daily Courier,* August 18, 1857.

44. Winslow, *Cosmography,* 13–14; *Boston Daily Courier,* August 18, 1857.

45. *Boston Daily Courier,* August 18, 1857.

46. *Boston Daily Journal,* October 26, 1857.

47. [Hill], "Scientific Intelligence," 152; emphasis in the original.

48. [Hill], "Scientific Intelligence," 152.

49. *Boston Traveler,* August 13, 1857, 4.

50. BP to JoLC, August 25, 1857, LFPB.

51. *Boston Daily Courier,* August 18, 1857 (emphasis in the original).

52. BP to JnLC, August 25, 1857, LFPB. One of these men, Chauncey Wright, published a paper the next year on the relationship between leaf arrangements and astronomy. He cites Peirce's work in the paper. See Wright, "On the Phyllotaxis."

53. *Boston Traveler,* August 13, 1857, 4; and *Rochester Union,* August 17, 1857.

54. *Boston Daily Courier,* August 20, 1857.

55. Ibid.

56. August 17, 1857.

57. Warner to Agassiz, December 1, 1856, John Warner Papers, APSL.

58. Ibid.

59. Much later, Warner made it clear that he was still obsessed with getting his paper back. In Warner to Kirkwood, December 19, 1857, John Warner Papers, APSL, he closed the letter with, "I want that paper or a reason why I [am] to rest satisfied with the explanation given." Warner's emphasis.

60. Warner to Agassiz, December 1, 1856, John Warner Papers, APSL.

61. Whitcomb to Sheafer, March 7, 1857, copy in John Warner Papers, APSL.

62. Whitcomb to Sheafer, March 24, 1857, copy in John Warner Papers, APSL.

63. Warner, *Studies.*

64. Warner to Agassiz, December 1, 1856, John Warner Papers, APSL.

65. Peirce, "Abstract of a Paper on Researches in Analytic Morphology."

66. JH to John Warner, December 19, 1853, Henry, *Papers,* vol. 8, 502–3.

67. Lurie, *Louis Agassiz,* 103.

68. See Peirce, "Mathematical Investigation."

69. BP to JnLC, July 9, 1852, LFPB.

70. Lesley to Warner, June 26, 1858, John Warner Papers, APSL.

71. Warner to Winslow, August 19, 1857, copy in John Warner Papers, APSL.

72. Ibid.

73. Since Warner and Winslow joined forces in their attacks on Peirce, the combined controversy is referred to as the Warner-Winslow affair. This incident has generally been viewed as an example of the clash between amateur scientists and the rising professional American scientific community. See, for example Kohlstedt, Formation of the American Scientific Community, 145–50, and *passim;* and James, "Elites in Conflict," 94–95. Unfortunately, I have been able to find only a few of Warner's letters.

74. Winslow's letter appeared in the morning edition of the *Boston Traveler,* August 13, 1857.

75. Winslow to Warner, November 21, 1857, John Warner Papers, APSL (Winslow's emphasis). Winslow consistently spelled Peirce's name as "Pierce."

76. Winslow to Warner, October 24, 1857, John Warner Papers, APSL.

77. "Professor Peirce and his use of the Labors of others," *Boston Daily Journal,* October 27, 1857.

78. Greeley to Warner, October 25, 1857, and Winslow to Warner October 24, 1857, both John Warner Papers, APSL.

79. Warner to Greeley (draft), January 30, 1858, John Warner Papers, APSL.

80. Winslow to Warner, November 21, 1857, John Warner Papers, APSL.

81. See *Boston Daily Journal,* "The Sun and the Continents," October 26, 1857, and Winslow to Warner, September 13, 1857, John Warner Papers, APSL.

82. Winslow to Warner, January 20, 1858, John Warner Papers, APSL (Winslow's emphasis).

83. Ibid., September 13, 1857, John Warner Papers, APSL.

84. Ibid., July 7, 1858, John Warner Papers, APSL.

85. Ibid., November 21, 1857, John Warner Papers, APSL.

86. Warner to Yardley, May 13, 1858, John Warner Papers, APSL.

87. Warner to Greeley (draft), January 30, 1858, John Warner Papers, APSL.

88. Winslow to Warner, November 5, 1857, John Warner Papers, APSL.

89. Ibid., June 22, 1858, John Warner Papers, APSL.

90. Ibid., September 3, 1857, John Warner Papers, APSL.

91. Ibid.

92. BP to JnLC, August 25, 1857, LFPB.

93. At least I have found none.

94. Winslow to Warner, August 1, 1858, John Warner Papers, APSL.

95. See for example, W. Parker Foulke to Warner, June 3, 1858, John Warner Papers, APSL.

96. Kirkwood to Warner, July 17, 1858, John Warner Papers, APSL (emphasis in the original).

97. Yardley to Warner, May 10, 1858, John Warner Papers, APSL.

98. J. P. Lesley to Warner, February 21, 1857, John Warner Papers, APSL.

99. The cousin of John Le Conte of Georgia and South Carolina.

100. J. P. Lesley to Warner, June 26, 1858, John Warner Papers, APSL (emphasis in the original).

101. Giovan Battista Donati discovered the comet named for him on June 2, 1858.

102. Winslow to Warner, July 7, 1860, John Warner Papers, APSL.

CHAPTER 7. AS THE TREE GROWS TOWARDS HEAVEN

1. In 1849 and 1850, Thomas Hart Benton, the Democratic senator from Missouri, attempted to put the Coast Survey under the navy. See Slotten, *Patronage*, 84–87.

2. BP to ADB, April 7, 1850, ADB file, PCHH.

3. See Shama, *Dead Certainties;* Sullivan, *Disappearance;* and the *ANB*, which gives additional references.

4. BP to ADB, April 7, 1850, ADB file, PCHH.

5. Ephraim Littlefield was the medical school janitor.

6. Edward D. Sohier and Pliny Merrick represented Webster. Although Merrick was senior council, Sohier actually managed the defense.

7. Daniel Treadwell, Rumford professor at Harvard, was Webster's close friend and neighbor.

8. Dr. Willard Codman testified at the trial.

9. BP to ADB, April 7, 1850, ADB file, PCHH.

10. Cooke introduced laboratory chemistry into the Harvard curriculum. See Miller, *Dollars for Research*, 83–86.

11. BP to JoLC, January 18, 1852, LFPB.

12. BP to SMP, August 18, 1851, SMP file, PCHH.

13. Ibid.

14. SMP to BP, August 21, 1851, SMP file, PCHH.

15. In BP to JoLC, September 13, 1858, LFPB. Peirce says "The chief is a true and [staunch] friend, but I have a truer and a stauncher one. The Queen [Josephine Le Conte] always believes in me for she knows me through and through." Even allowing for Peirce's usual piffle, it is a telling statement. Lupold, "From Physician to Physicist," 104, agrees with my assessment of the intensity of this friendship.

16. Stevens, "Sketch," 112.

17. Le Conte, "Biographical Memoir," 372.

18. Lupold, "From Physician to Physicist," iii. See also Stevens, "Sketch," 133.

19. Le Conte, "Biographical Memoir," 373.

20. Le Conte, *Autobiography,* 9. Le Conte inherited some portion of this estate. It is not clear how much. He shared the inheritance with his brother, Joseph.

21. Lupold, "From Physician to Physicist," 1.

22. Ibid., 100.

23. JnLC to BP, February 5, 1860, LFPB (Le Conte's emphasis); Le Conte also expressed aristocratic views in JnLC to BP, January 8, 1860, LFPB.

24. Lupold, "From Physician to Physicist," 101.

25. An undated letter, in LFPB, probably written shortly after the 1851 AAAS meeting in Albany, apparently to all of the members of the Lazzaroni, makes it clear that Le Conte was a member of this circle. Although John Le Conte is not always included in a list of Lazzaroni members, he belongs there. During the decade before the Civil War, Le Conte was intimate with both Peirce and Bache, and well acquainted with the other Lazzaroni members. He used the nicknames of the Lazzaroni in corresponding with these men. His biographer, Lupold, "From Physician to Physicist, 102, (see also 108–9) states that, "Le Conte could be considered a Lazzaroni." Kohlstedt, *Formation of the American Scientific Community,* 976, 163, and Stephens, *Joseph Le Conte,* 102, place him in the Lazzaroni as well; see also Lenzen, *Benjamin Peirce and the U. S. Coast Survey,* 44–45; BP to JoLC, July 7, 1857, March 3, 1858, and May 31, 1858, all in LFPB; and Le Conte, "Physical Geography of the Sea."

26. Stephens, *Joseph Le Conte,* 7.

27. Stevens, "Sketch," 114.

28. Stephens, *Joseph Le Conte*, 42.

29. CEP to her Salem aunt (Nichols), August 11, [1857], L 676, CSPP (emphasis in the original).

30. Le Conte, "Biographical Memoir," 375.

31. In BP to JoLC, January 18, 1852, LFPB, Peirce praised her wit; he often asked for her opinion.

32. BP to JoLC, December 4, 1853, LFPB.

33. Ibid.

34. Ibid., September 3, 1854, LFPB.

35. Ibid., October 23, 1854, LFPB.

36. Ibid., January 18, 1852, LFPB.

37. Norberg, "Early Astronomical Career," 213.

38. Bache, "Address on Retiring," xlvii–xlviii. Bache's emphasis.

39. Bache, "Address on Retiring," l.

40. Burton, *History of Mathematics*, 584–85; Rezneck, "Emergence," 230, 232

41. Madsen, *National University*, 15–16. Madsen notes that the possibility of an American university was mentioned in colonial times.

42. Fitzpatrick, *Writings of George Washington*, vol. 30, 494.

43. Washington to the vice president, November 15, 1794, in Fitzpatrick, *Writings of George Washington*, vol. 34, 23 (Washington's emphasis).

44. Washington to the commissioners of the District of Columbia, in Fitzpatrick, *Writings of George Washington*, vol. 34, 106. On Washington's thoughts on a national university see also: Washington to Thomas Jefferson, March 15, 1795, in Fitzpatrick, *Writings of George Washington*, vol. 34, 146–49; Washington to Alexander Hamilton, September 1, 1796, in Fitzpatrick, *Writings of George Washington*, vol. 35, 198–201; and Washington to the commissioners of the District of Columbia, October 21, 1796, in Fitzpatrick, *Writings of George Washington*, vol. 35, 248–50.

45. Fitzpatrick, *Writings of George Washington*, vol. 37, 280.

46. Dabney, *Jefferson's University*, 1–2.

47. Hawkins, *Pioneer*, 7.

48. Dabney, *Jefferson's University*, 1.

49. [Kirkland], "Literary Institutions." This article was actually written by Edward Everett; see McCaughey, *Josiah Quincy*, 244, fn. 3.

50. McCaughey, *Josiah Quincy*, 132–38; Bailyn, "Why Kirkland Failed."

51. Hofstadter and Smith, *American Higher Education*, 252–53.

52. Bruce, *Launching of Modern American Science*, 38–39.

53. Storr, *Graduate Education*, 68; Clarke, *James Hall of Albany*, 200–201; Bruce, *Launching of Modern American Science*, 226.

54. James, *Dudley Observatory*, 27–28.

55. Silverman and Beach, "A National University," 704, 709.

56. Hendrickson, *American Higher Education*, 367–68. This article also shows the general significance of the geological surveys.

57. Silverman and Beach, "A National University," 702.

58. *Daily Albany Argus*, February 6, 1851, cited by Silverman and Beach, "A National University," 703. See also Clarke, *James Hall of Albany*, 192.

59. Rezneck, "Emergence," 227–28.

60. Hofstadter and Smith, *American Higher Education*, 253, 334–75. See especially pp. 350–53. See also Silverman and Beach, "A National University," 703.

61. Agassiz to Hall, August 3, 1851; in Clarke, *James Hall of Albany*, 193.

62. Hall to Agassiz, October 27, 1851, in Clarke, *James Hall of Albany*, 196–97.

63. Clarke, *James Hall of Albany*, 193–94.

64. Bache, "Address on Retiring," xlv.

65. ADB to Alonzo Potter, August 2, 1852, in [Alonzo Potter] *Objections to a Reorganization Considered* (Philadelphia, 1853), as quoted by Storr, *Graduate Education*, 68.

66. Silverman and Beach, "A National University," 707.

67. BP to James Hall, October 19, 1851, Box 3, Series B0561, State Museum Director's, State Geologist's and State Paleontologist's Correspondence files, 1828–1944, New York State Archives.

68. The nine professors called for were Agassiz, Dana, Hall, Lovering, Mitchell, Norton, Peirce, Wright, and Wyman.

69. BP to James Hall, October 19, 1851, Box 3, Series B0561, State Museum Director's, State Geologist's and State Paleontologist's Correspondence files, 1828–1944, New York State Archives.

70. Ibid.

71. Ibid.

72. Silverman and Beach, "A National University," 788.

73. John P. Norton to BP, November 19, 1851, *54m-264, box 3 of 11 misc., PPHH.

74. Ibid., November 25, 1851, *54m-264, box 3 of 11 misc., PPHH.

75. Storr, *Graduate Education,* 70.

76. Silverman and Beach, "A National University," 709; Storr, *Graduate Education,* 70.

77. BP to JnLC and JoLC, February 19, 1852, LFPB (emphasis in the original).

78. *ANB;* see also Hofstadter and Smith, *American Higher Education,* 451; and Bruce, *Launching of Modern American Science,* 228.

79. Thompson, *Ruggles of New York; ANB;* Rezneck, "Emergence," 231; Silverman and Beach, "A National University," 711.

80. Ruggles et al., *Speeches,* 7.

81. BP to ADB, September 5, 1856, ADB file, PCHH.

82. Silverman and Beach, "A National University," 710.

83. Storr, *Graduate Education,* 70.

84. ADB to BP, February 24, 1853, ADB file, PCHH.

85. BP to SMP, no date, SMP file, PCHH.

86. BP to John V. Pruyn, February 6, 1853, John V. Pruyn file, PCHH; and James Hall to BP, March 28, 1853, James Hall file, PCHH. See also, Silverman and Beach, "A National University," 712.

87. BP to JoLC, February 20, 1853, LFPB.

88. Ruggles to BP, August 2, 1853, Ruggles file, PCHH (Ruggles emphasis).

89. Storr, *Graduate Education,* 69, 71.

90. Ibid., 74.

91. Silverman and Beach, "A National University," 709.

92. Ibid., 708–9.

93. Alonzo Potter to Joseph Henry, July 2, 1852, in Henry, *Papers,* vol. 8, 359–60.

94. Bruce, *Launching of Modern American Science,* 228.

95. BP to ADB, May 16, 1852, ADB file, PCHH.

96. Bruce, *Launching of Modern American Science,* 227–28; Silverman and Beach, "A National University," 711; and Storr, *Graduate Education,* 76–81.

97. BAG to BP, April 19, 1852, *54m-264, box 3 of 11 misc., PPHH.

98. Storr, *Graduate Education,* 94. BAG to BP, April 19, 1852, *54m-264, box 3 of 11 misc., PPHH, suggests Columbia as the site for a great university.

99. Bruce, *Launching of Modern American Science,* 228.

100. *The Diary of George Templeton Strong,* August 16, 1852 [Nevins and Thomas, *Diary,* vol. 2, 103]. James Renwick was Professor of Natural Philosophy and Experimental Chemistry at Columbia from 1820–53.

101. Storr, *Graduate Education*, 96; Bruce, *Launching of Modern American Science*, 228–29; Thompson, *Ruggles of New York*, 77–89.

102. JnLC to BP, December 4, 1853, JnLC file, PCHH (emphasis in the original).

103. BP to JoLC, [January 1853], LFPB.

104. BP to JnLC, December 11, 1853, LFPB.

105. Miller et al., *The Lazzaroni*, 39; *ANB;* Hofstadter and Smith, *American Higher Education*, 440.

106. *The Diary of George Templeton Strong*, January 9, 1854 [Nevins and Thomas, *Diary*, vol. 2, 146–47]; and Ruggles, *The Duty of Columbia College to the Community and its Right to Exclude Unitarians from its Professorship of Physical Science, Considered by One of its Trustees*, in Hofstadter and Smith, *American Higher Education*, 451–64.

107. *The Diary of George Templeton Strong*, January 9, 1854 [Nevins and Thomas, *Diary*, vol. 2, 146.]

108. BP to ADB, no date, ADB file, PCHH (emphasis in the original).

109. Hofstadter and Smith, *American Higher Education*, 452.

110. G. Ogden to H. Fish, February 12, 1854, Fish Mss, Columbia University Library, as quoted by Bruce, *Launching of Modern American Science*, 229.

111. S. Ruggles to BP, February 12, 1854, S. Ruggles file, PCHH (emphasis in the original).

112. Bruce, *Launching of Modern American Science*, 229.

113. Ibid., 230.

114. Chauvenet to BP, April 27, 1857, Chauvenet file, PCHH.

115. Davies to BP, May 24, 1857, C. Davies file, PCHH. Davies also apologized to Bache for the article, see Davies to ADB, August 11, 1857, BPLC.

116. Bache, "Remarks," 477.

117. Ibid., 478.

118. Ibid.

119. Peirce, "Remarks," 88.

120. Ibid.

121. Ibid., 90.

122. Ibid.

123. BP to ADB, December 30, 1855, ADB file, PCHH.

124. Bruce, *Launching of Modern American Science*, 230.

125. Peirce, "Remarks," 89. Peirce lists Cooper along with Smithson and Lawrence as benefactors of education.

126. Mack, *Peter Cooper*, 249.

127. Storr, *Graduate Education*, 85.

128. Henry P. Tappan to Samuel B. Ruggles, April 2, 1856, as quoted by Storr, *Graduate Education*, 88.

129. BP to Henry P. Tappan, as quoted in a letter from Henry P. Tappan to Samuel B. Ruggles, March 17, 1856, as quoted by Storr, *Graduate Education*, 88.

130. Tappan to BP, April 5, 1856, Tappan file, PCHH.

131. Ibid., June 19, 1856, Tappan file, PCHH.

132. Storr, *Graduate Education*, 88.

133. Ibid., 89. Storr cites S. Willis Rudy, *The College of the City of New York: A History 1847–1947, p. 40.* Rudy suggests the possibility that Wood was interested in the university only as a means of regaining a reputation for integrity.

134. *Peirce, "Working Plan,"* 4. There is a manuscript draft of this document in *54m-264, box 1 of 11, misc., PPHH, dated September 1, 1856.

135. BP to Armsby, September 3, 1856, Gratz Collection, Historical Society of Pennsylvania.

136. G[eorge] J. Alder to BP, December 10, 1857, Adler file, PCHH.

137. Armsby to BP, September 15, 1856, Armsby file, PCHH (emphasis in the original).

138. ADB to BP, October 15, 1856, ADB file, PCHH.

139. Alexander D. Bache, *Anniversary Address before the American Institute, of the City of New York, at the Tabernacle, October 28, 1856, during the Twenty-eighth Annual Fair* (New York, 1857), p. 13, as quoted by Storr, *Graduate Education,* 91.

140. ADB to BP, November 1, 1856, ADB file, PCHH.

141. BP to SMP, February 5, 1857, SMP file, PCHH.

142. Ibid., February 8, 1857, SMP file, PCHH.

143. Ibid., February 9, 1857, SMP file, PCHH.

144. Storr, *Graduate Education,* 92.

145. Mack, *Peter Cooper,* 250.

146. BP to JoLC, August 25, 1857, LFPB.

147. Bruce, *Launching of Modern American Science,* 231.

148. Ibid., 350.

149. William Elwood Byerly, "The Heat of the Sun" (PhD diss., Harvard University, 1873); HU 90.201, HUA; see also Morison, *Three Centuries of Harvard,* 334–35. Coolidge, "Mathematics at Harvard," 250, says, "Byerly received the first degree of Doctor of Philosophy ever awarded by the University." Morison says Byerly received the first PhD from Harvard, and mentions no other one. Joseph D. Zund, *ANB,* says Byerly's was one of the first two earned at Harvard. Mathematicians may be disappointed with the mathematical content of this thesis; see Birkhoff, "Mathematics at Harvard," 10. Byerly spoke well of Peirce later in his life (see, for example, Archibald et al., "Benjamin Peirce," 5–7), but no details of their relationship as student and mentor are known. Byerly's was the only PhD dissertation that Peirce supervised.

Chapter 8. No Idea of an Observatory

1. BP to ADB, June 20, 1855, ADB file, PCHH.

2. Ibid., August 1, 1855, ADB file, PCHH; see also, BP to ADB, July 23, 1855, ADB file, PCHH.

3. BP to ADB, July 30, 1855, ADB file, PCHH.

4. BP to SMP, August 24, [1855], SMP file, PCHH.

5. Ibid., undated [August 1855], SMP file, PCHH.

6. Ibid., Peirce often gave such monologues at scientific meetings. For another example see the *Boston Atlas and Daily Bee,* August 11, 1859, transcript in *54m-264, box 4 of 11, misc., PPHH.

7. [Harris], *Statement of the Trustees,* 3. This pamphlet as well as Scientific Council, *Defence of Dr. Gould,* and subsequent pamphlets from both sides are obviously propaganda and contain falsehoods. Hopefully I have been able to separate the wheat from the chaff. See also Miller, *Dollars for Research,* 40; and Boss, *History of the Dudley Observatory,* 2–3.

8. *Olson,* "The Gould Controversy," 267; McCormmach, "Sidereal Messenger," 35.

9. Olson, "The Gould Controversy," 268; and *ANB.*

10. Mitchel to Amos Dean, July 28, 1851, quoted by [Harris], *Statement of the Trustees,* 4 (emphasis in [Harris]).

11. [Harris], *Statement of the Trustees,* 4.

12. Ibid., 5.

13. Warner, "Astronomy," 62.

14. Hall to Agassiz, October 27, 1851, in Clarke, *James Hall,* 196–97.

15. [Harris], *Statement of the Trustees,* 5.

16. [Harris], *Statement of the Trustees,* 6.

17. Sears C. Walker to BAG, October 21, 1852, as quoted by Gould, *Reply,* 37–38.

18. Comstock, "Biographical Memoir," 156–57.

19. Ibid., 157–58. See also BAG to BP, May 11, 1867, BAG Jr. file, PCHH.

20. BAG to Baron von Humboldt, as quoted by Chandler, "Benjamin Apthorp Gould," 221. Baron von Humboldt was a distinguished German naturalist.

21. Herrmann, "B. A. Gould."

22. Chandler, "Benjamin Apthorp Gould," 222–23; Comstock, "Biographical Memoir," 157–58; McCormmach, "Sidereal Messenger," 35.

23. BP to ADB, October 16, 1853, ADB file, PCHH.

24. Ibid., September 21, 1851, ADB file, PCHH, in Reingold, *Science in Nineteenth-Century America,* 160–61.

25. Bruce, *Launching of Modern American Science,* 25. Rogers held a chair at the University of Glasgow from 1857 until his death. Nevertheless, about a decade earlier (see chapter 4), Peirce thought his science was too sloppy and lobbied against him when he applied for the Rumford professorship at Harvard.

26. BP to ADB, September 21, 1851, ADB file, PCHH, in Reingold, *Science in Nineteenth-Century America,* 160–61.

27. Gould, *Reply,* 37–40.

28. Peirce "Method of Determining Longitudes." [Harris], *Statement of the Trustees,* 7; claimed that C. H. F. Peters, an astronomer working for the Coast Survey, first suggested the advantages of a heliometer to Peirce. Peirce denied that the idea of installing a heliometer in the Dudley Observatory came from Peters: "The lie that Peters first suggested the use of the heliometer does not deserve contradiction." [BP to ADB, August 31, 58, Dudley Observatory file, PCHH]. This letter appears to be a draft, or the beginning of a letter never sent.

29. Boss, *History of the Dudley Observatory,* 12.

30. [Harris], *Statement of the Trustees,* 7.

31. Peters to ADB, Providence, August 22, 1855, BPLC (emphasis in the original).

32. BP to ADB, August 25, 1855, BPLC.

33. See Peirce, "Method of Determining Longitudes," 102; and [Harris], *Statement of the Trustees,* 9.

34. BP to Armsby, September 1, 1855, Gratz Collection, case 7, box 38, Historical Society of Pennsylvania.

35. Scientific Council of the Dudley Observatory to Anonymous, August 11, 1856, Armsby Collection, Dudley Observatory Archives, in Henry, *Papers,* vol. 9, 384–85; Peirce, "Method of Determining Longitudes," 102.

36. Cajori, *Chequered Career,* 71–73.

37. Stephens, "Astronomy as Public Utility," 26.

38. James, *Elites in Conflict,* 48. The sun and stars appear differently from different positions in the earth's orbit, although due to the extremely long distances, the differences are very slight. These changes in the appearance of the sun and stars are called solar and stellar parallax, respectively. As an example of Peirce's interest in solar parallax see, BP to ADB, August 9, 1863 and September 12, 1863, ADB file, PCHH.

39. Scientific Council, *Defence of Dr. Gould,* 18.

40. See James, *Elites in Conflict,* chapter five.

41. Armsby to BAG, September 15, 1855, copy in BPLC.

42. See, for example, *New York Times,* June 29, 1858; and the Scientific Council to Blandina Dudley, June 30, 1858, as printed in the *New York Times* July 5, 1858.

43. Armsby to BAG, September 15, 1855, copy in the BPLC. Gould discouraged the purchase of a prime vertical transit, primarily used for the accurate determination of time, because a meridian circle could do the same task. See James, "Dudley Observatory Controversy," 84–85.

44. Armsby to BAG, September 15, 1855, copy in BPLC.

45. BP to Wm B. Sprague Jr., September 22, 1855, Gratz Collection, case 7, box 38, Historical Society of Pennsylvania; BP to ADB, September 25, 1855, ADB file, PCHH.

46. BP to ADB, September 25, 1855, ADB file, PCHH.

47. James, "Dudley Observatory Controversy," 97.

48. BAG to Armsby, May 26, 1856, as quoted by Gould, *Garbling of Letters,* 7–9.

49. James, *Elites in Conflict,* 74.

50. Resolution of the Board, [January 9, 1857], copy in Dudley Observatory file, PCHH; Scientific Council, *Defence of Dr. Gould,* 23–24; [Harris], *Statement of the Trustees,* 20–21.

51. Scientific Council, *Defence of Dr. Gould,* 23.

52. BP to ADB, January 17, 1855, and BP to ADB, March 29, 1855, both ADB file, PCHH.

53. Bruce, *Launching of Modern American Science,* 236.

54. ADB, BP, and BAG to Thomas Olcott (copy), August 8, 1856, Dudley Observatory file, PCHH. This letter was apparently written by Peirce. A draft in his hand is in the Olcott file, PCHH.

55. James, "Dudley Observatory Controversy," 109, 120.

56. Gould, *Reply,* 322–25.

57. Gould, *Reply,* 13. See also BAG to BP, March 29, 1858, Dudley Observatory file, PCHH.

58. James, *Elites in Conflict,* 83.

59. BAG to John L. Le Conte, May 28, 1857, APSL.

60. [Harris], *Statement of the Trustees,* 24.

61. James, *Elites in Conflict,* 89. On the Dudley's inauguration, see Bartky et al., "Event of No Ordinary Interest."

62. BP to Armsby, September 3, 1856, Gratz Collection, Historical Society of Pennsylvania.

63. ADB to BP, November 1, 1856, ADB file, PCHH. Φ^n, which stood for Functionary, was Peirce's nickname within the Lazzaroni.

64. BAG to Peters, August 4, 1857, as quoted in Gould, *Garbling of Letters,* 14.

65. Scientific Council, *Defence of Dr. Gould,* 26–28; [Harris], *Statement of the Trustees,* 148–50.

66. McPherson, *Battle Cry of Freedom,* 189–90.

67. Scientific Council, *Defence of Dr. Gould,* 24–25.

68. James, "Dudley Observatory Controversy," 139.

69. Peters to ADB, November 11, 1857, copy in Dudley Observatory file, PCHH; ADB to Peters (copy), November 13, 1857, ADB file, PCHH; ADB to BP, November 13, 1857, ADB file, PCHH; and BP to Peters, November 17, 1857, Hamilton College Library Archives.

70. Gould, *Reply,* 56–60.

71. Ibid., 57–58.

72. [Harris], *Statement of the Trustees,* 163.

73. Ibid., 42.

74. *New York Times,* June 29, 1858.

75. [Harris], *Statement of the Trustees,* 43–45.

76. Scientific Council, *Defence of Dr. Gould,* 12; James, *Elites in Conflict,* 40–44; 77–78.

77. See, for example, ADB to BP, November 26, 1857; November 11, 1857; December 6, 1857; December 14, 1857; and December 22, 1857; all in ADB file, PCHH. See also [Harris], *Statement of the Trustees,* 43–45.

78. ADB to BP, December 6, [1857], ADB file, PCHH.

79. ADB to BP, December 6, [1857], ADB file, PCHH.

80. Ibid., December 14, 1857, ADB file, PCHH.

81. BAG to the Scientific Council, December 23, 1857, Dudley Observatory file, PCHH; James, *Elites in Conflict,* 104.

82. BP to ADB, December 23, 1857, HL. Peirce is apparently replying to ADB to BP, December 14, 1857, ADB file, PCHH. The correspondence does not specify the exact points upon which Bache and Peirce disagreed. Presumably Peirce was neither as anti-Peters, nor as pro-Gould as was Bache. Peirce also felt that the trustees' position was much stronger than Bache did.

83. James, "Dudley Observatory Controversy," 145.

84. See, for example, ADB to BP, December 14, 1857, ADB file, PCHH.

85. BAG to BP, January 14, 1858, Dudley Observatory file, PCHH.

86. Ibid., January 18, 1858, Dudley Observatory file, PCHH.

87. Ibid.

88. For Peters's career at Hamilton see Pilkington, *Hamilton College,* 197–98; see also *ANB* and *DSB.*

89. BP to JnLC, January 24, 1858, LFPB. John Van Schaick Lansing Pruyn, one of Albany's cultural elite, was a member of the New York Board of Regents.

90. BP to ADB, January 24, 1858, ADB file, PCHH.

91. Words of Mr. Barnard, given at a meeting to support the Scientific Council, as reported in the *New York Times,* July 14, 1858, p. 1.

92. BP to Frazer, February 2, 1858, APSL.

93. BP to SMP, February 9, 1858, and February 14, 1858, SMP file, PCHH.

94. BP to SMP, February 9, 1858, SMP file, PCHH.

95. James, *Elites in Conflict,* chapter 7.

96. BP to ADB, May 29, 1858, ADB file, PCHH.

97. ADB to BP, June 2, 1858. See also ADB to BP, May 21, 1858, and ADB to BP, May 30, 1858, all ADB file, PCHH.

98. BAG to BP, May 27, 1858, Dudley Observatory file, PCHH.

99. ADB and JH to BAG, May 28, 1858, Dudley Observatory file, PCHH.

100. BP to JoLC, June 6, 1858, LFPB (Peirce's emphasis). See also J. H. Toomer to BAG, May 20, 1858, Dudley Observatory file, PCHH; M. James Tilton to BAG, May 20, 1858, Dudley Observatory file, PCHH; Executive Committee of the Trustees to BAG (copy) May 22, 1858; Dudley Observatory file, PCHH; M. James Tilton to BAG, May 22, 1858, Dudley Observatory file, PCHH; BAG to Scientific Council, May 23, 1858, Dudley Observatory file, PCHH; J. H. Toomer to BAG, May 26, 1858, Dudley Observatory file, PCHH; and S. E. Winslow to BAG, May 29, 1858, Dudley Observatory file, PCHH.

101. ADB to BP, May 30, 1858, ADB file, PCHH. See also, BAG to BP, May 30, 1858, Dudley Observatory file, PCHH; undated newspaper clipping, *54m-264, box 1 of 11, misc., PPHH; and James, "Dudley Observatory Controversy," 217 and 503–4, footnote 43. The Scheutz calculating engine is now in the Smithsonian Institution.

102. BP to ADB, June 6, 1858, HL.

103. This resolution was dated June 12, 1858. There is a copy in the Dudley Observatory file, PCHH, and it was printed in the *New York Times*, June 29, 1859.

104. JH, ADB, and BP to Olcott, June 12, 1858, Dudley Observatory file, PCHH; *New York Times*, June 26, 1858; Scientific Council, *Defence of Dr. Gould*, 4; and Gould, *Garbling of Letters*, 6.

105. [Harris], *Statement of the Trustees*, 68, 152; Scientific Council, *Defence of Dr. Gould*, 2.

106. Scientific Council, *Defence of Dr. Gould*, 84.

107. Ibid., 4; *New York Times*, July 5, 1858. The Council was dismissed at a meeting of the trustees on July 3, 1858.

108. June 25, 1858, 2.

109. *New York Times*, June 29, 1858.

110. Scientific Council, *Defence of Dr. Gould*, 29–31, 73–74; see also *New York Times* July 12, 1858.

111. *New York Times*, June 29, 1858.

112. Mrs. Dudley to the Scientific Council, June 9, 1858, copy in the Dudley Observatory file, PCHH; also published in *New York Times*, July 3, 1858. James, "Dudley Observatory Controversy," 510, footnote 27, suggests that by this time Mrs. Dudley was senile, and the letter was written by her nephew, who was her heir. If the trustees could be shown to be incompetent, then the nephew might be able to get the money back that his aunt had given the observatory.

113. Bruce, *Launching of Modern American Science*, 225–26, 234; James, *Elites in Conflict*, 176.

114. BP to JnLC, June 27, 1858, LFPB. Peirce and Bache had a particularly low opinion of Matthew Maury, the superintendent of the Naval Observatory in Washington, D. C.

115. JH to ADB, September 18, 1858, Bache Papers, Smithsonian Archives, in Henry, *Papers*, vol. 10, 62–64.

116. *New York Times*, June 29, 1858.

117. BAG to George P. Bond (copy), June 29, 1858; Bond to BAG (copy), July 2, 1858; and BAG to Bond (copy), July 6, 1858; all in Dudley Observatory file, PCHH.

118. Bond to BAG, July 9, 1858 (copy), Dudley Observatory file, PCHH.

119. Scientific Council, *Defence of Dr. Gould*.

120. Ibid., 5.

121. James, *Elites in Conflict*, 175.

122. BP to JoLC, August 8, 1858, LFPB.

123. BP to JnLC, August 15, 1858, LFPB.

124. [Harris], *Statement of the Trustees*.

125. James, "Dudley Observatory Controversy," 279–80.

126. BP to ADB, October 3, 1858, and BP to ADB, October 8, 1858, both in HL.

127. JH to ADB, October 18, 1858, Bache Papers, Smithsonian Archives, in Henry, *Papers*, vol. 10, 73–74.

128. BP to JoLC, December 14, 1858, LFPB.

129. Ibid., January 16, 1859, LFPB.

130. James, "Dudley Observatory Controversy," 333. See also ADB to BP, September 16, 1859, ADB file, PCHH.

131. Bruce, *Launching of Modern American Science*, 238.

132. See, for example, BP to ADB, March 8, 1861, HL; and ADB to BP, December 9, 1862, ADB file, PCHH. See also Bruce, *Launching of Modern American Science*, 223–24.

133. See chapter 12.

134. See James, "Dudley Observatory Controversy," chapters 11 and 12.

135. BAG to BP, January 20, 1859, Dudley Observatory file, PCHH.

136. *New York Times,* January 14, 1859; BAG to BP, January 4, 1859, Dudley Observatory file, PCHH.

137. See chapter 12.

138. James, "Dudley Observatory Controversy," 373–87; Comstock, "Biographical Memoir;" and Chandler, "Benjamin Apthorp Gould."

139. Bruce, *Launching of Modern American Science,* 239.

140. James, *Elites in Conflict,* 225.

CHAPTER 9. A STRANGER IN MY OWN LAND

1. BP to JoLC, February 20, 1853, LFPB.

2. Ibid., April 30, 1856, LFPB.

3. BP to ADB, October 16, 1853, ADB file, PCHH; see also BP to JoLC, [January] 1853, LFPB.

4. BP to ADB, [1854?], ADB file, PCHH (emphasis in the original).

5. Ibid., October 16, 1853, ADB file, PCHH.

6. CEP to Aunt Nichols, February 28, 1857, L 676, CSPP.

7. BP to SMP, May 8, 1859, SMP file, PCHH.

8. BP to JoLC, November 26, 1857; BP to JnLC, January 26, 1858, both LFPB. It hadn't been printed by November 26, 1856, but was in print by January 26, 1857.

9. BP to ADB, October 16, 1853, October 18, 1853, ADB file, PCHH.

10. Birkhoff, "Mathematics at Harvard," 10, and Muir, *Theory of Determinants,* vol. 2, 251.

11. For example, J. P. Nichol, Professor of Astronomy at Glasgow, praised Peirce's *Mechanics,* in his *Cyclopaedia of the Physical Sciences,* published in 1860 [Crowe, *History of Vector Analysis,* 38]. Thomas Muir, [Muir, *Theory of Determinants,* vol. 2, 251] said Peirce's treatment was "free and masterful." Garrett Birkhoff [Birkhoff, "Mathematics at Harvard," 10] states, "Chapter X of his *Analytical Mechanics* (1855) contained a masterful chapter on 'functional determinants' of n × n matrices." V. F. Lenzen [Lenzen, *Benjamin Peirce and the U. S. Coast Survey,* 4] wrote, "Benjamin Peirce's treatment of mechanics was probably the highest level of any work in the field in English until the appearance of Whittaker's *Analytical Dynamics* [in 1904]." Cajori [Cajori, *Teaching and History of Mathematics,* 144] said that Peirce's *Mechanics* was regarded as the best work in the world on mechanics, even in Germany.

12. King, *Benjamin Peirce,* 10.

13. [Hill], "Peirce's Analytic Mechanics," 1.

14. Asa Gray to BP, March 26, 1848, Gray file, PCHH.

15. Dupree, *Asa Gray,* 141, 221, 226–28; Lurie, *Louis Agassiz,* 275–79. See also Royce, *"Some Relations."*

16. Late in his life, Peirce gave a series of lectures that were published as *Ideality in the Physical Sciences* [Peirce, *Ideality*].

17. Dupree, *Asa Gray,* 290–91.

18. Asa Gray to Charles Darwin, June [1863], as quoted by Dupree, *Asa Gray,* 291. See Peirce, "Mathematical Investigation," 444–47; and Gray, "Composition," 438–44.

19. L 666, CSPP contains a certificate saying Peirce was admitted to the American Academy. The date of admission is written on it in Peirce's hand.

20. BP to JoLC, August 28, 1858, LFPB; Kohlstedt, *Formation of the American Scientific Community,* 29–30.

21. BP to ADB, May 8, 1854, ADB file, PCHH.

22. Ibid., [no date, but probably September 1854], ADB file, PCHH.

23. Winslow to Warner, July 7, 1858, APSL. This is hardly an unprejudiced source, but I have no reason to doubt it in this case. See also BP to Josiah Quincy, May 26, 1858, Josiah Quincy file, PCHH.

24. BP to JoLC, May 31, 1858, LFPB.

25. Peirce mused about some of his options in an undated note to Agassiz, LA file, PCHH.

26. BP to ADB, May 29, 1858, HL.

27. Ibid.

28. Ibid.

29. ADB to BP, June 2, 1858, ADB file, PCHH.

30. BP to ADB, June 6, 1858, HL.

31. There is an undated note in Louis Agassiz file, PCHH, apparently to Louis Agassiz, in which Peirce discussed various scenarios for action the two of them might take.

32. BP to JoLC, first Sunday in October, 1858, LFPB. "Vox Populi Vox Diaboli" translates, "Voice of the people, voice of the devil."

33. BP to JoLC, November 7, 1858, and December 14, 1858 LFPB.

34. Ibid., October 17, 1857, LFPB.

35. BP to JnLC, January 24, 1858, LFPB.

36. Jones and Boyd, *Harvard College Observatory*, 54.

37. Bond, "Rings of Saturn" (*American Journal of Science*); Bond, "Rings of Saturn" (*Astronomical Journal*); Jones and Boyd, *Harvard College Observatory*, 99; and Holden, *Memorials*, 253–59.

38. Peirce, "Constitution of Saturn's Ring" (*American Journal of Science*); and Peirce, "Constitution of Saturn's Ring" (*Astronomical Journal*).

39. *Daily Evening Traveler*, May 13, 1851; a copy of the *Traveler* is in the Harvard Observatory Records, UAV 630.2.1, HUA.

40. *New York Daily Tribune*, May 12, 1851; a copy of the *Tribune* is in the Harvard University Records, UAV 630.2.1, HUA.

41. In 1857, James Clerk Maxwell corrected both Bond and Peirce, showing that the rings could not be fluid, but consisted of many small particles. Maxwell's results were confirmed by James Keeler in 1895. See Holden, *Memorials*, 257–58.

42. Stephens, "Astronomy as Public Utility," 31. Holden, *Memorials*, 254–59.

43. BP to O. M. Mitchel, August 3, 1851, Department of Special Collections, Kelvin Smith Library, Case Western Reserve University.

44. Kohlstedt, *Formation of the American Scientific Community*, 126, fn.

45. Stephens, "Astonomy as Public Utility," 26.

46. Obituary notice of W. C. Bond, *Proceedings of the American Academy of Arts and Sciences*, 4, 163–66.

47. BP to ADB, January 31, 1859, HL.

48. Ibid. See also BP to JoLC, February 13, 1859, LFPB.

49. Gibbs to BP, February 2, 1858, W. Gibbs file, PCHH.

50. BP to JnLC, February 6, 1859, LFPB; see also JH to ADB, February 13, 1859, Bache Papers, Smithsonian Archives, in Henry, *Papers*, vol. 10, 83–84.

51. BP to JnLC, February 6, 1859, LFPB.

52. Ibid.

53. BP to ADB, February 24, 1859, HL. Peirce gave a detailed account of his many attempts to work out an arrangement that would be acceptable to him. See also BP to JoLC, February 13, 1859, and September 24, 1859, LFPB; and JnLC to BP, March 6, 1859, JnLC file, PCHH.

54. See, for example, BP to JoLC, May 9, 1859, LFPB; and BP to SMP, May 20, 1859, SMP file, PCHH.

55. G. Bond to BP, March 12, 1859, G. Bond file, PCHH; also printed in Holden, *Memorials*, 163.

56. Holden, *Memorials*, 163; see also Jones and Boyd, *Harvard College Observatory*, 113. The Bond Papers are in the HUA.

57. BP [apparently] to G. Bond (perhaps a copy), undated, G. Bond file, PCHH. Peirce may, of course, have talked to Bond in person.

58. Stephens, "Astonomy as Public Utility," 31; Holden, *Memorials*, 91–93.

59. BP to ADB, January 15, 1864, ADB file, PCHH

60. BP to ADB, January 20, 1864, ADB file, PCHH.

61. BP to JnLC, January 18, 1859, LFPB.

62. BP to JoLC, March 27, 1859, LFPB.

63. BP to JnLC, April 8, 1859, LFPB.

64. BP to ADB, August 9, 1863, ADB file, PCHH.

65. BP to JoLC, February 28, 1860, LFPB.

66. BP to SMP, May 13, 1859, SMP file, PCHH. The French zoologist, Jean de Lamarck, proposed that the effort to deal with changing circumstances and bodily needs induced physical changes in individual species whose offspring then inherited them; thus producing better-adapted species. Lamarck published his principal work on evolution in 1807.

67. BP to JnLC, March 11, 1860, LFPB. Joe was Joseph Le Conte, John's brother. A geologist, he had studied under Agassiz at Harvard's Lawrence Scientific School. He later became a staunch supporter of Darwin.

68. Dupree, *Asa Gray*, chap. 15; Lurie, *Louis Agassiz*, 275–82, 293–95.

69. SMP to ADB, as a footnote on BP to ADB, February 6, 1861, ADB file, PCHH. Φ^n was the symbol for Peirce's nickname, "Functionary."

CHAPTER 10. THIS HOLY CAUSE

1. Morison, *Three Centuries of Harvard*, 287.

2. BP to SMP, September 22, 1844, SMP file, PCHH. Benjamin also wrote Sarah about his distaste for Clay a few years later; see BP to SMP, undated, but probably about February 1856, SMP file, PCHH. Peirce frequently spoke disparagingly of the abolitionists, for example, see, BP to ADB, April 7, 1850, ADB file, PCHH. Charles Phelps Huntington was the husband of Sarah Peirce's sister, Helen.

3. BP to SMP, February 20, 1847, SMP file, PCHH.

4. Charles Henry Peirce to BP, April 11, 1844, Charles Henry Peirce file, PCHH.

5. Le Conte, Joseph, "Biographical Memoir," 1895, 371.

6. Lupold, "From Physician to Physicist," 22–24; 68.

7. Ibid., 111.

8. Slotten, *Patronage*, 92–95.

9. BP to JnLC, February 19, 1860, LFPB.

10. BP to JoLC, February 27, 1858, LFPB; and Lupold, "From Physician to Physicist," 91. The Athenaeum was formed in Columbia for the exchange of scholarly papers.

11. BP to SMP, February 15, 1858, SMP file, PCHH.

12. BP to JoLC, March 3, 1858, LFPB (Peirce's emphasis).

13. Ibid., December 21, 1958, LFPB.

14. Ibid., March 3, 1858, LFPB.

15. Ibid., January 30, 1859, LFPB.

16. Ibid., December 14, 1858, LFPB.

17. Ibid., May 31, 1858, LFPB. This was not a new sentiment. Peirce had said the same thing a week earlier; see BP to JoLC, [mid-May, 1858], LFPB.

18. BP to JoLC, [mid-May, 1858], LFPB.

19. Ibid., January 16, 1859, LFPB.

20. JnLC to BP, October 9, 1859, LFPB.

21. BP to JoLC, November 6, 1859, LFPB.

22. Lupold, "From Physician to Physicist," October 14, 1835, 107.

23. SMP to JoLC, November 30, 1859, LFPB (emphasis in the original).

24. BP to JoLC, December 21, 1859, LFPB.

25. McPherson, *Battle Cry of Freedom*, 152–53.

26. Ibid., 204.

27. Ibid., 201–6.

28. Peirce is presumably referring to the "Secret Six."

29. BP to JoLC, October 31, 1859, LFPB.

30. Ibid., November 6, 1859, LFPB. Wendell Phillips was a social reformer and a leader in the antislavery crusade. William Henry Seward, then senator from New York, was the political leader of the antislavery movement.

31. BP to JoLC, November 13, 1859, LFPB (emphasis in the original). Ralph Waldo Emerson was the noted author, poet, and philosopher.

32. BP to JnLC, November 1, 1858, LFPB (emphasis in the original).

33. BP to ADB, May 29, 1858, HL.

34. BP to JoLC, November 7, 1858, LFPB. Anson Burlingame was a Republican congressman from Massachusetts. He achieved a national reputation as an antislavery orator.

35. JnLC to BP, November 14, 1858, LFPB.

36. BP to JoLC, December 21, 1859, LFPB.

37. BP to JnLC, December 11, 1859, LFPB.

38. Ibid.

39. BP to CSP, February 19, 1860, typescript copy in Pike County Historical Society of Milford, Pennsylvania.

40. BP to JoLC, January 8, 1860, LFPB. Peirce also gave lectures at the Lowell Institute in Boston in 1860.

41. BP to CSP, March 16, 1860, typescript copy in the Pike County Historical Society of Milford Pennsylvania.

42. Members of the Lazzaroni referred to Joseph Henry as Smithson.

43. BP to JoLC, January 21, 1860, LFPB.

44. Ibid. (emphasis in the original).

45. Ibid. (emphasis in the original).

46. BP to CSP, February 19, 1860, typescript, Pike County Historical Society.

47. Ibid.

48. Ibid.

49. CEP to Aunt [Nichols], April 1, 1860, L676 CSPP.

50. BP to JoLC, April 1, 1860, LFPB.

51. Ibid. (emphasis in the original). Φ^{na} stood for the female of functionary, Sarah's Lazzaroni nickname.

52. CEP to Aunt, Uncle, and Cousins, April 30, [1860?], L 686 CSPP.

53. CSP to JMP, April 23, 1860, L 339, CSPP.

54. BP to JnLC, May 3, 1860, LFPB.

55. See, for example, ADB to Charles Babbage, May 22, 1860, ADB file, PCHH.

56. BP to JoLC, May 26, 1860, LFPB.

57. BP 1860 Journal, June 10, 1860, HYG 1680 100, by permission of the Houghton Library, Harvard University.

58. Ibid.

59. BP 1860 Journal, May 24, 1860, HYG 1680 100, by permission of the Houghton Library, Harvard University. This is the date recorded by Peirce, but he must have written May mistakenly and meant June. Hereafter, this entry will be cited as [June] 24,. See also CEP to Aunt and Uncle Nichols, July 16, 1860, L 676, CSPP.

60. BP 1860 Journal, [June] 24, 1860, HYG 1680 100, by permission of the Houghton Library, Harvard University.

61. Ibid.

62. CEP to Aunt and Uncle Nichols, July 16, 1860, L 676, CSPP. See also, CEP to Aunt Nichols, December 30, 1860, L676, CSPP; and BP to ADB, July 30, 1860, ADB file, PCHH.

63. BP 1860 Journal, [June] 24, 1860, HYG 1680 100, by permission of the Houghton Library, Harvard University. See also CEP to Aunt and Uncle Nichols, July 16, 1860, L 676, CSPP.

64. BP 1860 Journal, July 6, and July 8 (sic), 1860, HYG 1680 100, by permission of the Houghton Library, Harvard University.

65. BP 1860 Journal, July 9, 1860, HYG 1680 100, by permission of the Houghton Library, Harvard University.

66. BP to JoLC, November 1, 1860, LFPB.

67. BP 1860 Journal, August 11, 1860, HYG 1680 100, by permission of the Houghton Library, Harvard University.

68. BP 1860 Journal, various dates, HYG 1680 100, by permission of the Houghton Library, Harvard University. The paper was published as "Lettre addressée à M. le président de L'Académie des Sciences sur la construction physique des cometes." *Comptes Rendus de l'Acadèm. D. Sc.* vol. 51, 174–76.

69. BP 1860 Journal, August 28, 1860, HYG 1680 100, by permission of the Houghton Library, Harvard University.

70. BP to JoLC, November 1, 1860, LFPB.

71. JnLC to BP, October 14, 1860, LFPB.

72. Ibid., November 25, 1860, LFPB (emphasis in the original).

73. BP to JoLC, November 25, 186[0], LFPB.

Chapter 11. Every Drop of Blood

1. JnLC to BP, May 6, 1860, LFPB.

2. Ibid., February 3, 1861, LFPB; and JnLC to BP, April 2, 1861, LFPB.

3. *ANB* (John Le Conte).

4. JnLC to Peirce, February 3, 1861, LFPB (emphasis in original).

5. Ibid. (emphasis in original).

6. JnLC to BP, April 2, 1861, LFPB (emphasis in original).

7. Ibid. (emphasis in original).

8. Emma Le Conte, *When the World Ended,* 86.

9. BP to ADB, April 14, 1861, ADB file, PCHH.

10. Ibid., April 15, 1861, HL.

11. ADB to BP, April 16, 1861, ADB file, PCHH.

12. There are several references to Peirce's retaining his Southern sympathies after the war, including: Lupold, "From Physician to Physicist," 101; Bruce, *Launching of Modern American Science,* 274, who quotes JH to BP, March 28, 1866, Henry Ms. Smithsonian Institution, "you had always been a democrat and that even in the midst of the war, you had not given countenance to the measures of the extreme view of the North." In addition, Ketner, *His Glassy Essence,* 241, cites a grandson of Peirce who said that the war wasn't popular in the Peirce house. While Peirce no doubt retained some sympathy for the plight of his southern friends, his correspondence certainly indicates a strong loyalty to the Union. The letters of his sister, Charlotte Elizabeth Peirce, which give a remarkably candid picture of the Peirces' domestic life, also describe a strongly pro-Union Peirce. As does Emerson, *Early Years of the Saturday Club,* 245, who says, "In spite of his having been a pro-slavery Democrat with close friendships with many Southerners, after the fall of Sumter Prof. Peirce was a strong Union man." Bowen, *Yankee from Olympus,* 140, also confirms that after the beginning of the war, Peirce was no longer a Southern sympathizer.

13. CEP to Aunt Nichols, June 2, 1861, L 676, CSPP.

14. BP to ADB, April 30, 1861, HL (Peirce's emphasis).

15. CEP to Aunt Nichols, June 9, 1861, L 676, CSPP (emphasis in original). Winfield Scott was seventy-five and in poor health when the war broke out. Lincoln quickly replaced him as commander of the Union army, but most of his strategic plans for winning the war were eventually carried out.

16. CEP to Aunt Nichols, August 23, 1861, L 676, CSPP.

17. Ibid., June 23, 1861, L 676, CSPP.

18. Ibid., August 14, 1861, L 676, CSPP (emphasis in original).

19. Ibid., September 6, 1861, L 676, CSPP (emphasis in original).

20. BP to ADB, May 4, 1861, HL.

21. For example, Lowell's great-grandfather had worded the preamble to the Massachusetts Bill of Rights in such a way as to make slavery forever void in that state [Emerson, *Life and Letters,* 369].

22. Cajori, *Teaching and History of Mathematics,* 143.

23. Bache said he would do everything in his power to aid Lowell (ADB to BP, May 10, 1861, ADB file, PCHH).

24. Charles Russell Lowell to Charles Sumner, April 23, 1861, in Emerson, *Life and Letters,* 201.

25. BP to ADB, June 11, 1861, BPLC.

26. ADB to BP, July 1, 1861, ADB file, PCHH.

27. CSP to ADB, August 11, 1862, National Archives Record Group 23, Records of the Coast Survey, Civil Assistants, F to Z, Roll 249, vol. III, NACP.

28. ADB to CSP, August 16, 1861, National Archives Record Group 23, Records of the Coast Survey, Civil Assistants, F to Z, Roll 249, vol. III, NACP.

29. CEP to Aunt Nichols, August 25–29, 1861, L 676, CSPP.

30. Ibid.

31. Ibid., September 30, 1861, L 676, CSPP.

32. BP to ADB, September 15, 1862, ADB file, PCHH. Probably the newspaper editor and radical Republican, Charles Anderson Dana. He was assistant secretary of war during the Civil War.

33. BP to ADB, September 15, 1862, ADB file, PCHH.

34. Quoted by Emerson, *Early Years of the Saturday Club.* 102.

35. ADB to C. Boutelle, January 3, 1861, Rhees Collection, as quoted by Bruce, *Launching of Modern American Science,* 298.

36. Slotten, *Patronage,* 110.

37. BP to ADB, January 8, 1863, BPLC.

38. Slotten, *Patronage*, 110.

39. *New York Times*, December 20, 1861, 4; see also *New York Times*, September 8, 1861, 4.

40. Bruce, *Launching of Modern American Science*, 298.

41. Ibid., 298–99, BP to ADB, December 23, 1861, BPLC.

42. BP to ADB, July 11, 1861, BPLC.

43. Wagenknecht, *Henry Wadsworth Longfellow*, 161, 172–77; Williams, *Henry Wadsworth Longfellow*, 90–92.

44. BP to ADB, March 27, 1862, BPLC; BP to ADB, May 17, 1863, ADB file, PCHH.

45. Hawkins, *Between Harvard and America*, 3.

46. Archibald et al., "Benjamin Peirce," 2.

47. Corporation Records, vol. 9, 1847–56, May 27, 1854, and June 24, 1854, UAI 530.2, HUA.

48. Hawkins, *Between Harvard and America*, 16, 22.

49. Lurie, *Louis Agassiz*, 177–78, 240.

50. Hawkins, *Between Harvard and America*, 24.

51. Dupree, *Asa Gray*, 315.

52. CEP to Aunt Nichols, April 9, 1862, L 676, CSPP.

53. BP to ADB, October 26, 1862, ADB file, PCHH.

54. Ibid., January 25, 1863, ADB file, PCHH (emphasis in original).

55. Morison, *Three Centuries of Harvard*, 305–6.

56. BP to ADB, December 23, 1862, ADB file, PCHH. William Chauvenet was a leading American mathematician.

57. BP to ADB, February 6, 1863, ADB file, PCHH.

58. Ibid., January 5, 1863, ADB file, PCHH.

59. Bruce, *Launching of Modern American Science*, 292.

60. BP to ADB, March 6, 1863 (two letters of this date), ADB file, PCHH.

61. Hawkins, *Between Harvard and America*, 26.

62. Ibid.,

63. BP to ADB, May 27, 1863, ADB file, PCHH.

64. Ibid., June 7, 1863, ADB file, PCHH.

65. Ibid., June 24, 1863, ADB file, PCHH.

66. Hawkins, *Between Harvard and America*, 26–27.

67. BP to ADB, June 24, 1863, ADB file, PCHH (Peirce's emphasis).

68. Hawkins, *Between Harvard and America*, 29, 35.

69. Bruce, *Launching of Modern American Science*, 301.

70. BP to ADB, May 8, 1854, ADB file, PCHH. It is not clear what "other society" Peirce is referring to, but he clearly saw the limitations of an all-inclusive organization such as the AAAS.

71. Agassiz to Frazer, July 12, 1858, copy in the National Academy of Sciences' Archives, as quoted by Cochrane, *National Academy*, 46.

72. Henry to Agassiz, August 13, 1864, JH file, PCHH. This letter is printed in full in Dupree, "Founding of the National Academy," 438–40, and in Reingold, *Science in Nineteenth-Century America*, 212–16.

73. Dupree, *Science in the Federal Government*, 137; Reingold, *Science in Nineteenth-Century America*, 201; and Reingold, "Science in the Civil War."

74. Cochrane, *National Academy*, 52.

75. Ibid., 53.

76. Quoted by Cochrane, *National Academy*, 53.

77. Cochrane, *National Academy*, 53–56.

78. Ibid.

79. Agassiz to ADB, March 6, 1863, Rhees Collection, HL, in Reingold, *Science in Nineteenth-Century America*, 203.

80. Agassiz to ADB, May 24, 1863, Hayden Papers, National Archives, in Reingold, *Science in Nineteenth-Century America*, 209–10.

81. JH to Stephen Alexander, March 9, 1863, family correspondence, Henry Papers, Smithsonian Archives, in Henry, *Papers*, vol. 10, 296.

82. Fairman Rogers was a civil engineer who worked for the Coast Survey. As evidence of his friendship with Peirce, see BP to ADB, June 24, 1863, ADB file, PCHH.

83. [Dana], "National Academy," 464.

84. BP to SMP, no date [January 1865], SMP file, PCHH.

85. BP to his mother and children April 25, 1863, L664, CSPP.

86. Reingold, *Science in Nineteenth-Century America*, 201.

87. Draper had just published the *Intellectual Development of Europe*, [Draper, *Intellectual Development*] which was decidedly anti-Catholic. In 1874, he published *History of the Conflict between Religion and Science* [Draper, *History*], also anti-Catholic. Draper was eventually elected a member of the National Academy of Sciences in 1877.

88. BP to ADB, January 3, 1863, ADB file, PCHH.

89. [Dana], "National Academy," 465; *New York Times*, May 21, 1863, 4.

90. BP to ADB, May 27, 1863, ADB file, PCHH.

91. Ibid., June 7, 1863, ADB file, PCHH.

92. Ibid., October 18, 1863, ADB file, PCHH.

93. Ibid., January 20, 1864, ADB file, PCHH.

94. Reingold, *Science in Nineteenth-Century America*, 202.

95. Henry to Agassiz, August 13, 1864, JH file, PCHH.

96. A. Guyot to Henry, February 13, 1865, J. P. Lesley Papers, American Philosophical Society, in Reingold, *Science in Nineteenth-Century America*, 221–22.

97. Reingold, *Science in Nineteenth-Century America*, 202.

98. Julius Erasmus Hilgard to William Adams Richardson, December 20, 1870, a copy made by Thomas Hill, March 15, 1871, is in Hill's copy of Peirce's *Linear Associative Algebra*, HUG 1680.138, HUA.

99. A copy of Bache's will is in the superintendents's file, 1866–1910, misc. letters 1867–92, NACP.

100. Dupree, *Science in the Federal Government*, 146–47; Miller, *Dollars for Research*, 125.

101. Emma Le Conte, *When the World Ended*, 28.

102. Joseph Le Conte, "Biographical Memoir," 376; *ANB* (John Le Conte).

103. Stephens, *Joseph Le Conte*, 79.

104. Emma Le Conte, *When the World Ended*, 21.

105. See Joseph Le Conte, *'Ware Sherman* (1937), 5, 8, 11,14, 24, 27; Stephens, *Joseph Le Conte*, 84–88.

106. McPherson, *Battle Cry of Freedom*, 825–29; Emma Le Conte, *When the World Ended*, xi.

107. Emma Le Conte, *When the World Ended*, 4, 6.

108. Stevens, "Sketch," 116; Stephens, *Joseph Le Conte*, 90.

109. Joseph Le Conte, "Biographical Memoir," 376.

110. Emma Le Conte, *When the World Ended*, 44–45.

111. Ibid., 61, (fn). The population of Columbia was about twelve thousand. Emma Le Conte, *When the World Ended*, 74.

112. Joseph Le Conte, *'Ware Sherman* (1999), xx.

113. Ibid., xix.

114. Emma Le Conte, *When the World Ended*, 105, 107; Joseph Le Conte, *'Ware Sherman* (1937), xv.

115. Emma Le Conte, *When the World Ended*, 90.

116. JoLC to BP, September 18, 1865, JoLC file, PCHH.

117. Emma Le Conte, *When the World Ended*, 112 (emphasis in original); Joseph Le Conte, *'Ware Sherman* (1999), xix, xx.

118. Joseph Le Conte, *'Ware Sherman* (1999), xxi.

119. Ibid.

120. Stephens, *Joseph Le Conte*, 99; Lupold, "From Physician to Physicist," 178.

121. Joseph Le Conte, *'Ware Sherman* (1999), xxi.

122. JoLC to BP, May 20, 1867, JoLC file, PCHH (emphasis in original).

123. Stephens, *Joseph Le Conte*, 107.

124. JnLC to P, March 18, 1879, LFPB. This letter indicates that a correspondence was still taking place; see also JnLC to BP, April 8, 1879, JnLC file, PCHH. Unfortunately most of the correspondence from this period is apparently lost.

125. Joseph Le Conte, "Biographical Memoir," 377; Ketner, *His Glassy Essence*, 315–16.

126. Quoted by Stephens, *Joseph Le Conte*, 206.

127. Morison, *Three Centuries of Harvard*, 302–3.

128. Ibid., 303.

129. CEP to Aunt [Nichols], April 10, [1859], L 676, CSPP.

130. Emerson, *Life and Letters*, 36; *ANB*.

131. Emerson, *Life and Letters*, 63.

132. Ibid., 3–71; Brown, *Harvard in the War*, 73–74.

133. Stewart, *Philanthropic Work*, 544; see also *ANB*.

Chapter 12. We Shall Always Be Boys Together

1. BP to ADB, April 12, 1864, ADB file, PCHH. JH to BP, February 25, 1867, SA, 18316, PLB, RU7001.

2. BP to ADB, July 5, 1863, ADB file, PCHH.

3. Ibid., May 25, 1863, ADB file, PCHH. Brent, *Charles Saunders Peirce*, 14–15 and 39–40, says that Peirce had trigeminal neuralgia, but I have found no evidence of this. BP to JoLC, May 3, 1860, LFPB, also indicates that Peirce was not suffering from trigeminal neuralgia. Peirce could, of course, have suffered from both diseases.

4. BP to ADB, January 15, 1864, ADB file, PCHH.

5. Ibid., April 12, 1864, ADB file, PCHH.

6. Ibid., May 2, 1864, and BP to ADB, May 18, 1864, both ADB file, PCHH.

7. Odgers, *Alexander Dallas Bache*, 210.

8. BP to ADB, May 18, 1864, ADB file, PCHH.

9. Ibid., September 11, 1864, ADB file, PCHH.

10. BP to SMP, [no date, probably January 1865], SMP file, PCHH.

11. Wife of the historian and senator, Henry Cabot Lodge.

12. BP to Anna Cabot Lodge, September 23, 1864, Anna Cabot Lodge file, PCHH.

13. JH to BP, July 5, 1866, JH file, PCHH.

14. NFB to BP, July 2, 1864, ADB file, PCHH.

15. Ibid., July 4, 1864, ADB file, PCHH.

16. Ibid., July 8, 1864, ADB file, PCHH.

17. Ibid., August 2, 1864, ADB file, PCHH.

18. Ibid., August 8, 1864, ADB file, PCHH.

19. Ibid., August 9, 1864, ADB file, PCHH. Charles Edward Brown-Séquard, a professor at the Harvard School of Medicine, was one of the world's leading physicians. The French doctor had come to the United States at the urging of Louis Agassiz.

20. NFB to BP, September 29, 1864, ADB file, PCHH.

21. ADB to BP, as dictated to NFB, December 27, 1864, ADB file, PCHH.

22. NFB to BP, January 3, 1865, ADB file, PCHH.

23. Ibid., August 11, 1865, ADB file, PCHH.

24. Ibid., March 2, 1866, ADB file, PCHH.

25. Ibid., April 2, 1866, ADB file, PCHH.

26. Ibid., September 14, 1866, ADB file, PCHH.

27. Ibid., October 14, 1866, ADB file, PCHH.

28. Ibid., November 4, 1866, ADB file, PCHH.

29. Ibid., [late 1866 or early 1867], ADB file, PCHH.

30. SMP to BMP, February 26, 1867, L 687, CSPP.

31. JH to BP, February 25, 1867, SA, 18316, PLB, RU7001; see also JH to Wolcott Gibbs, May 30, 1865, Wolcott Gibbs Papers, Franklin Institute Archives, in Henry, *Papers*, vol. 10, 517–18.

32. JH to William Petit Trowbridge, April 8, 1865, retained copy, Henry Papers, Smithsonian Archives, in Henry, *Papers*, vol. 10, 497.

33. JH to BP, March 28, 1866, JH file, PCHH.

34. Ibid. Peirce's salary at Harvard was $2,400. As superintendent of the Coast Survey, Peirce would have received a salary of $4,500 plus $1,500 for expenses. Henry, *Papers*, vol. 10, 523.

35. CEP to BMP, February 3, 1867, L 676, CSPP.

36. SMP to BMP, February 26, 1867, L 687, CSPP.

37. JH to BP, February 25, 1867, SA, 18316, PLB, RU7001.

38. Hilgard to BP, February 26, [1867], Hilgard file, PCHH.

39. John L. Le Conte, of Philadelphia, was a cousin of Peirce's friend John Le Conte, of Columbia and Berkeley.

40. BP to John L. Le Conte, March 4, 1867, APSL.

41. SMP to BMP, March 1, 1867, L 687, CSPP.

42. CEP to BMP, July 26, 1868, L 676, CSPP.

43. SMP to BMP, March 19, 1867, L 687, CSPP.

44. Angle, *The Lincoln Reader*, 318; the description is by Illinois congressman, Albert G. Riddle.

45. Peirce, H., "The Externals of Washington," 701. Zina's husband, Charles, did not like Washington either; see CSP to BP, September 17, 1872, L333, CSPP.

46. SMP to her children, December 4, 1867, L 687, CSPP.

47. SMP to BMP, March 27, [1870?], L 687, CSPP.

48. See, for example, *Coast Survey Report for 1867*, 42; and *Coast Survey Report for 1868*, 39.

49. BP to JH, December 8, 1867. [Appears to be a draft], JH file, PCHH.

50. Ibid., Peirce's emphasis.

51. BP to JH, December 8, 1867, [appears to be a draft], JH file, PCHH (emphasis in the original).

52. Ibid.

53. BP to Rutherford, November 16, 1867, SCL.

54. CEP to BMP, April 4, 1867, L 676, CSPP.

55. SMP to BMP, March 19, 1877, L 687, CSPP.

56. BAG to BP, May 11, 1867, BAG Jr. file, PCHH.

57. James, "The Dudley Observatory Controversy," 380–81.

58. BP to ADB, December 23, 1857, HL.

59. James, "The Dudley Observatory Controversy," 381.

60. Ibid., 382.

61. BAG to BP, May 11, 1867, BAG Jr. file, PCHH.

62. Ibid.

63. BAG to BP, May 11 (2nd letter), 1867, BAG Jr. file, PCHH.

64. Ibid.

65. Muzzey to BAG, May 13, 1867 (copy); May 14, 1867 (copy); both BAG Jr. file, PCHH.

66. BAG to BP, May 15, 1867, BAG Jr. file, PCHH.

67. Ibid., May 17, 1867, BAG Jr. file, PCHH.

68. BP to BAG, May 18, 1867, BAG Jr. file, PCHH.

69. BP to JH, January 29, 1869, JH file, PCHH.

70. JH to BP, [no date, copy], JH file, PCHH.

71. Peabody to BP, December 4, 1868, A. Peabody file, PCHH.

72. See Corporation Records, vol. 11, 1866–73, December 8, 1868; and Corporation Records, vol. 11, 1866–73, December 26, 1868, UAI 530.2, HUA.

73. James, *Charles W. Eliot*, vol. 1, 184–94.

74. Eliot, "The New Education."

75. Lurie, *Louis Agassiz*, 361–63.

76. William James to Henry James, May 22, 1869, in Perry, *Thought and Character*, vol. 1, 296.

77. Lurie, *Louis Agassiz*, 363.

78. James, *Charles W. Eliot*, vol. 2, 12–13.

79. Brent, *Charles Saunders Peirce*, 106–11.

80. Warner, "Louis M. Rutherford;" *ANB*.

81. SMP to BMP, August 3, 1866, L687, CSPP.

Chapter 13. A Public Functionary

1. Archibald et al., "Benjamin Peirce," 3–4.

2. Lenzen, *Benjamin Peirce and the U. S. Coast Survey*, 26.

3. BP to Secretary [of the Treasury], Hugh McCulloch, August 10, 1870, LPBL.

4. BP to JH, December 8, 1867, JH file, PCHH.

5. JH to BP, December 10, 1867, JH file, PCHH.

6. Hugh McCulloch to Edward Loring, [March or September, probably September] 9, 1868, Hugh McCulloch file, PCHH.

7. JH to BP, December 10, 1867, JH file, PCHH.

8. BP to JH, January 15, 1868, HL.

9. Fisch, "Decisive Year," xxii.

10. Sherwood, "George Davidson," 253–54.

11. James, *First Scientific Exploration*.

12. BP to Davidson, May 27, 1867, Records of the Coast and Geodetic Survey, entry 22, "Superintendent's File," 1866–1905, George Davidson 1866–69, NACP.

13. *New York Herald*, April 29, 1867.

14. SMP to BMP, April 23, 1867, L 687, CSPP.

15. King, "George Davidson," 179–80; Lewis, *George Davidson*, 23–29, 35.

16. *Coast Survey Report for 1867*, 188.

17. BP to Davidson, May 5, 1867, DPBL.

18. Ibid.

19. Captain Bryant had been a whaler before becoming an influential member of the Massachusetts legislature. BP to Davidson, May 26, 1867, Records of the Coast and Geodetic Survey, entry 22, "Superintendent's File," 1866–1905, George Davidson 1866–69, NACP.

20. BP to Davidson, May 21, 1867, DPBL.

21. Ibid., June 9, 1867, LPBL.

22. King, "George Davidson," 193.

23. BP to Davidson, November 30, 1867, LPBL.

24. See, for example, Davidson to BP, May 1, 1865, May 6, 1865, and May 14, 1865, all George Davidson file, PCHH.

25. BP to Davidson, January 8, 1868, LPBL.

26. Ibid., February 7, 1868, LPBL.

27. *Russian America*, Executive Document No. 177, House of Representatives, 40th Congress, 2nd session (1868). Davidson's report comprises pp. 219–314. The document contains many other reports from the Coast Survey expedition.

28. BP to Davidson, June 1, 1867, DPBL.

29. *Congressional Globe*, 40th Congress, 2nd session 1868, vol. 39, app., 430. For similar comments see, for example, *Congressional Globe*, 40th Congress, 2nd session, 1868, vol. 39, app., 85; and *Congressional Globe*, 40th Congress, 2nd session 1868, vol. 39, app., 403.

30. Walker was an adept politician with strong expansionist views. He was hired by the Russian minister to the United States to lobby Congress to support the Alaskan purchase. He later advised President U. S. Grant to annex all of British North America.

31. Peirce, B. M., *Report*.

32. Fisch, "Decisive Year," xxii.

33. *Congressional Globe*, 40th Congress, 2nd session, 1868, part 5 and app., 399.

34. *Coast Survey Report for 1868*, app. no. 15, 260–61.

35. Now called San Juan del Norte.

36. *Coast Survey Report for the Year 1874*, 135.

37. *Coast Survey Report for the Year 1871*, 1.

38. *Coast Survey Report for 1870*, 229.

39. *Coast Survey Report for 1869*, 5.

40. Ibid., 6.

41. CEP to Cousin Sarah, August 13, 1869, L 676, CSPP.

42. Fisch, "Decisive Year," xxii–xxiii.

43. BP to Hon. John Bingham, House of Representatives, March 8, 1870, superintendent's file, 1866–1910, Congress, army, and navy, 1867–77, NACP.

44. *Congressional Globe*, 42, part 3, p. 2754. See also *Coast Survey Report for 1870*, 229.

45. BP to Hon John A Bingham, May 15, 1870, Coast Survey file, PCHH.

46. Ibid.

47. Bingham to BP, May 23, 1870, superintendent's file, 1866–1910, Congress, army, and navy, 1867–77, NACP.

48. Hilgard to John A. Bingham, May 23, 1870, Superintendent's file, 1866–1910, Congress, army, and navy, 1867–77, NACP.

49. BP to John A. Bingham, May 26, 1870, superintendent's file, 1866–1910, Congress, army, and navy, 1867–77, NACP.

50. Ibid.

51. Ibid.

52. Bingham to BP, August 29, 1870, superintendent's file, 1866–1910, Congress, army, and navy, 1867–77, NACP.

53. BP to John A. Bingham, March 8, 1870, 41st Congress, 2nd session, House miscellaneous doc. vol. 3, no. 127, Washington, 1870, NACP.

54. BP to Rutherford, September 18, 1870, SCL; see also, BP to Rutherford, September 11, 1870, SCL.

55. CSP to BP, July 12, 1870, L 333, CSPP; and Hilgard to BP, July 18, 1870, The Coast and Geodetic Survey, "Superintendents File" 1866–1910, assistant in charge, 1866–76, NACP.

56. Lenzen, *Benjamin Peirce and the U. S. Coast Survey*, 37.

57. BP to Lockyer, November 1870, in Lenzen, *Benjamin Peirce and the U. S. Coast Survey*, 37. Lockyer was later knighted, and so became Sir Norman Lockyer.

58. Lockyer and Lockyer, *Life and Work*, 56.

59. Ibid., 56–57; Newcomb, *Reminiscences*, 274–77.

60. BP to SMP, November 3, [1870], SMP file, PCHH.

61. Lockyer and Lockyer, *Life and Work*, 61; *Coast Survey Report for 1870*, 229.

62. *Coast Survey Report for 1870*, 144.

63. Ibid., 231.

64. Fisch, "Decisive Year," xxxv.

65. BP to Davidson, February 11, 1871, LPBL; see also, *Coast Survey Report for 1870*, 231.

66. CEP to BP, February 8, 1871, L 676, CSPP.

67. Whitney, "Surveys," 67.

68. Weber, *Coast and Geodetic Survey*, 8–9.

69. Burger, "Contributions," 82.

70. James Hall to BP, April 18, 1868; BP to James Hall, May 18, 1868, and June 6, 1868, all Box 8, Series B0561, State Museum Director's, State Geologist's, and State Paleontologist's Correspondence files, 1828–1944, in New York State Archives.

71. BP to LA, February 18, 1871, BMS Am 1419 (567), by permission of the Houghton Library, Harvard University.

72. LA to BP, February 24, 1871, by permission of the Ernst Mayr Library, Museum of Comparative Zoology Archives, Harvard University.

73. LA to BP, February 26, 1871, LA file, PCHH.

74. Elizabeth C. Agassiz described the trip in three articles in the *Atlantic Monthly*. See Agassiz, E., "Hassler Glacier," "Straits of Magellan," and "Galapagos." Louis de François Pouratlès recounted the voyage in the *Coast Survey Report for 1872*, 213–21.

75. Lurie, *Louis Agassiz*, 372–73.

76. LA to Carl Gegenbaur, quoted by Lurie, *Louis Agassiz*, 373.

77. Charles Darwin to Alexander Agassiz, June 1, 1871, in Agassiz, G., *Letters and Recollections*, 119.

78. LA to BP, December 2, 1871, in Agassiz, L., "Concerning Deep-Sea Dredgings."

79. Ibid., Agassiz, E., *Louis Agassiz*, 692–93.

80. Agassiz, E., *Louis Agassiz*, 689–764.

81. LA to BP, July 29, 1872, in Agassiz, E., *Louis Agassiz*, 762–63.

82. See, for example, LA to BP, April 27, 1872, and July 29, 1872, in Agassiz, E., *Louis Agassiz*, 751–53; 756–56; Lurie, *Louis Agassiz*, 375.

83. LA to BP, January 16, 1872, and July 29, 1872, in Agassiz, E., *Louis Agassiz*, 703–58.

84. LA to BP, October 22, 1872, LA file, PCHH.

85. Manning, *Coast Survey vs. Naval Hydrographic Office*, 40–41.

86. Ibid., 41.

87. Weber, *Coast and Geodetic Survey*, 2–4.

88. Dupree, *Science in the Federal Government*, 101–2; Weber, *Coast and Geodetic Survey*, 7.

89. Hilgard to BP, November 5, 1868, Records of The Coast and Geodetic Survey, Superintendents File, 1866–1910, assistant in charge, 1866–76, NACP.

90. BP to Patterson, July 10, 1868, Records of the Coast and Geodetic Survey, "Superintendent's File," 1866–1910, hydrographic Inspector, 1867–74, NACP.

91. Hilgard to BP, November 5, 1868, Records of the Coast and Geodetic Survey, "Superintendents File," 1866–1910, assistant in charge, 1866–76, NACP (emphasis in the original).

92. G. F. Edmonds to BP, August 12, 1870, Superintendent's File, 1866–1910, Congress, army, and navy, 1867–77, NACP; Manning, *Coast Survey vs. Naval Hydrographic Office*, 4.

93. Barry, *Rising Tide*, 32–91.

94. BP to Patterson, August 22, 1870, Records of the Coast and Geodetic Survey, "Superintendent's File," 1866–1910, hydrographic inspector, 1867–74, NACP (emphasis in the original).

95. *Boston Evening Journal*, January 24, 1872. The Lighthouse Board was established provisionally in 1851 and permanently a year later. It consisted of two civilians and six military officers. Henry and Bache were the original civilian members. Peirce replaced Bache, after his death. See Henry, *Papers*, vol. 6, 379 fn.

96. *National Republican* (Washington, D. C.), January 24, 1872.

97. CSP to BP, September 18, 1872, L 333, CSPP (emphasis in the original). William Harkness was a professor of mathematics with the U. S. Navy. Most of his career was at the Naval Observatory.

98. Agassiz to The Honorable Henry Wilson, February 8, 1873, Records of the Coast and Geodetic Survey, "Superintendent's File," 1866–1910, Congress, army, and navy, 1867–77, NACP.

99. Agassiz to the Honorable Charles Sumner, February 8, 1873, Records of the Coast and Geodetic Survey, "Superintendent's File," 1866–1910, Congress, army, and navy, 1867–77, NACP.

100. Agassiz to the Honorable Carl Schurz, February 8, 1873, Records of the Coast and Geodetic Survey, "Superintendent's File," 1866–1910, Congress, army, and navy, 1867–77, NACP.

101. Agassiz to Dr. Heck, February 8, 1873, Records of the Coast and Geodetic Survey, "Superintendent's File," 1866–1910, Congress, army, and navy, 1867–77, NACP. I have been unable to identify Dr. Heck.

102. LA to BP, February 9, 1873, LA file, PCHH.

103. BP to JH, February 10, 1873, SA, 17497, RU 7001.

104. BP to Josiah Whitney, February 10, 1873, William Dwight Whitney Family Papers, YUL.

105. Ibid., March 7, 1873, William Dwight Whitney Family Papers, YUL.

106. CSP to BP, June 18, 1873, L333, CSPP.

107. Bruce, *The Launching of Modern American Science*, 318.

108. BP to Davidson, May 31, 1870, LPBL (emphasis in the original).

109. BP to Simon Newcomb, October [1], 1872, Simon Newcomb Papers, container # 34, Manuscript Division, Library of Congress.

110. BP to JH, October 7, 1873, SA 17514, RU 7001 (emphasis in the original).

111. *Coast Survey Report for 1874*, 13–14.

112. BP to CEP, February 16, 1874, CEP file, PCHH (emphasis in the original). See also SMP to CEP, [February 1874], L 687, CSPP.

113. Patterson to BP, December 3, 1874, C. Patterson file, PCHH (emphasis in the original).

114. CSP to BP, June 19, 1872, L 333, CSPP.

115. SMP to CEP, [February 1874], L 687, CSPP.

CHAPTER 14. MY MOST PRECIOUS PEARLS

1. See, for example, BP to Wm. B. Sprague Jr., September 22, 1855, Gratz Collection, case 7, box 38, The Historical Society of Pennsylvania; BP to Bache, December 17, 1855, ADB file, PCHH; BP to ADB, September 25, 1855, ADB file, PCHH; BP to SMP, February 22, 1847, SMP file, PCHH.

2. BP to JoLC, December 22, 1854, LFPB.

3. The American Association for the Advancement of Science.

4. BP to ADB, October 25, 1854, ADB file, PCHH.

5. Peirce is referring to Joseph Henry, whose nickname among the Lazzaroni was Smithson, or Smithsonian.

6. BP to JoLC, December 14, 1856, LFPB.

7. JH to ADB, December 1, 1856, Bache Papers, Smithsonian Archives, in Henry, *Papers*, vol. 9, 417–18.

8. BP to ADB, December 3, 1856, ADB file, PCHH.

9. BP to JoLC, December 14, 1856, LFPB.

10. I have not found a copy of these lectures; Peirce spoke extemporaneously without notes. Peirce does mention that some transcriptions were made of them. Correspondence gives a good indication of their content.

11. ADB to SMP, [after January 16, 1857], copy in SMP file, PCHH. This letter contains a transcription of a favorable newspaper account.

12. BP to SMP, January 19, 1857, SMP file, PCHH.

13. Ibid., February 3, 1857, SMP file, PCHH.

14. Lurie, *Louis Agassiz*, 126–28.

15. [Hill], "Scientific Meeting."

16. Hitchcock, *Religion of Geology*. Hovenkamp, *Science and Religion*, and Numbers, *Creation by Natural Law*, discuss the relation between science and religion during Peirce's lifetime.

17. Peirce, *A System of Analytical Mechanics*, 477.

18. Peirce, "Working Plan," 11.

19. Peirce, *Ideality*, 133.

20. Peirce, "Working Plan," 14.

21. Archibald et al., "Benjamin Peirce," 5–6.

22. [Peirce and Palfrey], "Bowditch's *Translation*," 175–77.

23. Peirce, "Working Plan," 14; Peirce, "Remarks on a National University," 90.

24. Peirce, *A System of Analytical Mechanics*, 28, 31, 476–77.

25. BP to Quincy, October 14, 1835, Corporation Papers, 2nd series, 1835–44, box 3, UA I.5.130, HUA.

26. Peirce, *Ideality*, 10.

27. BP to ADB, December 30, 1855, ADB file, PCHH; BP to James Dwight Dana, July 11, 1856, Dana Family Papers, YUL; BP to JoLC, December 14, 1858, and January 9–10, 1859, LFPB.

28. Corporation Records, vol. 11, 1866–73, October 24, 1868, UAI 530.2, HUA.

29. Sargent, *Sketches and Reminiscences*, 243–50, 339–42, and 376–80; BP to JoLC,

October 9, 1859, and October 16, 1859, LFPB. Manuscript notes for this lecture, dated June 12, 1858, through May 1859, are in *54m-264, box 8 of 11, misc., PPHH.

30. Peterson, "Benjamin Peirce: Mathematician and Philosopher," gives a detailed analysis of Peirce's religious and philosophical views, and how his opinions changed over time.

31. BP to SMP, January 29, 1857, SMP file, PCHH.

32. BP to JoLC, December 14, 1858 and January 9–10, 1859, LFPB.

33. BP to SMP, January 30, 1857, SMP file, PCHH. Being introduced from the floor of the Senate was a great honor.

34. BP to SMP, February 6, 1857, SMP file, PCHH.

35. Ibid., January 17, 1857, SMP file, PCHH.

36. Ibid., January 28, 1857, SMP file, PCHH.

37. Ibid., January 24, 1857, SMP file, PCHH.

38. Archibald et al., "Benjamin Peirce," 12; Newcomb, *Popular Astronomy*, 350, 362, 363, and 403; discusses some of Peirce's astronomical work.

39. BP to Patterson, November 25, 1877, Records of the Coast and Geodetic Survey "Superintendent's File," 1866–1910, Private Correspondence, 1874–77, 1881–85, NACP.

40. Peirce, "Criterion;" Stigler, "Mathematical Statistics," 244–46; Stigler, "Simon Newcomb," 873; Sheynin, "Mathematical Treatment," 111; Anscombe, "Rejection of Outliers," 125–26; Archibald et al., "Benjamin Peirce," 13-14; and Rider, *Criteria*, 3–8.

41. One man who was not intimidated was quickly put in his place by the brilliant Chauncey Wright, who worked in the *Nautical Almanac* office. See [Wright], "Mathematics in Court."

42. Meier and Zabell, "Peirce and the Howland Will;" James, *Notable American Women*,, 81–83; Sparks and Moore, *Hetty Green;* Flynn, *Men of Wealth*, 215–61; and *54m-264, box 3 of 11, misc., PPHH.

43. BP to President Quincy, February 14, 1845, Josiah Quincy Papers, UAI 15.882, HUA. This letter is not in Peirce's hand, but appears to be in his wife's. A draft of the letter, parts of which are in Peirce's hand and parts of which are in his wife's, is in L 664, CSPP.

44. BP to James Dwight Dana, November 27, 1876, Dana Family Papers, YUL.

45. [Peirce and Palfrey], "Bowditch's *Translation*," 145.

46. Peirce, *Linear Associative Algebra* (1870), 1.

47. BP to George Bancroft, November 8, 1870, Benjamin Peirce Miscellaneous File, Manuscripts and Archives Division, New York Public Library, Astor, Lenox and Tilden Foundations. This letter was published in Ginsburg, "Hitherto Unpublished Letter," 280.

48. BP to Hugh McCulloch, August 10, 1870, LPBL.

49. Nagel, "Impossible Numbers," 432–35; Richards, "Art and Science," 346. Peirce himself referred to mathematics as the science of quantity; see [Peirce and Palfrey], "Bowditch's *Translation*," 144.

50. Pycior, "Benjamin Peirce's *Linear Associative Algebra*," 538.

51. Nagel, "Impossible Numbers," 435; Pycior, "George Peacock," 27–28.

52. Nagel, "Impossible Numbers," 432–35.

53. Peacock, *Treatise on Algebra, Vol. 1, Arithmetical Algebra*, iv-vi..

54. Peacock, *Treatise on Algebra* (1830), 71.

55. Peacock, *Treatise on Algebra, Vol. 2, On Symbolical Algebra*, 453.

56. Katz, *History of Mathematics*, 680; Nagel, "Impossible Numbers," 449, 455.

57. Richards, "Art and Science," 354.

58. Peirce, "Linear Associative Algebra (1881)," 107; BP to John L. Le Conte, October 23, 1856, Le Conte Papers, APSL.

59. Nagel, "Impossible Numbers," 464.

60. Burton, *History of Mathematics,* 297–99.

61. Parshall, "Joseph Wedderburn," 227. Others, including John Warner, John Wallis, Casper Wessel, and Karl F. Gauss had interpreted complex numbers as two-dimensional space.

62. Hankins, *Hamilton,* 263–64.

63. Ibid., *Hamilton,* 283–84.

64. Ibid., *Hamilton,* 291–93.

65. Katz, *History of Mathematics,* 686.

66. Nagel, "Impossible Numbers," 464.

67. Archibald et al., "Benjamin Peirce," 6. Peirce expressed his enthusiasm for quaternions many times; see, for example, Peirce, *A System of Analytical Mechanics,* 476–77.

68. Cajori, *Teaching and History of Mathematics,* 137.

69. Pycior, "Benjamin Peirce's *Linear Associative Algebra,*" 541.

70. Crowe, *History of Vector Analysis,* 127. Cajori, *Teaching and History of Mathematics;* George Salmon to W. R. Hamilton, June 16, 1865, in Graves, *Life of Hamilton,* vol. 3, 207–8.

71. [Newton], "Benjamin Peirce," 174.

72. Crowe, *History of Vector Analysis,* 126.

73. [Hill], "Imagination in Mathematics."

74. Peirce, *Ideality,* 28.

75. CEP to Aunt Nichols, [May?] 2, 1858, L 676, CSPP (emphasis in the original).

76. Peirce, *Ideality,* 29.

77. Peirce gives only the algebras of dimension six that contain an idempotent element.

78. For a more detailed description of Peirce's *Linear Associative Algebra,* see Parshall, "Joseph H. M. Wedderburn," 250–58.

79. BP to George Bancroft, November 8, 1870, Benjamin Peirce Miscellaneous File, Manuscripts and Archives Division, New York Public Library, Astor, Lenox and Tilden Foundations.

80. Peirce, *Linear Associative Algebra* (1870), 2. Peirce's definition is remarkably similar to one given by Alfred North Whitehead eighteen years later: "Mathematics in its widest signification is the development of all types of formal, necessary, deductive reasoning." Whitehead, *A Treatise on Universal Algebra,* vol. 1, vi. Whitehead may well have been influenced by Peirce's definition; he cites Peirce's *Linear Associative Algebra* in his preface. Whitehead, *A Treatise on Universal Algebra,* vol. 1, v.

81. Quoted by Lenzen, "Contributions," 239–40.

82. Archibald, "Benjamin Peirce's Linear Associative Algebra," 526. Peirce responded to this objection in Peirce, "Linear Associative Algebra (1881)," [105 fn.]. His son said his father's response was pure bosh.

83. Archibald et al., "Benjamin Peirce," 6. See also Lenzen, "Contributions," 239–40.

84. Peirce, "Uses and Transformations," 395–96; reprinted in Peirce, "Linear Associative Algebra (1881)," 216–17 (emphasis in the original).

85. Parshall, "Joseph Wedderburn," 253–54 fn. Clifford concurs with Peirce in a foot note.

86. An undated note in the National Academy file, PCHH, lists the topics Peirce presented at the National Academy of Sciences meeting in Hartford, Connecticut, on August 13, 1867.

87. Julius Erasmus Hilgard to William Adams Richardson, December 20, 1870, a copy made by Thomas Hill, March 15, 1871, is in Hill's copy of Peirce's *Linear Associative Algebra,* HUG 1680.138, HUA.

Others also found Peirce's presentations at scientific meetings unintelligible and were not as charitable as Agassiz. After attending the Providence meeting of the AAAS in 1855, William B. Rodgers wrote his brother, "We had as usual many scraps from the Coast Survey and many unintelligible things from Peirce." William B. Rodgers to Henry D. Rodgers, [August 1855], Rodgers Papers, Archives, Massachusetts Institute of Technology, as quoted in the Henry, *Papers*, vol. 9, 281.

88. Grattan-Guiness, "New Light," 606.

89. Julius Erasmus Hilgard to William Adams Richardson, December 20, 1870, copy made by Thomas Hill, March 15, 1871, is in Hill's copy of Peirce's *Linear Associative Algebra*, HUG 1680.138, HUA. *Linear Associative Algebra* was published posthumously in the *American Journal of Mathematics* (Peirce, "Linear Associative Algebra [1881]") and as a monograph, (Peirce, *Linear Associative Algebra* [1882]), both edited by Charles S. Peirce. It was also reprinted, along with other material related to Peirce, in Cohen, *Benjamin Peirce*.

90. CSP to BP, July 12, 1870, L 333, CSPP.

91. Anon., *Proceedings*, 220; Brent, *Charles Saunders Peirce*, 79.

92. Spottiswoode, "Remarks," 150–51.

93. Cayley, "On Multiple Algebra," 459, 465.

94. BP to George Bancroft, November 8, 1870, Benjamin Peirce Miscellaneous File, Manuscripts and Archives Division, New York Public Library, Astor, Lenox and Tilden Foundations.

95. Ginsburg, "Hitherto Unpublished Letter," 2278–80.

96. Peirce, C., "Algebras" and "Relative Forms;" see also, Pycior, "Benjamin Peirce's *Linear Associative Algebra*," 547,

97. Peirce, "Linear Associative Algebra (1881)," 97.

98. CSP to JMP, May 22, 1881, L 339, CSPP.

99. Ginsburg, "Hitherto Unpublished Letter," 279.

100. Cayley, "On Multiple Algebras," 459.

101. [Newton], "Benjamin Peirce," 175.

102. Hawkes, "Estimate."

103. Hawkes, "Hypercomplex Number Systems," 313.

104. Grattan-Guinness, "New Light," 599 gives a detailed listing of this work.

105. Whitehead, Treatise on Universal Algebra, vol. 1, v.

106. Pycior, "Benjamin Peirce's *Linear Associative Algebra*," 547.

107. Wedderburn, "Hypercomplex Numbers," 78, 92; Parshall, "Joseph Wedderburn," 256. Wedderburn, however, questioned the importance of Peirce's *Linear Associative Algebra*. See Wedderburn to Flexner, February 13, 1932, L 726, CSPP.

108. Pycior, "Benjamin Peirce's *Linear Associative Algebra*," 551; Bourbaki, *History*, 118; Novy, "Concept of Linear Algebra," 212; Bell, *Development of Mathematics*, 246; Schlote, "Zur Geschichte der Algebrentheorie," 16; Harkin, "Development of Modern Algebra," 21; Artin, "Influence," 65; Archibald et al., "Benjamin Peirce," 15; Shaw, *Synopsis*, 6; Lenzen, "Contributions," 240; and Grattan-Guiness, "New Light," 599–600.

109. Rudolph, *The American College*, 244, 270–72; Hawkins, *Pioneer*, especially chapters 1 and 2; Birkhoff, "Fifty Years," 272–73; Cordasco, *Daniel Coit Gilman;* and French, *Johns Hopkins*.

110. Rice, "Mathematics in the Metropolis," 383–84.

111. Parshall and Rowe, *Emergence*, 65, esp. fn. 44.

112. Sylvester to Henry, April 12, 1846, in Parshall, *James Joseph Sylvester*, 15–18.

113. Parshall, "Joseph Wedderburn," 240; Sylvester to Lord Brougham, July 10, 1855, and September 16, 1855, in Parshall, *James Joseph Sylvester*, 86–87, 89–90.

114. Rice, "Mathematics in the Metropolis," 398, 402.

115. Parshall, *James Joseph Sylvester*, 83.

116. Parshall and Rowe, *"American Science Comes of Age,"* 7.

117. Sylvester to Barbara Bodichon, September 19, 1875, in Parshall, *James Joseph Sylvester*, 144–45.

118. BP to Gilman, September 11, 1875, Daniel Coit Gilman Papers, ms. 1, Special Collections and Archives, Johns Hopkins University.

119. BP to Gilman, [no date], Daniel Coit Gilman Papers, ms. 1, Special Collections and Archives, Johns Hopkins University. "1878?" has been written on the letter, but 1878 seems too late. The letter was most likely written right after Sylvester's appointment in 1876.

120. Hawkins, *Pioneer*, 42–43.

121. Sylvester to Barbara Bodichon, August 21, 1876, in Parshall, *James Joseph Sylvester*, 154–55.

122. Parshall and Rowe, *"American Science Comes of Age,"* 10.

123. Sylvester to BP, July 14, 1876, James Joseph Sylvester Papers, Brown University Library.

124. JEM to CSP, undated fragment, [summer 1876], L 339, CSPP. Sylvester was an avid poet. See Sylvester to SMP, March 25, 1880, Sylvester file, PCHH; and Smith, "Sylvester as a Poet."

125. Sylvester to Gilman, June 12, 1876, in Parshall, *James Joseph Sylvester*, 152–54. The title of the position was "associate," but it was equivalent to a modern assistant professorship; see Cooke and Rickey, "W. E. Story," 35.

126. Story to JEM, August 1, 1876, William Story file, PCHH.

127. Cooke and Rickey, "W. E. Story," 34, and Hawkins, *Pioneer*, 44–45.

128. Story to JEM, August 1, 1876, William Story file, PCHH.

129. Cooke and Rickey, "W. E. Story," 35, 49.

130. Parshall, "America's First School."

131. Eliot, "Address;" see also Rudolph, *The American College*, 449.

132. Gilman, *Launching*, 115–16.

133. Cordasco, *Daniel Coit Gilman*, 107.

134. BP to Sylvester, November 19, 1876, Cambridge, by permission of the master and fellows of St. Johns College, Cambridge.

135. Sylvester to BP, November 19, 1877, Sylvester file, PCHH. Simon Newcomb and George William Hill were both mathematical astronomers; both had also worked under Peirce at the *Nautical Almanac*.

136. Story to BP, November 24, 1877, William Story file, PCHH.

137. Sylvester to BP, March 17, 1878, Sylvester file, PCHH.

138. Story to BP, October 18, 1878, William Story file, PCHH.

139. Peirce's *Linear Associative Algebra* was published in the *American Journal of Mathematics*, after Peirce's death in 1881.

CHAPTER 15. MY BEST LOVE

1. BP to JnLC, April 8, 1859, LFPB.

2. Emerson, *Early Years of the Saturday Club*. See also Dana, *Journal*, vol. 2, 830.

3. Story was the son of Justice Joseph Story, Lydia Nichols's suitor.

4. Peirce's son Charles lists many of his father's friends in Hardwick, *Semiotic and Significs*, 113–14; ms 619:03ff; and ms 296:41ff, CSPP. See also Howe, *Later Years of the Saturday Club*, 107–16; and Peabody, *Harvard Reminiscences*, 180–86.

5. CEP to Aunt Nichols, July 25, 1858, L 676, CSPP. *New York Times*, October 30, 1857, gives a description of Lady Napier.

6. Ibid., November 23, 1861, L 676, CSPP (emphasis in the original).

7. CEP to BMP, July 21, 1868, L 676, CSPP.

8. CEP on Aunt Saunders's will, April 10, 1872, L 676, CSPP (emphasis in the original).

9. Fragment undated and unsigned, perhaps by HPE's (Helen Peirce Ellis) husband, William Ellis, CSPP, as quoted by Brent, *Charles Saunders Peirce.*

10. Lodge, *Early Memories,* 55–56. Peirce was often at the Lodge's home.

11. Sargent, *Sketches and Reminiscences,* 370.

12. Emerson, *Early Years of the Saturday Club,* 102.

13. Archibald et al., "Benjamin Peirce," 7.

14. SMP to BMP, October 9, 1866, L 687, CSPP. For another example of Sarah's troubles with servants, see SMP to BMP, March 6, 1870, L 687, CSPP.

15. SMP to BMP, March 14, 1870, L 687, CSPP (emphasis in the original).

16. CEP to Aunt [Nichols], July 13, 1857, L 676, CSPP.

17. CEP to Aunt Nichols, October 4, [1857?], L 676, CSPP.

18. C. H. Davis to BP, July 20, 1848, C. H. Davis file, PCHH.

19. SMP to CEP, [letter fragment, no date], L 687, CSPP (emphasis in the original).

20. BP to SMP, August 24, [1855], SMP file, PCHH.

21. JoLC to BP, [no date, probably 1858], JoLC file, PCHH.

22. CEP to Aunt [Nichols], March 16, [1856?], L 676, CSPP.

23. BP to JoLC, October 23, 1854, LFPB (emphasis in the original).

24. Ibid., August 21, 1859, LFPB.

25. BP to SMP, February 24, 1858, SMP file, PCHH.

26. JnLC to BP, March 14, 1858, LFPB; see also Stevens, "Sketch," 114. Henry described the relationship between Benjamin and Josephine as a "platonic flirtation." JH to Harriet Henry, August 1[3], [1859], in Henry, *Papers,* vol. 10, 98–102.

27. JnLC to BP, July 4, 1858, LFPB.

28. BP to JnLC, April 1, 1857, LFPB.

29. BP to ADB, January 29, 1846, ADB file, PCHH.

30. CEP to Aunt [Nichols], April 1, 1860, L 676, CSPP.

31. BP to Loomis , December 24, 1845, YB. See also BP to Loomis, January 22, 1846, YB.

32. BP to C. P. Huntington, April 2, 1844, L 664, CSPP.

33. CEP to Aunt Nichols, January 13, [1856], L 676, CSPP.

34. SMP to BMP, April 30, 1867, L 687, CSPP.

35. Ibid., July 16, 1867, L 687, CSPP (emphasis in the original).

36. Ibid., October 17, [1869?], L 687, CSPP.

37. Ibid.

38. BP to CEP, June 25, 1874, CEP file, PCHH.

39. Herbert was also called Berts.

40. SMP to HPE, [no date], L 687, CSPP.

41. SMP to BMP, September 11, [1866?], L 687, CSPP.

42. Ibid., July 16, 1867, L 687, CSPP.

43. Harvard Faculty Records, vol. 16, 1860–65, p. 40, UAIII 5 5.2, HUA.

44. BP to BMP, July 29, 1869, L 664, CSPP.

45. SMP to BMP, October 17, [1869?], L 687, CSPP.

46. Ibid., September 11, 1869, L 687, CSPP (emphasis in the original).

47. Ibid., April 30, 1869, L 687, CSPP (emphasis in the original).

48. *National Cyclopedia of American Biography,* vol. 9, 539–40. New York: James T. White and Co. (1907); Ketner, *His Glassy Essence,* 16; and Brent, *Charles Saunders Peirce,* 77.

49. Whittemore, "James Mills Peirce;" Byerly, "James Mills Peirce;" and Kennedy, "Towards a Biography."

50. CEP to SMP, August 13, 1840, L 676, CSPP.

51. Corporation Records, vol. 9, 1847–56, May 27, 1854 and June 24, 1854, UAI 530.2, HUA.

52. BP to JoLC, October 2, 1859, LFPB.

53. CEP to Aunt Nichols, November 13, 1859, L 676, CSPP.

54. BP to JoLC, January 8, 1860, LFPB.

55. JnLC toBP, February 5, 1860, LFPB (emphasis in the original).

56. See JMP to CSP, May 2, 1860, as cited by Ketner, *His Glassy Essence,* 196.

57. SMP to BMP, June 19, [1869], L 687, CSPP.

58. Ibid., June 27, 1869, L 687, CSPP.

59. Byerly, "James Mills Peirce," 574.

60. See, for example, JMP to BP, August 17, 1869, James Mills Peirce file, PCHH.

61. JMP to BP, August 17, 1869, L 667, CSPP.

62. Coolidge, "The Story of Mathematics at Harvard," 374.

63. BP to ADB, May 17, 1863, ADB file, PCHH; see also, Ketner, *His Glassy Essence,* 254; and Brent, *Charles Saunders Peirce,* 78.

64. BP to JoLC, August 27, 1857, LFPB.

65. Ibid., May 22, 1859; BP to JnLC, August 28, 1859, LFPB.

66. BP to JoLC, August 25, 1857, LFPB.

67. BP to ADB, February 10, 1856, ADB file, PCHH. Theophilus Parsons was Dane Professor of Law at Harvard.

68. BP to JoLC, March 3, 1858, LFPB.

69. CEP to Aunt Nichols, January 13, [1856], L 676, CSPP.

70. Morison, *Three Centuries of Harvard,* 177. ZP to BMP, March 10, [1863], L 687, CSPP.

71. BP to BMP, June 7, 1863, L 668, CSPP.

72. Benjamin Mills Peirce's offenses are listed in the Harvard Faculty Records, vol. 16, 1860–65, pp. 176, 225, 238, 245, 261, 264, 273, 285, 386, 441, 461, and 506; UA III 5.5.2, HUA.

73. The *National Cyclopedia of American Biography,* s. v. "Benjamin Mills Peirce," says that Benjamin Mills Peirce studied at the School of Mines in Paris.

74. While employed by the Coast Survey, Charles went to Europe to study gravimetric, but not at any single institution; see Brent, *Charles Saunders Peirce,* 95.

75. CEP to BMP, [November?] 18, 1866, L 676, CSPP.

76. Lodge, *Early Memories,* 159–61.

77. *National Cyclopedia of American Biography,* vol. 10, 449. New York: James T. White and Company.

78. BP to BMP, June 18, 1868, L 664, CSPP.

79. Ibid., note added to a letter dated September 8, 1869, Herbert to BMP, L 664, CSPP.

80. SMP to BMP, September 11, 1869, L 687, CSPP.

81. See, for example, SMP to BMP, September 16, 1868, and July 3, [1869?], L 687, CSPP.

82. SMP to BMP, January 13, 1870, L 687, CSPP.

83. Ibid., February 21, 1870, written on back of BP to BMP, March 1, 1870, L 664, CSPP.

84. BP to BMP, March 1, 1870, L 664, CSPP.

85. SMP to BMP, March 14, 1870, L 687, CSPP.

86. Brent, *Charles Saunders Peirce*, 340.

87. BP to JMP, April 25, 1870, telegram, L 664, CSPP.

88. This is the opinion of Max H. Fisch, who studied Charles Saunders Peirce extensively, but failed to write a planned biography of him. Fisch wrote, "Who is the most original and the most versatile intellect that the Americas have so far produced? The answer 'Charles S. Peirce' is uncontested, because any second would be so far behind as not to be worth nominating." See Fisch's "Introductory Note," in Sebeok, *The Play of Musement*, 17.

89. BP to LRP, September 10, 1839, LRP file, PCHH.

90. BP to Charles Henry Peirce and CEP, September 10, 1839, Charles Henry Peirce file, PCHH.

91. Eisele, *Studies*, 1.

92. BP to SMP, July 30, 1855, SMP file, PCHH.

93. Obituary prepared by [HPE and HDP], *Boston Evening Transcript*, May 16, 1914.

94. Ms 1608, CSPP.

95. Ms 619, CSPP.

96. Ms 1606: 02, CSPP.

97. Fisch, "Introduction," xvii–xviii. Leibig developed an experimental approach to teaching chemistry.

98. Ms 619, CSPP.

99. BP to ADB, March 27, 1862, BPLC.

100. CEP to Aunt Nichols, April 9, 1862, L 676, CSPP.

101. BP to ADB, March 27, 1862, BPLC; see also, ADB to BP, April 3, 1862, ADB file, PCHH.

102. BP to ADB, October 26,1862, ADB file, PCHH.

103. Atkinson, "Life and Thought," 136–39; Brent, *Charles Saunders Peirce*, 62, 95, 111.

104. Brent, *Charles Saunders Peirce*, 74.

105. Ketner suggests that Eliot's antipathy towards Charles stemmed from the tiff Peirce and Eliot had over the Rumford professorship. While this incident may have left some bad feelings on Eliot's part, they were not so strong as to preclude Eliot's being a pallbearer at Peirce's funeral, or his saying generous things about Peirce later in his life. (See Archibald et al., "Benjamin Peirce," 1–5.) It also did not prevent Eliot from working closely and well with another of Peirce's sons, James. Charles seemed quite capable of engendering ill will towards himself without help from his father.

106. Murphey, *Development of Peirce's Philosophy*, 19.

107. Houser, "Introduction," xxv.

108. Ibid., xxxvii–xxxviii.

109. Ibid., xxxviii

110. BP to SMP, January 21, 1880, SMP file, PCHH; see also Brent, *Charles Saunders Peirce*, 144.

111. CSP to Gilman, December 25, 1879, as quoted by Houser, "Introduction," xxvi.

112. Brent, *Charles Saunders Peirce*, 14–15 and 39–40, makes this diagnosis, obviously a problematic thing to do without a patient and over a hundred years later. Nevertheless, his diagnosis is convincing.

113. BP to JoLC, June 27, 1859, LFPB; ZP to BMP, March 10, 1863, L 668, CSPP; Atkinson, "Life and Thought," 129–33; and Brent, *Charles Saunders Peirce*, 40, 149.

114. Brent, *Charles Saunders Peirce*, 128, 139.

115. Ketner, *His Glassy Essence*, 281–84.

116. CEP to HPE, December 18, 1883, L 676, CSPP.

117. Ketner, *His Glassy Essence*, suggests that Juliette's ethnic background made it unlikely they had an affair, and that they only appeared to be intimate. The attitude of Charles's own family, however, suggests otherwise.

118. This is by Ketner's reckoning; Ketner, *His Glassy Essence*, 291. Brent points out that the wedding certificate said she was twenty-six, but according to the age she claimed in Charles's obituary, she would have been sixteen. Brent, *Charles Saunders Peirce*, 142.

119. Ketner, *His Glassy Essence*, 279–92; and Brent, *Charles Saunders Peirce*, 141–47. Ketner's and Brent's accounts of Juliette differ substantially.

120. SMP to HPE, September 19, 1883, L 687, CSPP.

121. CEP to HPE, [November 20, 1883], CSPP, as quoted by Brent, *Charles Saunders Peirce*, 146 (emphasis in the original).

122. CEP to HPE, March 18, [1884], L 676, CSPP (emphasis in the original).

123. Brent, *Charles Saunders Peirce*, 147.

124. Ibid., 169–70.

125. Ibid., 166–71.

126. Houser, "Introduction," lxvi.

127. Moyer, *Scientist's Voice*, xi; Norberg, "Early Astronomical Career," 209.

128. JH to John Rodgers, as quoted by Norberg, "Astronomical Career," 211–12.

129. Norberg, "Astronomical Career," 213.

130. Ibid.

131. Newcomb to BP, July 12, 1855, Newcomb file, PCHH.

132. Newcomb, *Reminiscences*, chapter 3.

133. Ibid., 95.

134. Norberg, "Astronomical Career," 215.

135. Newcomb, *Reminiscences*, 75.

136. Eisele, *Studies*, 74–76; Parshall and Rowe, *Emergence*, 114, fn. 139.

137. Brent, *Charles Saunders Peirce*, 153; Menand, *Metaphysical Club*. On Wright, see Madden, *Chauncey Wright*.

138. Brent, *Charles Saunders Peirce*, 151.

139. Ibid., 152.

140. Newcomb to Mendehall, March 30, 1888, National Archives, as quoted by Eisele, *Studies*, 54–55.

141. Lenzen, "Unpublished Scientific Monograph," 20. See also Houser, "Introduction," xxvii–xxxvi.

142. Brent, *Charles Saunders Peirce*, 198–200.

143. Ibid., 151–52.

144. Ibid., 152.

145. In the *Iliad*, Arisbe was the home of Axylus, who welcomed all passersby into his house.

146. Ketner, *His Glassy Essence*, 37, 44–53;

147. BP to SMP, January 21, 1880, SMP file, PCHH.

148. Ibid., January 22, 1880, SMP file, PCHH.

149. Hill to SMP, July 15, 1880, Hill file, PCHH.

150. CEP to CSP, September 26, [1880], L 336, CSPP (emphasis in the original).

151. Matz, "Benjamin Peirce," 178–79. Jonathan Ingersoll Bowditch was a successful Boston merchant.

Bibliography

Ackerberg-Hastings, Amy K. "Mathematics Is a Gentleman's Art: Analysis and Synthesis in American College Geometry Teaching 1790–1840." PhD diss., Iowa State University, 2000.

Agassiz, Elizabeth C. "A Cruise through the Galapagos." *Atlantic Monthly* 31 (1873): 579–84.

———. "The Hassler Glacier in the Straits of Magellan." *Atlantic Monthly* 30 (1872): 472–78.

———. "In the Straits of Magellan." *Atlantic Monthly* 31 (1873): 89–95.

Agassiz, Elizabeth C., ed. *Louis Agassiz, His Life and Work.* Boston: Houghton Mifflin, 1885.

Agassiz, G. R., ed. *Letters and Recollections of Alexander Agassiz.* Boston: Houghton Mifflin, 1913.

Agassiz, Louis. "Concerning Deep-Sea Dredgings." *American Naturalist* 6 (1872): 1–6.

Angle, Paul M., ed. *Lincoln Reader.* New Brunswick, N.J.: Rutgers University Press, 1947.

Anonymous. "Benjamin Peirce." *Proceedings of the American Academy of Arts and Sciences* 8 (1881): 443–54.

———. "Cambridge Course of Mathematics." *American Journal of Science* 6 (1823): 283–302.

———. "Cambridge Course of Mathematics." *Southern Review* 3 (1829): 289–307.

———. "Day's Algebra." *Christian Examiner* 35 (1844): 398.

———. "Geometry and Calculus." *Southern Review* 1 (1828): 107–34.

———. "An Introduction to Algebra upon the Inductive Method of Instruction." *United States Literary Gazette* 3 (1826): 241–50.

———. "The Late Benjamin Peirce." *Nation,* no. 798 (October 14, 1880): 268.

———. *Memorial of Joseph Lovering.* (American Academy of Arts and Sciences), Cambridge, Mass.: John Wilson and Son University Press, 1892.

———. *Proceedings of the London Mathematical Society* 3 (1869–71): 220.

———. "Review of the Cambridge Course of Mathematics." *American Journal of Science* 5 (1822): 304–26.

———. "Sketch of Joseph Lovering." *Popular Science Monthly* (September 1889) copy in HUA, HUG 1531.71.

———. *Spiritualism Shown As It Is! Boston Courier Report of the Proceedings of Professed Spiritual Agents and Mediums.* Boston: Office of the Boston Courier, 1859.

Anscombe, F. J. "Rejection of Outliers." *Technometrics* 2 (1960): 123–47.

402

Archibald, Raymond Clare. "Benjamin Peirce's Linear Associative Algebra and C. S. Peirce." *American Mathematical Monthly* 34 (1927): 525–27.

———. "Unpublished Letters of Sylvester." *Osiris* 1 (1936): 85–154.

Archibald, Raymond Clare, Charles W. Eliot, A. Lawrence Lowell, William Elwood Byerly, and Arnold B. Chase. "Benjamin Peirce." *American Mathematical Monthly* 32 (1925): 1–30. This article contains a bibliography of Peirce's work.

———. *Benjamin Peirce 1809–1880*. Oberlin, Ohio: Mathematical Association of America , 1925. This is a reprint of Archibald et al., "Benjamin Peirce."

Artin, Emil. "The Influence of J. H. M. Wedderburn on the Development of Modern Algebra." *Bulletin of the American Mathematical Society* 56 (1950): 65–72.

Atkinson, Norma P. "An Examination of the Life and Thought of Zina Fay Peirce an American Reformer and Feminist." PhD diss., Ball State University, 1983.

Bache, Alexander Dallas. "Address on Retiring from the Duties of President." *Proceedings of the American Association for the Advancement of Science* 6 (1851): xli–lx.

———. "Remarks at the Opening of the Fifth Session of the American Association for the Advancement of Education." *American Journal of Education* 1 (1855): 477–79.

Bailey, Sarah Loring. *Historical Sketches of Andover (Comprising the Present Towns of North Andover and Andover), Essex County Massachusetts*. Boston: Houghton Mifflin, 1880.

Bailyn, Bernard. "Why Kirkland Failed." *Glimpses of the Harvard Past*, 19–44. Cambridge: Harvard University Press, 1986.

Baker, Ruth, Dorothy K. Smith, and Elizabeth S. Versailles. *History and Genealogy of the Families of Chesterfiled Massachusetts*. Northampton, Mass.: Gazette Printing, 1962.

Barry, John M. *Rising Tide, The Great Mississippi Flood of 1927 and How it Changed America*. New York: Simon & Schuster, 1997.

Bartky, Ian R., Norman S. Rice, and Christine H. Bain. "'An Event of No Ordinary Interest'—The Inauguration of Albany's Dudley Observatory." *Journal of Astronomical History and Heritage* 2 (1999): 1–20.

Bassett, John Spencer. *The Round Hill School*. Worcester, Mass.: Davis Press, 1917.

Beach, Mark. "Was There a Scientific Lazzaroni?" *Nineteenth-Century American Science, A Reappraisal*. Edited by George H. Daniels, 115–32. Evanston, Illinois: Northwestern University Press, 1972.

Bell, Eric Temple. *The Development of Mathematics*. 2nd ed. New York: McGraw-Hill, 1945.

Bellows, Henry W. "The Round Hill School." *Harvard Register* 3 (1881): 3–7.

Berry, Robert Elton. *Yankee Stargazer The Life of Nathaniel Bowditch*. New York: Whittlesey House, 1941.

Birkhoff, Garrett. "Mathematics at Harvard, 1836–1944." *A Century of Mathematics in America* 2. Edited by Peter Duren, Richard A. Askey, and Uta C. Merzbach, 3–58. Providence: American Mathematical Society, 1989.

Birkhoff, George D. "Fifty Years of American Mathematics." *Semicentennial Addresses of the American Mathematical Society* 2, 270–315. New York: American Mathematical Society, 1938.

Bond, George. P. "On the Rings of Saturn." *American Journal of Science* 12 (1851): 97–105.

———. "On the Rings of Saturn." *Astronomical Journal* 2 (1851): 5–10.

Bonnycastle, Charles. *Inductive Geometry or an Analysis of the Relations of Form and Magnitude*. Charlottesville, Va.: Clement P. M'Kennie, 1834.

Boss, Benjamin. *History of the Dudley Observatory 1852–1956*. Albany, N.Y.: The Dudley Observatory, 1968.

Bourbaki, Nicholas. *Elements of the History of Mathematics*. Translated by John Meldrum. Berlin: Springer-Verlag, 1991.

[Bowditch, Henry Ingersoll]. "Memoir of Nathaniel Bowditch." Nathaniel Bowditch, *Celestial Mechanics*, vol. 1, 9–168. Bronx, N.Y.: Chelsea Publishing, 1839.

Bowen, Catherine Drinker. *Yankee from Olympus, Justice Holmes and His Family*. Boston: Little, Brown, 1944.

Brent, Joseph. *Charles Saunders Peirce: A Life*. Bloomington and Indianapolis: Indiana University Press, 1993.

Brown, Francis H. *Harvard University in the War of 1861–1865*. Boston: Cupples, Upham, 1886.

Bruce, Robert V. *The Launching of Modern American Science*. New York: Alfred A. Knopf, 1987.

Burger, William Henry. "The Contributions of the United States Coast and Goedetic Survey to Geodesy." *Centennial Celebration of the United States Coast and Geodetic Survey*, 81–85. Washington, D.C.: Government Printing Office, 1916.

Burton, David M. *The History of Mathematics*. 3rd ed. Dubuque, Iowa: Wm. C. Brown, 1995.

Byerly, William Elwood. "James Mills Peirce." *Harvard Graduates Magazine* 14 (1906): 573–77.

Cajori, Florian. "Attempts Made During the Eighteenth and Nineteenth Centuries to Reform the Teaching of Geometry." *American Mathematical Monthly* 17 (1910): 181–201.

———. *The Chequered Career of Ferdinand Rudolph Hassler*. Boston: The Christopher Publishing House, 1929.

———. *The Teaching and History of Mathematics in the United States*. Washington, D. C.: U. S. Government Printing Office, 1890.

Cayley, Arthur. "On Multiple Algebra." (From *Quarterly Journal of Pure and Applied Mathematics* 12, 270–308.) *Collected Mathematical Papers of Arthur Cayley* vol. 11, 459–89, Cambridge, Mass.: The University Press, 1889–97.

Chandler, S. C. "Benjamin Apthorp Gould." *Monthly Notices of the Royal Astronomical Society* 57 (1897): 218–22.

Clarke, James Freeman. *Autobiography, Diary and Correspondence*. Edited by Edward Everett Hale. Boston: Houghton Mifflin, 1891.

Clarke, John M. *James Hall of Albany, Geologist and Paleontologist 1811–1898*. 1923. Reprint, New York: Arno Press, 1978.

Clerke, Agnes M. *A Popular History of Astronomy*. London: Adam and Charles Black, 1908.

Cochrane, Rexmond C. *The National Academy of Sciences: The First Hundred Years 1863–1964*. Washington, D. C.: Academy, 1978.

Cohen, I. Bernard, ed. *Benjamin Peirce: Father of Pure Mathematics in America*. New York: Arno Press, 1980.

Colaw, J. M. "Dr. Joel E. Hendricks, A. M." *American Mathematical Monthly* 1 (1894): 65–67.

Comstock, George C. "Biographical Memoir of Benjamin Gould 1824–1846." *Memoirs of the National Academy of Sciences* 17 (1924): 153–80

Cooke, Roger, and V. Frederick Rickey. "W. E. Story of Hopkins and Clark." *A Century of Mathematics in America*. Edited by Peter Duren, Richard A. Askey, and Uta Merzbach, vol. 3, 29–76. Providence, R.I.: American Mathematical Society, 1989.

Coolidge, Julian Lowell. "Mathematics." *The Development of Harvard University Since the Inauguration of President Eliot, 1869–1929*. Edited by Samuel Eliot Morison, 248–57, Cambridge: Harvard University Press, 1930.

———. "The Story of Mathematics at Harvard." *Harvard Bulletin*, 26 (1924): 372–76.

Cordasco, Francesco. *Daniel Coit Gilman and the Protean Ph. D. The shaping of American Graduate Education*. Leiden: E. J. Brill, 1960.

Coulson, Thomas. *Joseph Henry His Life and Work*. Princeton: Princeton University Press, 1950.

Crowe, Michael J. *A History of Vector Analysis*. Notre Dame, Ind.: University of Notre Dame Press, 1967.

Dabney, Virginius. *Mr. Jefferson's University*. Charlottesville, Va.: University of Virginia Press, 1981.

[Dana, James Dwight]. "National Academy of Sciences." *American Journal of Science and Arts* 35 (1863): 462–65.

Dana, Richard Henry Jr. *The Journal of Richard Henry Dana, Jr.* Edited by Robert F. Lucid. 3 vols. Cambridge: The Belknap Press of Harvard University, 1968.

Daniels, George H. *American Science in the Age of Jackson*. New York: Columbia University Press, 1968.

———. "The Process of Professionalization in American Science." *Isis* 58 (1967): 151–66.

———. *Science in American Society*. New York: Alfred A. Knopf, 1971.

Davis, Captain Charles H. *Life of Charles Henry Davis Rear Admiral 1807–1877*. Boston: Houghton, Mifflin, 1899.

Day, Gardiner M. *The Biography of a Church*. Cambridge, Mass.: Riverside Press, 1951.

Day, Jeremiah, and James L. Kingsley. "Original Papers in Relation to a Course of Liberal Education." *American Journal of Science and Arts* 15 (1829): 197–351.

De Hegermann-Lindencrone, L. *In the Courts of Memory 1858–1875*. Garden City, N.Y.: Garden City Publishing Company, 1925.

———. *The Sunny Side of Diplomatic Life*. New York: Harper & Brothers, 1914.

Devlin, Keith. *Mathematics: The New Golden Age*. New York: Penguin Books, 1990.

Dickson, Leonard Eugene. *History of the Theory of Numbers*. 3 vols. New York: Chelsea Publishing Company, 1966.

Draper, John William. *History of the Conflict between Science and Religion*. New York: D. Appleton, 1875.

———. *History of the Intellectual Development of Europe*. 2 vols. New York: Harper& Brothers, 1876.

Dunne, Gerald T. *Justice Joseph Story and the Rise of the Supreme Court*. New York: Simon and Schuster, 1970.

Dupree, A. Hunter. *Asa Gray*. Cambridge, Mass.: Belknap Press of Harvard University, 1959.

———. "The Founding of the National Academy of Sciences—A Reinterpretation." *Proceedings of the American Philosophical Society* 44 (1957): 434–40.

———. *Science in the Federal Government.* Cambridge, Mass.: The Belknap Press of Harvard University, 1957.

Eisele, Carolyn. *Studies in the Scientific and Mathematical Philosophy of Charles S. Peirce.* Edited by R. M. Martin. The Hague: Mouton Publishers, 1979.

Eliot, Charles. "The New Education." *Atlantic Monthly* 23 (1869): 203–20, 358–67.

———. "President Eliot's Address." *Johns Hopkins University Celebration of the Twenty-fifth Anniversary of Founding of the University and Inauguration of Ira Remson,, LL. D. as President of the University,* 104–8. Baltimore: Johns Hopkins Press, 1902.

Elliott, Clark A. *Biographical Dictionary of American Science.* Westport, Conn.: Greenwood Press, 1979.

Elliott, Maud Howe. *Uncle Sam Ward and His Circle.* New York: MacMillian., 1938.

Emerson, Edward Waldo. *The Early Years of the Saturday Club: 1855–1870.* Boston: Houghton Mifflin, 1918.

———. *Life and Letters of Charles Russell Lowell.* 1907. Reprint, Port Washington, N.Y.: Kennikat Press, 1971.

Emerson, George. "Course of Mathematics." *North American Review* 13 (1821): 363–80.

Emerson, George B. *Reminiscences of an Old Teacher.* Boston: Alfred Mudges & Son, 1878.

Everett, Edward. *Orations and Speeches on Various Occasions.* Boston: Little, Brown, 1856.

Fairchild, Herman L. "The History of the American Association for the Advancement of Science." *Science* 59 (1924): 365–69, 384–90, 410–15.

Feuer, L. S. "America's First Jewish Professor: James Joseph Sylvester at the University of Virginia." *American Jewish Archives* 36 (1984): 152–201.

Fisch, Max H. "The Decisive Year and Its Consequences." *Writings of Charles S. Peirce A Chronological Edition.* Edited by Edward C. Moore, vol. 2, xxi–xxxvi. Bloomington, Ind.: Indiana University Press, 1984.

———. "Introduction." *Writings of Charles S. Peirce A Chronological Edition.* Edited by Max H. Fisch, vol. 1, xv–xxxv. Bloomington, Ind.: Indiana University Press, 1982.

Fite, Gilbert Courtland, & Jim E. Reese. *An Economic History of the United States (2nd ed.).* Boston: Houghton Mifflin, 1965.

Fitzpatrick, John C., ed. *The Writings of George Washington from Original Manuscript Sources 1745–1799.* Washington, D.C.: Government Printing Office, 1939.

Flynn, John T. *Men of Wealth.* New York: Books for Libraries Press, 1971.

French, John C. *A History of the University Founded by Johns Hopkins.* Baltimore: Johns Hopkins Press, 1946.

Frothingham, Paul Revere. *Edward Everett Orator and Statesman.* 1925. Reprint, Port Washington, N.Y.: Kennikat Press, 1971.

Genuth, Sara Schechner. "From Heaven's Alarm to Public Appeal." *Science at Harvard University, Historical Perspectives.* Edited by Clark A. Elliott and Margaret W. Rossiter, 28–54. Bethlehem, Pa.: Lehigh University Press, 1992.

Gilman, Daniel Coit. *The Launching of a University and Other Papers. A Sheaf of Remembrances.* New York: Dodd, Mead, 1906.

Ginsburg, Jekuthiel. "A Hitherto Unpublished Letter by Benjamin Peirce." *Scripta Mathematica* 2 (1934): 278–82.

Gould, Benjamin Apthorp Jr. *Reply to the "Statement of the Trustees" of the Dudley Observatory*. Albany, N.Y.: Charles Van Benthuysen, 1859.

———. *Report on the History of the Discovery of Neptune*. Washington City: Smithsonian Institution, 1850.

———. *Specimens of the Garbling of Letters by the Majority of the Trustees of the Dudley Observatory*. Albany, N.Y.: Charles Van Benthuysen, 1858.

Grattan-Guinness, Ivor. "Benjamin Peirce's *Linear Associative Algebra* (1870): New Light on its Preparation and 'Publication.'" *Annuals of Science* 54 (1997): 597–606.

Graves, Robert Perceval. *The Life of Sir William Rowan Hamilton*. 3 vols. New York: Arno Press, 1975.

Gray, Asa. "On the Composition of Plant by Phytons, and Some Applications of Phyllotaxis." *American Association for the Advancement of Science, Proceedings* 2 (1850): 438–44.

Grosser, Morton. *The Discovery of Neptune*. Cambridge: Harvard University Press, 1962.

Groves, Don. "The Unsung Sailor of Science." *Naval Engineers Journal* 81 (1969): 27–34.

Guralnick, Stanley. *Science and the Ante-Bellum American College*. Philadelphia: American Philosophical Society, 1975.

Hale, Edward E. *James Russell Lowell and His Friends*. New York: AMS Press, 1965.

———. "My College Days." *Atlantic Monthly* 71 (1893): 355–63.

———. *A New England Boyhood*. New York: Cassell Publishing Company, 1893.

———. "T Is Sixty Years Since At Harvard." *Atlantic Monthly* 78 (1896): 496–505.

Hankins, Thomas L. *Sir William Rowan Hamilton*. Baltimore: Johns Hopkins University Press, 1980.

Hanson, Norwood Russell. "Leverrier: The Zenith and Nadir of Newtonian Mechanics." *Isis* 53 (1962): 359–78.

Hardwick, Charles S., ed. *Semiotic and Significs, The Correspondence between Charles S. Peirce and Victoria Lady Welby*. Bloomington, Ind.: Indiana University Press, 1977.

Harkin, Duncan. "The Development of Modern Algebra." *Norsk Matematisk Tidsskrift* 33 (1951): 17–26.

[Harris, Ira]. *The Dudley Observatory and the Scientific Council, Statement of the Trustees*. Albany, N.Y.: Van Benthuysen, 1858.

Hassler, Ferdinand R. *Elements of Analytic Trigonometry, Plane and Spherical*. New York: James Bloomfield, 1826.

Hawkes, Herbert Edwin. "Estimate of Peirce's Linear Associative Algebra." *American Journal of Mathematics* 24 (1902): 88–95.

———. "On Hypercomplex Number Systems." *Transactions of the American Mathematical Society* 3 (1902): 312–30.

Hawkins, Hugh. *Between Harvard and America*. New York: Oxford University Press, 1972.

———. *Pioneer: A History of the Johns Hopkins University, 1974–1889*. Ithaca, N.Y.: Cornell University Press, 1960.

Hendrickson, Walter B. "Nineteenth-Century State Geological Surveys: Early Government Support of Science." *Isis* 52 (1961): 357–71.

Henry, Joseph. *The Papers of Joseph Henry*. Vols. 1–5 edited by Nathan Reingold, Stuart Pierson, Arthur Molella, James M. Hobbins, and John R. Kerwood; vols. 6–10 edited by Marc Rothenberg, Kathleen Dorman, Frank R. Milikan, and Deborah Y. Jefferies. Washington, D.C.: Smithsonian Institution Press.

Herrmann, D. B. "B. A. Gould and His Astronomical Journal." *Journal for the History of Astronomy* 2 (1971): 98–108.

Hilen, Andrew, ed. *The Letters of Henry Wadsworth Longfellow.* 6 vols. Cambridge, Mass.: Belknap Press of Harvard University, 1966.

[Hill, Thomas]. "The Imagination in Mathematics." *North American Review* 85 (1857): 223–37.

———. "Peirce's *Analytic Mechanics.*" *North American Review* 87 (1858): 1–21.

———. "Scientific Intelligence." *The Christian Examiner* 63 (1857): 151–52. (Winslow identifies Hill as the author in the *Boston Daily Journal,* October 26, 1857.)

———. "The Scientific Meeting at Cambridge." *The Christian Examiner* 47 (1849): 325–337.

Hill, Thomas. "Benjamin Peirce." *Harvard Register* 1 (1880): 91–92.

Hindle, Brooke. *The Pursuit of Science in Revolutionary America 1735–1789.* Chapel Hill, N.C.: University of North Carolina Press, 1956.

Hitchcock, Edward. *The Religion of Geology and Its Connected Sciences.* Boston: Phillips, Sampson, 1851.

Hoar, George F. *Autobiography of Seventy Years.* 2 vols. New York: Charles Scribner's Sons, 1903.

Hofstadter, Richard, and Wilson Smith. *American Higher Education, A Documentary History.* 2 vols. Chicago: University of Chicago Press, 1961.

Hogan, Edward R. "The Mathematical Miscellany (1836—1839)." *Historia Mathematica* 12 (1985): 245–57.

———. "'A Proper Spirit Is Abroad:' Peirce, Sylvester, Ward, and American Mathematics, 1829–1843." *Historia Mathematica* 18 (1991): 158–72.

———. "Robert Adrain: American Mathematician." *Historia Mathematica* 4 (1977): 157–72.

Holden, Edward S. *Memorials of William Cranch Bond and his Son George Phillips Bond.* 1897. Reprint, New York: Arno Press, 1980.

Houser, Nathan 1989. "Introduction." *Writings of Charles S. Peirce A Chronological Edition.* Edited by Christian J. W. Kloesel, vol. 4, xix–lxx. Bloomington, Ind.: Indiana University Press, 1989.

Hovenkamp, Herbert. *Science and Religion in America 1800–1860.* Philadelphia: University of Pennsylvania Press, 1978.

Howe, M. A. DeWolfe, ed. *Later Years of the Saturday Club, 1870–1920.* 1927. Reprint, New York: Books for Libraries Press, 1972.

Hubbell, John G., and Robert W. Smith. "Neptune in America: Negotiating a Discovery." *Journal for the History of Astronomy* 23 (1992): 261–91.

Hutton, Charles. *A Course of Mathematics.* Edited by Robert Adrain. New York: Samuel Campbell et al., 1812.

Jahns, Patricia. *Matthew Fontaine Maury & Joseph Henry, Scientists of the Civil War.* New York: Hastings House, 1961.

James, Alton James. *The First Scientific Exploration of Russian America and the Purchase of Alaska.* Northwestern University Studies in the Social Sciences, Number 4. Evanston, Ill.: Northwestern University, 1942

James, Edward T. *Notable American Women 1607–1950.* Cambridge, Mass.: Belknap Press of Harvard University, 1971.

James, Henry. *Charles W. Eliot.* 2 vols. Boston: Houghton Mifflin, 1930.

James, Henry Rosher. *Mary Wollstonecraft, A Sketch*. London: Oxford University Press, 1932.

James, Mary Ann. "The Dudley Observatory Controversy." PhD diss., Rice University, 1980.

———. *Elites in Conflict, The Antebellum Clash over the Dudley Observatory*. New Brunswick, Conn.: Rutgers University Press, 1987.

———. "Engineering an Environment of Change Bigelow, Peirce, and Early Nineteenth-Century Education at Harvard." *Science at Harvard University Historical Perspectives*, edited by Clark A. Elliot and Margaret W. Rossiter, 55–75. Bethlehem, Pa.: Lehigh University Press, 1992.

Johnson, Claudia L., ed. *The Cambridge Companion to Mary Wollstonecraft*. Cambridge, Mass.: Cambridge University Press, 2002.

Jones, Bessie Z. "Diary of the Two Bonds: 1846–1849 First Directors of the Harvard College Observatory." *Harvard Library Bulletin* 15 (1967): 368–86 and 16 (1968): 49–71 and 178–207.

Jones, Bessie Zaban, and Lyle Gifford Boyd. *The Harvard College Observatory: The First Four Directorships, 1839–1919*. Cambridge, Mass.: Belknap Press of Harvard University Press, 1971.

Karpinski, Louis C. *Bibliography of Mathematical Works Printed in America Through 1850*. Ann Arbor, Mich.: The University of Michigan Press, 1940.

Katz, Victor J. *A History of Mathematics, An Introduction*. Reading, Mass.: Addison-Wesley, 1998.

Kennedy, Hubert. "Towards a Biography of James Mills Peirce." *Historia Mathematica* 6 (1979): 195–201.

Ketner, Kenneth Laine. *His Glassy Essence*. Nashville, Tenn.: Vanderbilt University Press, 1998.

King, Moses. *Benjamin Peirce: A Memorial Collection*. Cambridge, Mass.: Rand, Avery 1881.

King, William F. "George Davidson: Pacific Coast Scientist for the U. S. Coast and Geodetic Survey, 1845–1895." PhD diss., Claremont, 1973.

[Kirkland, John Thornton]. "Literary Institutions—University." *North American Review* 7 (1818): 270–78.

Kohlstedt, Sally Gregory. *The Formation of the American Scientific Community: The American Association for the Advancement of Science, 1846–60*. Urbana, Ill.: University of Illinois Press, 1976.

Le Conte, Emma. *When the World Ended: The Diary of Emma LeConte*. Edited by Earl Schenck Miers. New York: Oxford University Press, 1957.

[Le Conte, John]. "The Physical Geography of the Sea." *Southern Review* 1 (1856): 151–67.

Le Conte, Joseph. *The Autobiography of Joseph Le Conte*. Edited by William Dallam Armes. New York: D. Appleton, 1903.

———. "Biographical Memoir of John Le Conte 1818–1891." *National Academy of Sciences: Biographical Memoirs* 3 (1895): 371–89.

———. '*Ware Sherman, A Journal of Three Months' Personal Experience in the Last Days of the Confederacy*. Introduction by Caroline Le Conte. Berkeley, Calif.: University of California Press, 1937.

———. '*Ware Sherman, A Journal of Three Months' Personal Experience in the Last Days of the Confederacy.* Introduction by William Blair. Baton Rouge, La.: Louisiana University Press, 1999.

Lenzen, Victor Fritz. *Benjamin Peirce and the U. S. Coast Survey.* San Francisco: San Francisco Press, 1968.

———. "The Contributions of Charles S. Peirce to Linear Algebra." *Phenomenology and Natural Existence, Essays in Honor of Marvin Farber.* Edited by Dale Riepe. Albany, N.Y.: State University of New York Press, 1973.

———. "An Unpublished Scientific Monograph by C. S. Peirce." *Transactions of the Charles S. Peirce Society: A Quarterly Journal in American Philosophy* 5 (1969): 5–24.

Leverrier, Urbain Jean Joseph . "Le Verrier's Further Vindication of His Predicted Theory of Neptune." *American Journal of Science and Arts,* 2nd ser., 7 (1849): 442–47.

———. "Le Verrier's Remarks on the Planet Neptune." *American Journal of Science and Arts,* 2nd ser., 7 (1849): 118–22.

Lewis, Charles Lee. *Matthew Fontaine Maury, the Pathfinder of the Seas.* Annapolis, Md.: United States Naval Institute, 1927.

Lewis, Oscar. *George Davison Pioneer West Coast Scientist.* Berkeley, Calif.: University of Los Angeles Press, 1954.

Lockyer, Mary T., and Winifred L. Lockyer. *Life and Work of Sir Norman Lockyer.* London: Macmillan, 1928.

Lodge, Henry Cabot. *Early Memories.* New York: Charles Scribner's Sons, 1913.

Loomis, Elias. "Historical Notice of the Discovery of the Planet Neptune." *American Journal of Science and Arts,* 2nd ser., 5 (1848): 187–205.

———. *The Recent Progress of Astronomy; Especially in the United States.* New York: Harper & Brothers, 1850.

Lovering, Joseph. "Communication Presented by Joseph Lovering at 380th Meeting." *Proceedings of the American Academy of Arts and Sciences* 3 (1852–57): 38–40.

Lupold, John Samuel. "From Physician to Physicist: The Scientific Career of John LeConte, 1818–1891." PhD diss., University of South Carolina, 1970.

Lurie, Edward. *Louis Agassiz, A Life in Science.* Chicago: University of Chicago Press, 1960.

Mack, Edward C. *Peter Cooper Citizen of New York.* New York: Duell, Sloan, and Pearce, 1949.

Madden, Edward H. *Chauncey Wright and the Foundations of Pragmatism.* Seattle: University of Washington Press, 1963.

Madsen, David. *The National University Enduring Dream of the USA.* Detroit: Wayne State University Press, 1966.

Manning, Thomas, G. *U. S. Coast Survey vs. Naval Hydrographic Office, A 19th-Century Rivalry in Science and Politics.* Tuscaloosa, Ala.: University of Alabama Press, 1988.

Marvin, Winthrop L. *The American Merchant Marine: Its History and Romance from 1620 to 1902.* New York: Charles Scribner's Sons, 1902.

Matz, F. P. "Benjamin Peirce." *American Mathematical Monthly* 2 (1895): 173–79.

McCaughey, Robert A. *Josiah Quincy, 1772–1864, The Last Federalist.* Cambridge: Harvard University Press, 1974.

McClintock, Emory. "Charles Gill: The First Actuary in America." *Actuarial Society of America, Transactions* 14 (1913): 9–16, 211–38; 15 (1913): 11–40, 227–70.

McCormmach, Russell 1966. "Ormsby MacKnight Mitchel's Sidereal Messenger 1846–1848." *Proceedings of the American Philosophical Society* 10, 35–47.

McPherson, James M. *Battle Cry of Freedom.* New York: Oxford University Press, 1988.

Meier, Paul, and Sandy Zabell. "Benjamin Peirce and the Howland Will." *Journal of the American Statistical Association* 75 (1980): 497–506.

Menand, Louis. *The Metaphysical Club.* New York: Farrar, Straus and Giroux, 2001.

Miller, Howard S. *Dollars for Research: Science and Its Patrons in Nineteenth-Century America.* Seattle: University of Washington Press, 1970.

Miller, Lillian B., Frederoick Voss, and Jeannette M. Hussey. *The Lazzaroni.* Washington, D.C.: Smithsonian Institution Press, 1972.

Mitchel, Ormsby MacKnight. "Le Verrier's Planet." *Sidereal Messenger* 1 (1846): 42–44.

———. *The Orbs of Heaven.* London: Office of the National Illustrated Library, 1851.

Morison, Samuel Eliot. "The Great Rebellion in Harvard College." *Publications of the Colonial Society of Massachusetts,* Boston. (April 1928): 54–112.

———. *The Maritime History of Massachusetts: 1783–1866.* Boston: Houghton Mifflin, 1921.

———. *Three Centuries of Harvard: 1636–1936.* Cambridge, Mass.: The Belknap Press of Harvard University, 1965.

Moyer, Albert E. *A Scientist's Voice in American Culture, Simon Newcomb and the Rhetoric of Scientific Method.* Berkeley, Calif.: University of California Press, 1992.

Muir, Thomas. *The Theory of Determinants.* Vol. 2. New York: Dover Publications, 1811.

Murphey, Murray G. *The Development of Peirce's Philosophy.* Cambridge: Harvard University Press, 1961.

Nagel, Ernest. " 'Impossible Numbers': A Chapter in the History of Modern Logic." *Studies in the History of Ideas.* Edited by the Dept. of Philosophy of Columbia University, 3, 429–74. New York: Columbia University Press, 1935.

Nevins, Allan, and Milton Halsey Thomas. *The Diary of George Templeton Strong.* New York: Macmillan, 1952.

Newcomb, Simon. *Popular Astronomy.* New York: Harper & Brothers, 1878.

———. *The Reminiscences of an Astronomer.* Boston: Houghton, Mifflin, 1903.

Newmyer, R. Kent. *Supreme Court Justice Joseph Story Statesman of the Old Republic.* Chapel Hill, N.C.: The University of North Carolina Press, 1985.

[Newton, Hubert Anson.] "Benjamin Peirce." *American Journal of Science,* 22, 3rd ser. (1881): 167–78.

Nichols, Thomas Low. *A Biography of the Brothers Davenport.* London: Saunders, Otley, 1864.

Norberg, Arthur. "Simon Newcomb's Early Astronomical Career." *Isis* 69 (1978): 209–25.

Novy, Lubos. "Benjamin Peirce's Concept of Linear Algebra." *Acta historiae rerum naturalium necnon technicarum* 7 (1974): 211–30.

Numbers, Ronald L. *Creation by Natural Law.* Seattle: University of Washington Press, 1977.

Odgers, Merle M. *Alexander Dallas Bache Scientist and Educator 1806–1867.* Philadelphia: University of Pennsylvania Press, 1947.

Olson, Richard G. "The Gould Controversy at Dudley Observatory: Public and Professional Values in Conflict." *Annals of Science* 27 (1971): 265–76.

Palfrey, John Gorham. "Notice of Professor Farrar." *Christian Examiner* (July 1853). Boston: Crosby, Nichols, 1853.

Pannekoek, A. "The Discovery of Neptune." *Centarus* 12 (1953): 126–37.

[Parkman, F.] "Review of *Catalogue of the Library of Harvard University in Cambridge, Massachusetts.*" *Christian Examiner* 8 (1830): 321–25.

Parshall, Karen Hunger. "America's First School of Mathematical Research: James Sylvester at the Johns Hopkins University 1876–1883." *Archive for History of Exact Sciences* 38 (1988): 153–96

———. *James Joseph Sylvester: Life and Work in Letters.* Oxford: Clarendon Press, 1998.

———. "Joseph H. M. Wedderburn and the Structure Theory of Algebras." *Archive for History of Exact Sciences* 32 (1985): 223–94.

Parshall, Karen, and David E. Rowe. "American Mathematics Comes of Age: 1875–1900." *A Century of Mathematics in America.* Edited by Peter Duren, Richard A. Askey, and Uta C. Merzbach. Vol. 3, 2–28. Providence: American Mathematical Society, 1989.

———. *The Emergence of the American Mathematical Community, 1876–1900: J. J. Sylvester, Felix Klein and E. H. Moore.* Providence: American Mathematical Society, 1994.

Parsons, Theophilus. *The Report of the Committee to Whom was Referred the Report and Resolutions of the President and Fellows of Harvard University Respecting the Introduction of the Voluntary System in the Studies of the Mathematics, Latin, and Greek.* Cambridge, Mass.: Folsom, Wells, and Thurston, Printers to the University, 1841. (UA II 10 7.5.872, HUA.)

Peabody, Andrew P. *Harvard Reminiscences.* 1888. Reprint, Freeport, N.Y: Books for Libraries Press, 1972.

Peacock, George. *A Treatise on Algebra.* Cambridge, Mass.: J & J. Deighton, 1830.

———. *A Treatise on Algebra. Vol 1. Arithmetical Algebra.* 1842. Reprint, New York: Scripta Mathematica, 1940.

———. *A Treatise on Algebra. Vol 2. On Symbolical Algebra.* 1845. Reprint, New York: Scripta Mathematica, 1940.

Pierce, Benjamin. "Abstract of a Paper on Researches in Analytic Morphology." *AAAS, Proceedings* 9 (1856): 67–69.

———. "Criterion for the Rejection of Doubtful Observations." *Astronomical Journal* 2 (1852): 161–63.

———. *An Elementary Treatise on Algebra.* Boston: James Munroe, 1837.

———. *An Elementary Treatise on Curves, Functions and Forces.* Cambridge, Mass.: James Munroe, 1852.

———. *An Elementary Treatise on Plane and Spherical Trigonometry.* (New ed.) Boston: James Munroe, 1852.

———. *An Elementary Treatise on Sound.* Boston: James Munroe, 1836.

———. *First Part of an Elementary Treatise on Spherical Trigonometry.* Boston: James Munroe, 1836.

———. *Ideality in the Physical Sciences.* Boston: Little, Brown, 1881.

———. "The Intellectual Organization of Harvard University." *Harvard Register* 1 (1880): 77.

———. "Linear Associative Algebra." *American Journal of Mathematics* 4 (1881): 97–229.

———. *Linear Associative Algebra. New Edition, with Addenda and Notes by C. S. Peirce, Son of the Author.* New York: Van Nostrand, 1882.

———. *Linear Associative Algebra.* Washington, D.C.: 1870. (100 lithographed copies.)

———. "Mathematical Investigation of the Fractions which Occur in Phyllotaxis." *AAAS, Proceedings* 2 (1850): 444–47.

———. "The Mathematical Miscellany." *American Journal of Science*. 31 (1836): 184–85.

———. "Method of Determining Longitudes by Occultations of the Pleiades." *AAAS Proceedings* 9 (1856): 97–102.

———. "On Perfect Numbers." *Mathematical Diary* 2 (1832): 267–77.

———. "On the Constitution of Saturn's Ring." *American Journal of Science* 12 (1851): 106–8.

———. "On the Constitution of Saturn's Ring." *Astronomical Journal* 2 (1851): 17–19.

———. "On the Uses and Transformations of Linear Algebra." *Proceedings of the American Academy of Arts and Sciences* 10 (1875): 395–400.

———. "Opening Address by Professor Benjamin Peirce, President of the Association." *AAAS, Proceedings* 7 (1856): xvii–xx. (This address was delivered at the July 1853 meeting of the AAAS.)

———. *Plane and Solid Geometry*. Boston: James Munroe, 1847.

———. "Presidential Address." *AAAS, Proceedings* 8 (1855): 1–17.

———. "Remarks on a National University." *American Journal of Education* 2 (1856): 86–92.

———. "Report to the American Academy on 4 April 1848." *Proceedings of the American Academy of Arts and Sciences* 1 (1846–48): 332–42.

———. *A System of Analytical Mechanics*. Boston: Little, Brown, 1855.

———. "Two Hundred and Eighty-eighth Meeting." *Proceedings of the American Academy of Arts and Sciences* 1 (1846–48): 39–43.

———. "Two Hundred and Ninety-fifth Meeting." *Proceedings of the American Academy of Arts and Sciences* 1 (1846–48): 144–49.

———. "Two Hundred and Ninety-third Meeting." *Proceedings of the American Academy of Arts and Sciences* 1 (1846–48): 57–71.

———. Working Plan for the Foundation of a University. Cambridge, Mass.: 1856. 184–85.

[Peirce, Benjamin]. "*The American Almanac and Repository of Useful Knowledge*." *North American Review* 44 (1837): 267–70.

———."American Astronomical and Magnetic Observations." *Cambridge Miscellany* 1 (1842): 25–28.

———. "Bartlett's *Elementary Treatise on Optics*." *North American Review* 48 (1839): 540–41.

———. "Magnetic Observations." *North American Review* 53 (1841): 520–21.

———. "Meteors." *North American Review* 56 (1843): 409–35.

———. "Nichol's *Contemplations on the Solar System*." *North American Review* 66 (1848): 253–55.

———. "Norton's *Elementary Treatise on Astronomy*." *North American Review* 48 (1839): 539–40.

———. "On the New Planet Neptune." *Astronomische Nachrichten* 25 (1846–47): columns 375–88.

———. "The Mathematical Miscellany." *American Journal of Science* 31 (1837): 184–85.

———. "Worcester's *Geography*." *North American Review* 50 (1840): 272–73.

[Peirce, Benjamin, and John Gorham Palfrey.] "Bowditch's Translation of the *Mécanique Céleste.*" *North American Review* 48 (1839): 144–80.

Peirce, Benjamin Mills. *A Report on the Resources of Iceland and Greenland.* Washington, D.C.: Government Printing Office, 1868.

Peirce, Benjamin Osgood. "Biographical Memoir of Joseph Lovering." *National Academy of Sciences: Biographical Memoirs* 6 (1909): 329–44.

Peirce, Benjamin, Sr. *A History of Harvard University from Its Foundation, in the Year 1636 to the Period of the American Revolution.* Cambridge, Mass.: Brown, Shattuck, 1833.

———. *An Oration, Delivered at Salem, on the Fourth of July, 1812.* Salem, Mass.: Thomas C. Cushing, 1812.

Peirce, Charles Saunders. "The Century's Great Men in Science." *Annual Report of the Smithsonian Institution for 1900.* Washington, D. C.: Government Printing Office, 1901.

———. "On the Algebras in which Division is Unambiguous." *American Journal of Mathematics* 4 (1881): 225–29.

———. "On the Relative Forms of the Algebras." *American Journal of Mathematics* 4 (1881): 221–25.

Peirce, Harriet Melusina Fay. "The Externals of Washington." *Atlantic Monthly,* 32 (1873): 701–16.

Perry, Ralph Barton. *The Thought and Character of William James.* 2 vols. Boston: Little, Brown, 1935.

Peterson, Sven R. "Benjamin Peirce: Mathematician and Philosopher." *Journal for the History of Ideas* 16 (1955): 89–112.

Phillips, James Duncan. *Salem and the Indies, The Story of the Great Commercial Era of the City.* Boston: Houghton Mifflin, 1947.

———. *When Salem Sailed the Seven Seas.* Princeton: Princeton University Press, 1946.

Pickering, J. "Eulogy on Dr. Bowditch, President of the American Academy." *Memoirs of the American Academy of Arts and Sciences* 2 (1846): i–lxxvi.

Pilkington, Walter. *Hamilton College: 1812/1862.* Deposit, N.Y.: Courier Printing, 1962.

Pulsifer, Susan Farley (Nichols). *Supplement to Witch's Breed: the Peirce-Nichols Family of Salem.* Cambridge, Mass.: Dresser, Chapman and Grimes, 1967.

———. *Witch's Breed: the Peirce-Nichols Family of Salem.* Cambridge, Mass.: Dresser, Chapman and Grimes, 1967.

Pycior, Helena M. "Benjamin Peirce's *Linear Associative Algebra.*" *Isis* 70 (1979): 537–51.

———. "British Synthetic vs. French Analytic Styles of Algebra in the Early American Republic." *History of Modern Mathematics.* Edited by David E. Rowe and John McCleary. Vol. 1, 124–54. Boston: Academic Press.

———. "George Peacock and the British Origins of Symbolical Algebra." *Historia Mathematica* 8 (1981): 23–45.

Rantoul, Robert S. "Memoir of Benjamin Peirce." *Historical Collections of the Essex Institute* 18: 161–76.

Rawlins, D. "Some Simple Results Regarding Gravitational Disturbance by Exterior Planets—with Historical Applications." *Monthly Notices of the Royal Astronomical Society* 147 (1970): 177–86.

Redman, George A. *Mystic Hours; or Spiritual Experiences.* New York: Charles Partridge, 1859.

Reingold, Nathan. *Science in Nineteenth-Century America, A Documentary History.* New York: Hill and Wang, 1964.

———. "Science in the Civil War, the Permanent Commission of the Navy Department." *Isis* 49 (1958): 307–18.

Rendall, Jane. *The Origins of Modern Feminism: Women in Britain, France and the United States, 1780–1860.* New York: Schocken Brooks, 1984.

Rezneck, Samuel. "The Emergence of a Scientific Community in New York State A Century Ago." *New York History* 43 (1962): 211–38.

Rice, Adrian. "Mathematics in the Metropolis: A Survey of Victorian London." *Historia Mathematica* 23 (1996): 376–417.

Richards, Joan L. "The Art and the Science of British Algebra: A Study in the Perception of Mathematical Truth." *Historia Mathematica* 7 (1980): 343–65.

Rider, Paul R. *Criteria for Rejection of Observations.* Saint Louis, Mo.: Washington University Studies—New Series, Science, and Technology—No.8, 1933.

Rosenstein, George M. "The Best Method. American Calculus Textbooks of the Nineteenth Century." *A Century of Mathematics in America.* Edited by Peter Duren, Richard A. Askey, and Uta C. Merzbach. Vol. 3, 77–109. Providence: 1989.

Rossiter, Margaret W. "The Emergence of Agricultural Science, Justice Liebig and the Americans, 1840–1880." New Haven: Yale University Press, 1975.

Rothenberg, Marc. "Patronage of Harvard College Observatory, 1839–1851." *Journal for the History of Astronomy* 21 (1990): 36–46.

Royce, Josiah. "Some Relations between Philosophy and Science in the First Half of the Nineteenth Century in Germany." *Science,* n. s., 38 (1913): 567–84.

Rudolph, Fredrick. *The American College and University A History.* New York: Alfred A. Knopf, 1962.

Ruggles, Samuel B., Duncan Kennedy, Azor Taber, and Ray Palmer. *Speeches in Behalf of the University of Albany.* Albany, N.Y.: C. Van Benthuysen, Printer, 1852.

Sargent, Mrs. John T. *Sketches and Reminiscences of the Radical Club of Chestnut Street, Boston.* Boston: James R. Osgood, 1880.

Schlote, Karl-Heinz. "Zur Geschichte der Algebrentheorie—Peirce's 'Linear Associative Algebra.'" *Schriftenreihe der Geschichte der Naturwissenschaften, Technik und Medizin* 20 (1983): 1–20.

Scientific Council. *Defence [sic] of Dr. Gould by the Scientific Council of the Dudley Observatory.* Albany, N.Y.: Weed, Parson, 1858.

Sebeok, Thomas A. *The Play of Musement.* Bloomington, Ind.: Indiana University Press, 1981.

Shaler, Nathaniel Southgate. *The Autobiography of Nathaniel Southgate Shaler.* Boston: Houghton Mifflin, 1909.

Shama, Simon. *Dead Certainties.* New York: Alfred A. Knopf, 1991.

Shaw, James Byrnie. *Synopsis of Linear Associative Algebra.* Washington, D. C.: Carnegie Institution of Washington, 1907.

Sherwood, Morgan B. "George Davidson and the Acquisition of Alaska." *Alaska and Its History.* Edited by Morgan B. Sherwood. Seattle: University of Washington Press, 1967.

Sheynin, O. B. "Mathematical Treatment of Astronomical Observations (A Historical Essay)." *Archive for History of the Exact Sciences* 11 (1973): 97–126.

Silverman, Robert, and Mark Beach. "A National University for Upstate New York." *American Quarterly* 22, no. 3 (1970): 701–13.

Slotten, Richard Hugh. *Patronage, Practice, and the Culture of American Science, Alexander Dallas Bache and the U. S. Coast Survey*. Cambridge, Mass.: Cambridge University Press, 1994.

Smith, David Eugene. "Sylvester as a Poet." *American Mathematical Monthly* 29 (1922): 14–15.

Smith, David Eugene, and Jekuthiel Ginsburg. *A History of Mathematics in America before* 1900. Chicago: Open Court, 1934.

Sparks, Boyden, and Samuel Taylor Moore. *Hetty Green A Woman Who Loved Money*. New York: Doubleday, Doran, 1930.

Spottiswoode, William. "Remarks on Some Recent Generalizations of Algebra." *Proceedings of the London Mathematical* 4, (1873): 147–64.

Stephens, Carlene E. "Astronomy as Public Utility: The Bond Years at the Harvard College Observatory." *Journal for the History of Astronomy* 21 (1990): 21–35.

Stephens, Lester D. *Joseph LeConte, Gentle Prophet of Evolution*. Baton Rouge, La.: Louisiana State University Press, 1982.

Stevens, W. Le Conte. "Sketch of Professor John Le Conte." *Popular Science Monthly* 36 (1889): 112–20.

Stewart, William Rhinelander. *The Philanthropic Work of Josephine Shaw Lowell*. 1911. Reprint, Montclair, N.J.: Patterson Smith, 1974.

Stigler, Stephen M. "Simon Newcomb, Percy Daniell and the History of Robust Estimation." *Journal of the American Statistical Association* 63 (1973): 872–79.

———. "Mathematical Statistics in the Early States." *The Annals of Statistics* 6 (1978): 239–65.

Stillwell, John. *Mathematics and Its History*. 2nd ed. New York: Springer-Verlag, 2002.

Storr, Richard J. *The Beginnings of Graduate Education in America*. Chicago: University of Chicago Press, 1953.

Struik, Dirk J. *Yankee Science in the Making*. New rev. ed. New York: Collier Books, 1962.

Sullivan, Robert. *The Disappearance of Dr. Parkman*. Boston: Little, Brown, 1971.

Sunstein, Emily W. *A Different Face, the Life of Mary Wollstonecraft*. New York: Harper & Row, 1975.

Tharp, Louise Hall. *Three Saints and a Sinner*. Boston: Little, Brown. Co., 1956

Thomas, Lately. *Sam Ward King of the Lobby*. Boston: Houghton Mifflin, 1965.

Thompson, D. G. Brinton. *Ruggles of New York*. New York: AMS Press, 1968.

Ticknor, Anna Eliot. *Life of Joseph Green Cogswell as Sketched in his Letters*. Cambridge, Mass.: Riverside Press, 1874.

Tyler H. W. "John Daniel Runkle." *American Mathematical Monthly* 10 (1903): 183–85.

Waff, Craig B. "Charles Henry Davis, the Foundation of the American Nautical Almanac, and the Establishment of an American Prime Meridian." *Vistas in Astronomy* 28 (1985): 61–66.

Wagenknect, Edward. *Henry Wadsworth Longfellow: Portrait of an American Humanist*. New York: Oxford University Press, 1966.

Walker, Sears Cook. "Astronomy." *American Journal of Science*, n. s., 4 (1847): 132–34.

———. "Researches Relative to the Planet Neptune." *Smithsonian Contributions to Knowledge* 2 (1851): 3–60.

[Walker, T.] "Farrar's Mathematics." *North American Review* 27 (1828): 191–214.

Walsh, Alison. "Relations Between Logic and Mathematics in the Work of Benjamin and Charles S. Peirce." PhD diss., Middlesex University, 1999.

Warner, Deborah Jean. "Astronomy in Antebellum America." *Sciences in the American Context.* Edited by Nathan Reingold. Washington, D. C.: Smithsonian Institution Press, 1979.

———. "Lewis M. Rutherford: Pioneer Astronomical Photographer and Spectroscopist." *Technology and Culture* 12 (1971): 190–216.

Warner, John. *Studies in Organic Morphology, an Abstract of Lectures Delivered before the Pottsville Scientific Association, in 1855 and 1856.* Philadelphia: J. B. Lippincollt, Printed by Order of the Association, 1857.

Weber, Gustavus A. *The Coast and Geodetic Survey Its History, Activities and Organization.* Baltimore: Johns Hopkins Press, 1923.

Wedderburn, Joseph Henry Maclagan. "On Hypercomplex Numbers." *Proceedings of the London Mathematical Society* 6 (1907): 77–118.

Whitehead, Alfred North. *A Treatise on Universal Algebra, with Applications.* Vol. 1. Cambridge, Mass.: University Press, 1898. (The planned vol. 2 never appeared.)

Whitney, J. D. "Geographical and Geological Surveys." *North American Review* 121 (1875): 37–85.

Whittemore, J. K. "James Mills Peirce." *Science*, n. s., 24, no. 602 (1906): 40–48.

Williams, Cecil B. *Henry Wadsworth Longfellow.* New York: Twayne Publishers, 1964.

Williams, Francis L. *Matthew Fontaine Maury: Scientist of the Sea.* New Brunswick, N.J.: Rutgers University Press, 1963.

Winslow, Charles F. *Cosmography, or Philosophical Views of the Universe.* Boston: Crosby, Nichols, 1853.

Wollstonecraft, Mary. *A Vindication of the Rights of Woman.* 1792. Reprint, 2nd ed. Edited by Carol H. Poston. New York: W. W. Norton, 1988.

Wright, Chauncey. "On the Phyllotaxis." *Astronomical Journal* 5 (1858): 22–24.

[Wright, Chauncey]. "Mathematics in Court." *Nation* 5 (1857): 238.

Wright, Nathalia. *Horatio Greenough The First American Sculptor.* Philadelphia: University of Pennsylvania Press, 1963.

Yates, R. C. "Sylvester at the University of Virginia." *American Mathematical Monthly* 44 (1937): 194–201.

Index

abolitionists: in Massachusetts, 203; John Brown, 208–9; Le Conte's dislike of, 146; Charles Henry Peirce, 203; Peirce's attitude toward, 205, 221, 381 n. 2

Academy of Sciences (France), 215, 230

Adams, John Couch, 15–22, 24, 27

Adams, John Quincy, 107

Adler, George J., 162

Adrain, Robert, 46, 47, 354 n. 90, 356 n. 47

Agassiz, Alexander, 272

Agassiz, Elizabeth, 84, 271

Agassiz, Louis: AAAS, 130; comes to America, 83–86; American Academy of Arts and Sciences, 192; Civil War, 222; Coast Survey, 222, 257, 275–76; Darwinism, 192, 272; Dudley Observatory, 167; contrasted with Gray, 191; Harvard University, 79, 227; *Hassler* (voyage of), 270–73; Howland will case, 287; Lazzaroni, 123–24; medical school attendance, 191; National Academy of Sciences, 230–31; relationship with Peirce, 110, 306; science and religion, 283; spiritualism, 130; teaching of, 146, 226; Warner-Winslow affair, 135–36

Airy, George Biddle, 27, 167, 267

Alaska: purchase of, 259–64

Albany: University at, 301

Allison Commission, 335

amateur scientists, 98–99

American Academy of Arts and Sciences, 347 n. 3, 350 n. 80; Agassiz-Gray debates, 201; Gray's leadership in, 79; Harvard Observatory, 116; National Academy of Sciences, 232; Peirce's frustration with, 192–94, 206,

233; Peirce's membership in, 379 n. 19; Peirce-Leverrier controversy, 15, 17–22; Rutherford recognized by, 255

American Association for the Advancement of Education, 87, 159, 170, 284

American Association for the Advancement of Science, 195, 230, 283; Albany 1851 meeting, 144; Albany 1856 meeting, 132; Civil War, 218; founding of, 129–30; Montreal 1857 meeting, 131, 134; Providence 1855 meeting, 136, 165

American Association of Geologists and Naturalists, 129

American Ephemeris and Nautical Almanac. See *Nautical Almanac*

American Journal of Mathematics: Charles Saunders Peirce, 337; first mathematical research journal, 95; founding of the, 304–6; *Linear Associative Algebra*, 299, 300; William E. Story, 303

American Journal of Science, 26, 98, 195, 232, 365 n. 88

American Philosophical Society, 103, 106, 192, 347 n. 3, 350 n. 80

American Practical Navigator, 42

American science: inferiority of, 25

Amherst College, 129

Analyst (Joel Hendricks), 95

Analyst (Robert Adrain), 46

Analytical Mechanics, 189–91, 379 n. 11

Argelander, Friedrich W. A., 168

Arisbe, 338, 401 n. 145

Armsby, James: Dudley Observatory, 166, 169, 170–72; relationship with Gould, 175; Peirce's proposal for a

Armsby, James (*continued*)
university, 163; scientific council of
the Dudley Observatory (conflict
with), 180–83, 186
Army Corps of Engineers, 274–75
Astor Library, 52, 161
Astor, William, 161
Astronomical Journal, 168–69, 185, 195
Astronomische Nachrichten, 23, 100, 169
Atlantic cable, 251
Atlantic Monthly, 254

Babinet, Jacques, 27
Bache (Coast Survey ship), 271
Bache, Alexander Dallas: AAAS, 129;
Albany (observatory at), 167, 170;
American Philosophical Society, 192;
American research university, 160;
C. H. F. Peters, 178; Civil War, 219,
223; Coast Survey, 107, 273; congress,
259; Davidson's teacher, 263;
Jefferson Davis, 204; death of,
241–44; Dudley Observatory, 169–74,
176–89; early career of, 103–4;
Lazzaroni, 123–24; National
Academy of Sciences, 149, 231, 234;
Neptune (discovery of), 25; as a
peacemaker, 121; Peirce (friendship
with), 124, 306; Royal Society of
London, 27; Smithsonian Institution,
126; William and George Bond
(frustration with), 171
Bache, Nancy Fowler: death of
Alexander Bache, 241–44
Baconianism (Baconian philosophy),
101, 364n. 40
Baird, Spencer, 232–33
Bancroft, George: Coast Survey, 107;
European education of, 52, 68;
Linear Associative Algebra, 296, 299;
Naval Observatory, 111; Peirce family
friendship, 307; Peirce visits, 163;
Round Hill School, 52–53, 57
Banks, Nathaniel Prentiss, 128
Bartlett, John, 307
Beagle: voyage of, 272
Beck, Charles, 74, 76
Berlin Observatory, 15
Bingham, John Armor, 265–68
Bode's Law, 17
Bombelli, Rafael, 293

Bond, George: American Academy of
Arts and Sciences, 193; Harvard
Observatory, 79, 196–98; Leverrier
visit, 23; National Academy of
Sciences, 198, 232; Neptune
(discovery of), 16, 19; Saturn's rings,
194–95
Bond, William Cranch: Harvard
Observatory, 79, 113; Neptune
(discovery of), 16, 24; Peirce
eulogizes, 196; supported by Peirce,
123
Bonnycastle, Charles, 70, 358n. 41
Boston, 28
Bowditch, Ingersoll, 40, 193, 222, 340
Bowditch, Jonathan Ingersoll, 401n.
151
Bowditch, Nathaniel: accomplishments
of, 42–43; *American Practical
Navigator,* 42; Benjamin Peirce Sr., 51;
John Thornton Kirkland, 45–46;
Mécanique céleste, 43; Peirce's mentor,
28, 40, 52; Peirce's regard for, 100,
356n. 47; Royal Society of London,
27; Samuel Ward, 95
Bowen, Francis, 307
Brattle, Thomas, 113
British Association for the
Advancement of Science, 129–30,
213–14
Brown, John, 208–10
Brown-Séquard, Charles Edward, 388n.
19
Brünnow, Francis, 161–62
Buchanan, James, 204–5, 286
Bull Run, first battle of, 220
Byerly, William Elwood, 164, 297, 303

Cambridge High School, 318, 324, 329,
331
Cambridge Miscellany, 99–100
Cambridge, Mass., 49
Cardano, Jerome, 293; formula of, 293
Carnegie Institution, 338
Cary, Elizabeth Cabot. *See* Agassiz,
Elizabeth
Cauchy, Augustin Louis, 86
Cayley, Arthur, 299; *Linear Associative
Algebra,* 300
Challis, James, 16
Chauvenet, William, 70, 159, 385n. 56

Choate, Rufus, 107, 307; Smithsonian Institution, 127–28

Christian Examiner, 132

Cincinnati Observatory, 16, 113, 167–68, 173

Civil War, 164, 273

Clark University, 304

Clarke, James Freeman, 41, 307, 340

Clifford., William Kingdon, 298

Coast Survey. *See* United States Coast Survey

cocaine, 333

Cogswell, Joseph, 52–57, 68, 161, 163

Colburn, Warren 357n. 7

Columbia Athenaeum, 204

Columbia University, 161, 251

Columbia, S.C.: burning of, 236; description of, 217

complex numbers, 292–93, 297

Confederate States of America, 217

Cooke, John P., 144, 370n. 10

Cooper, Peter, 161–63

Count Rumford. *See* Thompson, Benjamin

Cranch, Joseph, 115

Cuvier, Georges, 84

Dallas, George M., 103

Dana, Charles Anderson, 384n. 32

Dana, James Dwight: Lazzaroni, 130; National Academy of Sciences, 231–33; Peirce-Leverrier controversy, 24, 27

Darwin, Charles, 192, 272

Davidson, George, 277; heads Coast Survey expedition to Alaska, 260–63; transit of Venus, 278

Davis, Charles Henry: Coast Survey, 107; *Nautical Almanac,* 118–19; Peirce (relationship with), 118, 122, 306; scientific council of the Dudley Observatory, 171; transit of Venus, 277

Davis, Jefferson, 204, 237

Day, Jeremiah, 72

De Hegermann-Lindencorne, Lillie. *See* Greenough, Lillie

de Lamarck, Jean, 381n. 66

De Morgan, Augustus, 101, 292, 296, 299, 301

Del Ferro, Scipione, 293

Democratic Party, 210

Depot of Charts and Instruments, 111

determinants, 190

Dewey, John, 333

Dixwell, E. S., 329

Dorpat Observatory, 115, 366n. 112

Draper, John William: National Academy of Sciences, 232, 386n. 87

Dudley, Blandina, 167, 170, 174, 184, 378n. 112

Dudley Heliometer, 171

Dudley Observatory, 124, 169–89, 251; inauguration of, 175; scientific council of, 171, 174–75, 177–79, 181–84, 187–89; trustees of, 167, 169, 174–88

École Polytechnique, 104, 150, 264, 301, 325, 327

education, college, 66–68

Eliot, Charles W.: administrative ability of, 226; European travel, 229; Harvard University, 224, 229, 254; James Mills Peirce, 321, 323; Lawrence Scientific School, 226; Massachusetts Institute of Technology, 229; Charles Saunders Peirce, 254, 400n. 105; evaluations of Peirce, 90–91, 93, 256; pallbearer at Peirce's funeral, 340; Rumford Chair, 224–28

Eliot, Samuel, 117, 366n. 124

Ellis, William (son-in-law), 316

Emerson, George, 41, 80

Emerson, Ralph Waldo, 210; Saturday Club, 306

Encke, Johann, 16, 26, 173

ether, 213, 240, 333

Eustis, Henry L., 86, 226

Everett, Edward: Dudley Observatory inauguration (speaks at), 175; European education of, 68; Harvard, president of, 71, 76, 79, 81–82; Lawrence Scientific School, 83; Neptune (discovery of), 19–20; political career of, 81

Exeter College, 214

Farrar, John, 46; declining health of, 66; Peirce's teacher, 41–42

Fay, Melusina. *See* Peirce, Harriet Melusina Fay

Felton, Cornelius C., 222
Ferguson, James, 109
Ferrell, William: *Nautical Almanac,* 119;
  response to Maury's *Physical
  Geography of the Sea,* 113
Flagg, George A.: Peirce as a teacher, 93
Florentine Academy. *See* Lazzaroni
Fort Sumter, 217–21
Fox, Kate, 130
Frankenstein, 37
Franklin, Benjamin, 103
Franklin, Fabian, 333
Frederick William III, 150
Frend, William, 291
fugitive slave law, 216
Fuller, Margaret, 307

Galapagos Islands, 272
Galle, J. G., 15
Garcia, Manuel, 90
Gardener, H. F., 130
Gauss, Karl Frederick: complex
  numbers, 395 n. 61; Benjamin
  Apthorp Gould, 168–69; Neptune,
  15, 26–27; C. H. F. Peters, 173
German universities, 150
Gibbs, Josiah Willard, 164, 294
Gibbs, Oliver Wolcott, 251; Columbia
  University, 157–59; National
  Academy of Sciences, 231; Rumford
  Chair, 228–29; scientific training of,
  123
Gill, Charles, 97–99
Gilliss, James M., 111–12, 234
Gilman, Daniel Coit, 301–5, 332–40
Girard College, 104
glaciation, 272
Gladstone, William Ewart, 269
Goodwin, William, 37
Göttingen, University of, 52, 104, 159,
  168–69
Gould, Benjamin Apthorp: American
  science (dedication to), 168–69;
  Armsby (relationship with), 175;
  Astronomical Journal, 169; George
  Bond, 185; career of, 188; Civil War,
  222; Coast Survey, 168, 249–53;
  Dudley Observatory, 170–89; William
  Ferrell, 113; University of Göttingen,
  169; Harvard Observatory, 149,
  196–97; mental illness of, 175;

Neptune (discovery of), 21, 26;
  student of Peirce, 306; Saturn's rings
  controversy, 195; scientific education
  of, 123, 168; spiritualism, 130; Henry
  P. Tappan, 162
Gray, Asa: American Academy of Arts
  and Sciences, 191–92; debates on
  Darwinism, 192; Harvard University,
  79; Joseph Henry, 81; Neptune
  (discovery of), 25; teaching of, 93
Greeley, Horace, 307; Warner-Winslow
  affair, 137
Green, Hetty, 286–88
Greenough, Horatio, 307; statue of
  Washington, 110
Greenough, Lillie, 87–90, 215; Peirce as
  a teacher, 88
Grund, Francis, 355 n. 17
Guyot, Arnold, 234

Hackley, Charles, 109
Hale, Edward Everett: Peirce as a
  teacher, 93
Hall, James, 151, 153, 270; Dudley
  Observatory, 167, 174
Halleck Henry W., 86
Halley, Edward, 115
Hamilton College, 173, 180
Hamilton, William Rowan, 86, 296;
  quaternions, 292–95
Harkness, William, 275, 392 n. 97
Harper's Ferry, 208–10
Harris, Ira, 187
Harvard Medical School, 142, 306
Harvard Observatory: George Bond,
  185, 193, 196–98; Coast Survey, 171;
  endowment of, 173; founding of,
  114–17; Benjamin Apthorp Gould,
  251; *Nautical Almanac,* 119; Neptune
  (discovery of), 16; Charles Saunders
  Peirce, 255, 332; refractor at, 167;
  solar eclipse of 1870, 268; success of,
  118; Joseph Winlock, 149, 259
Harvard University: astronomy at, 113;
  electives at, 74–77; engineering at,
  77–79, 82; graduate education at,
  164; Lawrence Scientific School (*see*
  Lawrence Scientific School; master of
  arts degree, 66); Rumford
  Professorship, 79–80, 82, 315
Hassler (Coast Survey ship), 271

Hassler, Ferdinand Rudolph, 70, 104, 171, 273

Hawks, Herbert: *Linear Associative Algebra*, 300

Hawthorne, Nathaniel: Saturday Club, 306

Heaviside, Oliver, 294

heliometer, 169

Hendricks, Joel, 95, 362n. 2

Henry, Joseph: AAAS, 129; Albany (observatory at), 167; Spencer Baird, 233; Coast Survey, 244–46, 257, 259, 275; Dudley Observatory, 171, 174, 180, 183–88; education, of, 123; Benjamin Gould, 253; John Le Conte, 146; Lazzaroni, 123–24; Matthew Fontaine Maury, 113; Samuel Morse, 187; National Academy of Sciences, 230, 234; Simon Newcomb, 336; as a peacemaker, 253; Peirce and Bache (disagreement with), 187; Peirce (visits home of), 306; Peirce-Leverrier controversy, 20, 26; Royal Society of London, 27; Rumford Professorship, 81; scientific accomplishments of, 103–4, 127; Smithsonian Institution, 126–29; James J. Sylvester, 302; Warner-Winslow affair, 136

Herrick, Edward, 17–18, 123, 348n. 17

Hilgard, Julius, 267, 273, 280; Coast Survey, 248; drunkenness of, 335; *Linear Associative Algebra*, 298

Hill, George William, 305–6; *Nautical Almanac*, 119

Hill, Thomas: *Analytical Mechanics*, 190; Harvard presidency, 93, 227–28, 253; *Hassler* (voyage of), 271; pallbearer at Peirce's funeral, 340; Peirce's textbooks defense, 70; Peirce's student, 306; quaternions, 294; science and religion, 283

Hitchcock, Edward: and AAAS founding, 129; science and religion, 283

Hoar, Ebenezer Rockwood, 257; Saturday Club, 306

Hoar, Leonard, 92

Holmes, Oliver Wendell, 210, 340; Howland will case, 287; Saturday Club, 306

Hooker, Joseph:, 302

Horsford, Eben, 82–83, 130, 224, 228, 331

Howland will case, 286–88

Humboldt, Alexander von, 84, 168

Humphreys, Andrew A., 274–75

Huntington, Charles Phelps, 381n. 2

Huntington, Fredrick Dan, 307

Hutton, Robert, 301

Hydrographic office of the navy, 273, 275

Ice Age, 84

Ideality in the Physical Sciences, 345, 379n. 16

imaginary numbers, 291

inverse perturbation problem, 16

Jahrbuch über die Fortschritte der Mathematik, 299

James, William, 254, 332, 337; Charles Saunders Peirce, 339

Jastrow, Joseph, 333

Jefferson, Thomas: Coast Survey, 106; University of Virginia, 150

Jewett, Charles, 128

Johns Hopkins, 86, 164, 301–4, 332–35

Jones, Judge Samuel, 64–65

Kepler, Johannes, 295

Kingsley, James L., 72

Kirkland, John Thornton, 45–46, 151

Kirkwood, Daniel, 140

Ladd, Christine, 333

Lagrange, Joseph-Louis, 48

laudanum, 213, 240, 333

Lawrence Scientific School: Louis Agassizx, 86; Charles Eliot, 226, 254; founding of, 83, 151; Joseph Le Conte, 146; *Nautical Almanac* Office relationship with, 119; Herbert Henry Davis Peirce, 320; Charles Saunders Peirce, 331; quaternions, 294

Lawrence, Abbott, 83–84, 163; endows scientific school at Harvard, 83

Lazzaroni: AAAS (leadership in), 130; American Academy of Arts and Sciences, 192, 206; American research university, 157; George

Lazzaroni (*continued*)
　Bond, 193, 198; Columbia University,
　157, 159; Jefferson Davis, 204;
　description of, 122–24; disintegration
　of, 187; Dudley Observatory, 176,
　180–81, 184; Oliver Wolcott Gibbs,
　229; Benjamin Apthorp Gould, 197;
　great American University, 162;
　Harvard University, 254; Thomas
　Hill, 228; John Le Conte, 237;
　meeting in Albany, 144; National
　Academy of Sciences, 149, 230–33;
　New York University, 161; opposition
　to, 140; Samuel B. Ruggles, 155;
　scientific quacks, 140; Smithsonian
　Institution, 126; spiritualism, 130;
　Henry P. Tappan, 162; University of
　California, 238; Washington (annual
　dinner at), 285
Le Conte, Emma: Fort Sumter, firing
　on, 218; life during Civil War,
　235
Le Conte, Guillaume, 146
Le Conte, John: abolitionists, 146;
　aristocratic views of, 146; Civil War,
　235–36; Confederate States, 218;
　death of, 238; Lazzaroni, 146, 370n.
　25; medical education of, 191; Nitre
　and Mining Corps., 235; Northern
　ties, 204; Peirce (friendship with),
　281, 306; *Physical Geography of the Sea*,
　113, 146; Secession of South, 217;
　Southern viewpoint of, 215, 219;
　spiritualism, 130; University of
　California, 237, 238
Le Conte, John Eatton, 146
Le Conte, John L., 140, 146
Le Conte, Joseph: Louis Agassiz, 145;
　Civil War, 235–36; Nitre and Mining
　Corps., 235; University of California,
　237
Le Conte, Josephine: Civil War, 236;
　death of, 238; description of, 146;
　Lazzaroni circle, 147; James Mills
　Peirce, 323; meets Peirce, 144;
　relationship with Peirce, 313–15
Le Conte, Julian, 235
Le Conte, Louis, 146
Lee, John, 98
Lee, Robert E., 208
Lesley, J. Peter, 140

Leverrier, Urbain Jean Joseph, 15–27,
　112, 192
Liebig, Justus, 80, 331
Lighthouse Board, 274, 392n. 95
Lincoln (steam cutter), 260–61
Lincoln, Abraham, 222, 231, 238
Lind, Jennie, 90
Linear Associative Algebra, 288–301;
　description of, 296; influence of, 301;
　publication of, 298–99
Lockyer, J. Norman, 268, 269
Lodge, Anna Cabot, 241
Lodge, Henry Cabot, 332; Benjamin
　Mills Peirce, 325; Peirce's domestic
　life, 309
London Mathematical Society, 299
Longfellow, Fanny Appleton: death of,
　224
Longfellow, Henry Wadsworth:
　Saturday Club, 306; wife's death, 224
Loomis, Elias, 26, 116–17, 306, 316
Lovering, Joseph: American Academy
　of Arts and Sciences, 192, 233;
　Cambridge Miscellany, 100, 363n. 35;
　Eliot, Charles, 229; Harvard
　University, 360n. 91; pallbearer at
　Peirce's funeral, 340; proposed
　engineering school, 79; Warner-
　Winslow affair, 132, 136–37
Lowell Institute, 84, 285; Peirce's
　lectures at, 202, 382n. 40
Lowell, A. Lawrence: Peirce as a
　teacher, 91; quaternions, 294
Lowell, Charles Russell: record of Civil
　War, 238; seeks commission in Union
　army, 220
Lowell, James Russell: Saturday Club,
　306
Lyell, Charles, 84, 152

Marchesi, Mathilde, 90
Marquand, Allan, 333
Maseres, Francis, 291
Massachusetts Institute of Technology,
　229
Mathematical Correspondent, 46
Mathematical Diary, 46–48, 54, 57, 95–96
Mathematical Miscellany, 97–98
mathematics, definition of, 291
Maury, Matthew Fontaine: career of,
　111; Dudley Observatory, 185;

Lazzaroni, 140; Naval Observatory, 112, 118; oceanography, 112; Peirce (relationship with), 112; Peirce-Leverrier controversy, 23–24; Peirce's assessment of, 110, 378n. 114; *Physical Geography of the Sea*, 113, 146; Sears Cook Walker, 16, 18

Maxwell, James Clerk, 294

McCulloch, Hugh, 248, 257

Mécanique céleste, 42–43, 100, 221, 284

Mendenhall, Thomas Corwin, 337–38

Metaphysical Club, 337

Milford, Pennsylvania, 338

Millerites, 115

Mills, Charles (brother-in-law), 306

Mills, Harriette Blake (sister-in-law), 118

Mitchel, Ormsby MacKnight, 113; Dudley Observatory, 167–69, 177; Neptune, discovery of, 16, 20, 23–24

Mitchell, Oscar Howard, 333

Mitchell, William, 111, 123

Monge, Gaspard, 47, 86

Morse, Samuel: controversy with Henry, 187

Moulton, Charles, 215

Moulton, Lillie. *See* Greenough, Lillie

Muzzey, Artemus B., 252

Napier, Lady, 307

Napoleon, 150

Nation, 306

National Academy of Sciences: George Davidson, 263; founding, 230–35; Benjamin Apthorp Gould, 188; and the Lazzaroni, 124, 187; *Linear Associative Algebra*, 298; transit of Venus, 277

National Institute for the Promotion of Science, 129

natural theology, 283

Naturphilosophie, 191

Nautical Almanac: Cambridge headquarters of, 119; employees of, 119; William Ferrell, 113; founding of, 118; funding for cut, 188; Simon Newcomb, 336; Peirce employed by, 194; purpose of, 119; Joseph Winlock, 149, 251; workload of employees, 336

Naval Observatory, 348n. 26; and William Bond, 117; Coast Survey, competition with the, 277; failure to become the national observatory, 113; Maury made superintendent of, 112; and the *Nautical Almanac*, 118; Neptune (discovery of), 17; Peirce-Leverrier controversy, 23–24; Sears Cook Walker, 18, 112

negative numbers, 291

Neptune: discovery of, 15–28

New American Practical Navigator, 111

New Theoretical and Practical Treatise on Navigation, 111

New York Geological Survey, 151

New York University, 151, 156, 162, 232

Newcomb, Simon: *American Journal of Mathematics*, 304; Coast Survey, 275; *Nautical Almanac* (purpose of), 119; Charles Saunders Peirce, 335–38; Peirce's funeral (pallbearer at), 340; Peirce's student, 306; transit of Venus, 278

Newton, H. A.: *Linear Associative Algebra*, 300

Nicaragua, 264

Nichols, Ichabod (grandfather), 28, 31, 65

Nichols, Ichabod (uncle), 40

Nichols, Lydia Ropes. *See* Peirce, Lydia Nichols

North American Review, 81, 100, 306–7, 363–64n. 38

Northampton, 53, 57, 59

Norton, Andrew, 224

Norton, Charles E.: Saturday Club, 306

Olcott, Thomas, 171, 174, 176–81, 183–86

Oliver, James Edward, 306; *Nautical Almanac*, 119

opium, 240, 333

organic morphology, 136–38, 140, 298

Origin of the Species, 192; Agassiz-Gray debates on, 201–2

Otis, Mrs. Harrison Gray, 307

Oxford, 214

Panama, 264

Parkman, George: murder of, 142–40

Patterson, Carlile, 278, 280, 340

Peabody, Andrew, 41, 67, 307, 340; acting president of Harvard, 253

Peacock, George, 291–92, 295
Peirce Elizabeth (aunt): early
  education, 35
Peirce, Benjamin: AAAS, 130;
  abolitionists, 203, 209, 221; Agassiz
  (relationship with), 110; aging
  (thoughts on), 198–99; American
  Academy of Arts and Sciences,
  192–94; *Analytical Mechanics*, 189–91;
  Alexander Dallas Bache, 107, 124;
  birth of, 28; William Bond, 117, 196;
  brother's death, 165; *Cambridge
  Miscellany*, 99–100; children (love of),
  310; Civil War, 218–22, 224, 384n. 12;
  Coast Survey, 107–10, 222, 264, 270,
  280; Coast Survey (superintendent
  of), 244–46, 256–81; Peter Cooper,
  163; courtship, 57, 59–60; Charles
  Henry Davis (relationship with), 122;
  death of, 340; description of, 15;
  domestic life, 313; Dudley
  Observatory, 172; education of,
  40–42, 43–46, 45–46, 52, 357n. 27;
  educational reform, 68–69, 74–77;
  Charles Eliot, 229; European trip,
  213–15; financial problems of, 194;
  funeral of, 340; Benjamin Apthorp
  Gould, 249–53; Gray (relationship
  with), 191–92; Harvard Observatory,
  114–17, 196–98; Harvard University,
  59, 62, 66, 161; Harvard
  (engineering at), 77–79, 82;
  heliometer, 170; Joseph Henry, 187;
  humor of, 165–66; illness of, 213,
  240; *Ideality in the Physical Sciences*,
  345, 379n. 16; imaginary numbers,
  297; introduced on floor of the
  Senate, 212, 257; Lazzaroni, 123–24;
  Le Contes (friendship with), 149;
  lectures at Johns Hopkins, 339;
  Linear Associative Algebra, 297,
  288–301; marriage, 62; *Mathematical
  Diary*, 46–48; mathematical work, 96,
  286–89; mathematics, 288;
  mathematics (definition of), 296;
  mathematics (nature of), 101;
  Matthew Maury, 110; Mitchel-Bond
  dispute, 195; National Academy of
  Sciences, 231, 233; national
  university, 160; national university
  (plan for), 162–63; *North American*

Review articles, 363–64n. 38; old age
  (comments on), 189; Sarah Peirce,
  144, 212; Perkins Professorship, 66;
  PhD dissertation (supervision of),
  164; political activity of, 210–11, 222;
  as a public speaker, 281; quaternions,
  294; racism of, 211–12; religious
  views of, 281–85; Benjamin Rogers,
  375n. 25; Round Hill School, 52,
  54–55; Royal Society of London, 27;
  Saturn's rings, 194–95; science
  (conception of), 21, 125; scientific
  patriot, 25; slavery, 203–5, 209;
  Smithsonian Institution (lectures at),
  282; solar eclipse of 1870, 269;
  Southern sympathies, 204, 206, 384n.
  12; students of, 306, 359n. 56; James
  J. Sylvester, 102, 302; teacher, 56,
  87–95; temper of, 309; textbooks of,
  69–72; Transit of Venus Commission,
  278; Warner-Winslow affair
  (supported by Winlock, Wright, and
  Newcomb), 133, 134
Peirce, Benjamin Mills (son), 147, 221,
  324–28; death of, 328; mining
  engineer, 327; poor health of, 328;
  report on Iceland and Greenland,
  264, 327; rustication of, 325; student
  in Paris, 325
Peirce, Benjamin Senior (father):
  business failure, 49, 354–55n. 1;
  courtship, 28–34; death of, 63;
  description of, 28; early career, 34;
  Harvard (history of), 63–64;
  Librarian of Harvard College, 51;
  marriage, 34, 40; mathematical work
  of, 40; political career, 34;
  valedictorian, 40
Peirce, Charles Henry (brother), 57,
  63–64, 331; abolitionists, 203;
  medical practice, 65
Peirce, Charles Saunders (son), 144,
  275, 277, 280, 328–39; *American
  Journal of Mathematics*, 305; birth of,
  328; brilliance of, 400n. 88; Civil War,
  221; Coast Survey, 332, 338; educated
  by father, 331; Charles W. Eliot, 255,
  400n. 105; Lillie Greenough
  (romance with), 90; Harvard
  Observatory, 259; Howland will case,
  287; Johns Hopkins, 332; *Linear*

Associative Algebra, 297, 299; photometric research of, 332; quaternions, 294; solar eclipse of 1870 expedition, 268; students of, 333

Peirce, Charlotte Elizabeth (sister), 308–9, 331; AAAS 1857 Montreal meeting, 131; African American servant, 249; Alexander Dallas Bache, 104; Cambridge (moves to), 49; Charles Henry Davis, 119; Lillie Greenough, 89; Lady Napier, 307; Josephine Le Conte, 147; life at Rhinebeck, 65; Peirce's *Analytical Mechanics,* 190; racism of, 220; slavery (hopes for the end of), 220; views on teaching mathematics, 63; works as a governess, 63

Peirce, Harriet Melusina Fay (Zina) (daughter-in-law), 90, 331; divorce of, 334; solar eclipse of 1870 expedition, 268–69

Peirce, Helen (daughter), 147, 316–18; birth, 316

Peirce, Herbert Henry Davis (son), 221, 318–21; career in the Foreign Service, 320

Peirce, James Mills (son), 221, 321–24; career as a minister, 321; dean of Arts and Sciences at Harvard, 323; dean of Harvard graduate school, 323; Harvard Divinity School, 321; homosexuality of, 323; Perkins Professor of Mathematics and Astronomy, 323; quaternions, 295; solar eclipse of 1870 expedition, 268

Peirce, Jerathmiel (grandfather), 28

Peirce, Juliette Froissy Pourtalais (wife of Charles Saunders Peirce), 334–35

Peirce, Lydia Nichols (mother), 62; courtship, 28–34; feminist views, 35–39; governess and housekeeper, 64–65; marriage, 34, 40; move to Cambridge, 49, 50

Peirce, Sarah Mills (wife), 57, 118, 310–13; Civil War, 221; Coast Survey, 246; domestic duties, 310; illness of, 212, 315; jealously of, 314; Peirce (relationship with), 144; Peirce's devotion to, 315; religion, 311; Rutherfords, 255

Peirce's Criterion, 286

perfect numbers, 96, 286, 362 n. 10

Permanent Commission of the Navy, 223, 230

Peters, Christian H. F., 174, 186; Dudley Observatory, 173, 176–81; heliometer, 170

Philadelphia High School, 17, 116

Phillips, Wendell, 209, 382 n. 30

Phyllotaxy, 79, 191

Physical Geography of the Sea, 112, 146

Pickering, John, 50, 64, 107, 357 n. 9

Pierce, Franklin, 285

Pourtalès, Louis François de, 271

Pruyn, J. V. L., 257

Pruyn, J. V. L., Mrs., 180

Pulkovo Observatory, 116, 366 n. 112

Pythagoreans, 96

quaternions, 292–95

Quincy, Josiah, 68, 72, 77, 91, 192; Harvard Observatory, 113; Harvard (electives at), 76

Radcliff College, 86

Radical Club: Peirce's lectures at, 285

recitation: instruction by, 67

Religion of Geology and Its Connected Sciences, The, 283

Rensselaer Polytechnic Institute, 156

Republican Party, 210, 216

Rodgers, Fairman: National Academy of Sciences, 231

Rogers, Henry Darwin, 80, 169, 315; AAAS, 130

Rogers, William Barton: National Academy of Sciences, 232

Round Hill School, 53, 55, 59, 95, 307

Royal Military Academy at Woolwich, 301

Royal Society of London, 101

Ruggles, Samuel B., 157; Albany (university at), 155–56; Columbia University, 157–59; Lazzaroni, 157; New York City (university in), 161, 163

Rumford Professorship. *See* Harvard University, Rumford Professorship

Runkle, John D., 306; *Nautical Almanac,* 119

Rush, Benjamin, 150

Rutherford, Lewis, 123, 249, 255, 268
Ryan, James, 95–96

Safford, Truman: *Nautical Almanac,* 119
Salem Grammar School, 40
Salem, Mass., 28, 50
Sands, Benjamin F., 277
Sanitary Commission, 188, 223–24
Saturday Club, 306; Peirce resigns
    from, 205
Saunders, Charles (uncle), 65, 309;
    death of, 240; Harvard University
    (endows Civil War memorial at),
    238
Saunders, Charlotte (aunt), 65
Scheutz calculating engine, 182, 377 n.
    101
Schurz, Carl, 275
science and religion, 281–85
scientific council of the Dudley
    Observatory. *See* Dudley Observatory,
    scientific council of
Scott, General Winfield, 219, 384 n. 15
Seward, William Henry, 209, 382 n. 30;
    Alaska (purchase of), 259; Iceland
    and Greenland, 264; St. Thomas and
    St. John (acquisition of), 264
Shaler, Nathaniel, 110
Shaw, Josephine, 239
Shaw, Robert, 238
Sheafer, P. W., 135
Sheffers, George, 300
Sheffield Scientific School, 151
Shelley, Percy, 37
Sherman, William Tecumseh, 235
Sidereal Messenger, 167
Simpson, Robert, 291
Simpson, Thomas, 301
Smith, Edwin, 278
Smithson, James, 126
Smithsonian Institution: founding of,
    126–29; Henry as secretary of, 81;
    Peirce's lectures at, 281
solar eclipse of 1869, 265
solar eclipse of 1870, 265
solar parallax, 171, 278, 375 n. 38
Southern university, 207
Sparks, Jared, 20, 77, 221
Spencer, Charles A., 173
spiritualism, 130–31, 368 n. 38
Spottiswoode, William, 299

St. Thomas and St. John: United States
    purchases from Denmark, 264
Steindachner, Franz, 271
stellar parallax, 375 n. 38
Story, Joseph, 51, 78, 107; career of, 34;
    courtship of Lydia Nichols, 29–34
Story, William E., 303, 305
Story, William W., 307
Study, Eduard, 300
Sumner, Charles, 257, 259, 275, 277;
    Saturday Club, 306
Sylvester, James J.: *American Journal of
    Mathematics,* 337; anti-Semitism, 301;
    career of, 301–2; in the United
    States, 102–3; Johns Hopkins
    University, 164, 301–6; *Linear
    Associative Algebra,* 300; Peirce,
    friendship with, 102, 214, 306, 335;
    Peirce's funeral (pallbearer at), 340;
    University of Virginia, 101–2

Tait, Peter Guthrie, 294
Talcott, Andrew, 109
Tappan, Henry P., 160–63
telegraphic transmissions, 251
Thomas Hart Benton, 204
Thompson, Benjamin, 79–80, 83
Ticknor, George: accomplishments of,
    355 n. 21; collegiate instruction, 67;
    Charles Eliot (uncle of), 224;
    European education of, 52, 68;
    Harvard (professor at), 151; modern
    languages, study of, 72
Torrey, John, 191
transit of Venus, 113, 277, 278
Transit of Venus Commission: Peirce
    serves on, 278
Treadwell, Daniel 360 n. 91
trigeminal neuralgia, 333, 387 n. 3
Troy, New York, 144

Union College, 156
Unitarianism, 30, 81, 93, 158, 307, 321
United State Military Academy at West
    Point, 42, 86, 97, 103, 109, 276
United States Coast and Geodetic
    Survey, 270
United States Coast Survey, 104, 142;
    Alaska (role in purchase of), 259–64;
    army (competition with), 274–77;
    Civil War (importance to the),

222–23; Congress (investigated by), 335; early years of, 104–7; expansion of under Peirce, 270; navy (competition with), 273–74; observatories (unable to have), 171; Charles Saunders Peirce, 338; solar eclipse of 1869, 265; training ground for American scientists, 244

United States navy: and the Coast Survey, 106

United States Supreme Court: Peirce dines with justices of, 285

University College, London, 101, 301

University of Berlin, 150, 152

University of Bern, 104

University of California, 237–38, 323

University of Chicago, 304

University of Michigan, 160, 162

University of Pennsylvania, 104, 156, 161

University of Virginia, 42, 101

Uranus, 16, 19, 20, 22–23, 27, 348 n. 32

Van Rensselaer, Stephen, 167

Veblen, Thorstein, 333

vector analysis, 294

Vestiges of Creation, 80

Walker, Robert J., 109, 264, 390 n. 30

Walker, Sears Cook: Dudley Observatory, 168; fired by Maury, 112; illness of, 168; *Nautical Almanac*, 119; Neptune (discovery of), 16–20, 22, 25

Walker, Timothy, 51

Walsh, John, 40

Ward, Samuel, 95, 96–97

Warner, John, 135–38

Warner-Winslow affair, 131–42, 369 n. 73

Wartmann, Louis François, 17–18

Washington, D.C.: descriptions of, 247

Washington, George: and a national university, 150

Webster, Daniel, 59, 81, 307

Webster, John W., 79, 82; murder trial of, 144

Wedderburn, Joseph, 300

Western Messenger, 307

Whig Party, *210*

Whitehead, Alfred North, 300, 395 n. 80

Whitney, Josiah, 276

Whittier, John Greenleaf: Saturday Club, 306

Wilkes expedition, 111, 129

Williams College, 59

Wilson, Henry, 230–31, 257, 275, 277

Winlock, Joseph: Coast Survey, 275; Harvard Observatory, 196, 251, 259; meets Peirce, 149; *Nautical Almanac,* 119; Newcomb, Simon, 337; Charles Saunders Peirce, 332; solar eclipse of 1870, 268–69

Winslow, Charles F., 122, 131–35

Winthrop, John, 113

Wollstonecraft, Mary, 36–39

Wood, Fernando, 162

Wright, Chauncey: Howland will case, 394 n. 41; Metaphysical Club, 337; organic morphology, 368 n. 52; *Nautical Almanac,* 119; Peirce's student, 306

Wyman, Jeffries, 80, 208

Yale Report of 1828, 72

Yale University, 72, 79; graduate education at, 164

Yardley, J. W., 140